T0213809

Texts in Computer Science

Series Editors

David Gries, Department of Computer Science, Cornell University, Ithaca, NY, USA

Orit Hazzan⬤, Faculty of Education in Technology and Science, Technion—Israel Institute of Technology, Haifa, Israel

More information about this series at http://www.springer.com/series/3191

R. M. R. Lewis

Guide to Graph Colouring

Algorithms and Applications

Second Edition

 Springer

R. M. R. Lewis
Cardiff School of Mathematics
Cardiff University
Cardiff, UK

ISSN 1868-0941   ISSN 1868-095X   (electronic)
Texts in Computer Science
ISBN 978-3-030-81056-6   ISBN 978-3-030-81054-2   (eBook)
https://doi.org/10.1007/978-3-030-81054-2

This Springer imprint is published by the registered company Springer Nature Switzerland AG
The registered company address is: Gewerbestrasse 11, 6330 Cham, Switzerland

For Fifi, Maiwen, Aoibh, and Maccy
Gyda cariad

Preface

Graph colouring is one of those rare examples in the mathematical sciences of a problem that, while easy to state and visualise, has many aspects that are exceptionally difficult to solve.

In this book, our goal is to examine graph colouring as an *algorithmic* problem, with a strong emphasis on implementation and practical application. To these ends, in addition to providing a theoretical treatment of the problem, we also dedicate individual chapters to real-world problems that can be tackled via graph colouring techniques. These include designing sports schedules, solving Sudoku puzzles, timetabling lectures at a university, and creating seating plans.

Portable source code for all of the algorithms considered in this book is also available for free download.

Organisation and Features

The first chapter of this book is kept deliberately light; it gives a brief tour of the graph colouring problem, avoids jargon, and gives plenty of illustrated examples.

In Chap. 2, we discover that graph colouring is a type of "intractable" problem, meaning that it usually needs to be tackled using inexact heuristic algorithms. To reach this conclusion, we introduce the topics of problem complexity, polynomial transformations, and $\mathcal{NP}$-completeness. We also review several graph topologies that are easy to colour optimally.

Chapter 3 of this book starts with some theory and uses various techniques to derive bounds on the chromatic number of graphs. It then looks at five well-established constructive heuristics for graph colouring (including the Greedy, DSatur, and RLF algorithms) and analyses their relative performance. Source code for these algorithms is provided.

The intention of Chap. 4 is to give the reader an overview of the different strategies available for graph colouring, including both exact and heuristic methods. Techniques considered include backtracking, integer programming, column generation, and various metaheuristics. No prior knowledge of these techniques is assumed. We also describe ways in which graph colouring problems can be reduced in size and broken into smaller parts, helping to improve algorithm performance.

Chapter 5 gives an in-depth analysis of six high-performance algorithms for the graph colouring problem. The performance of these algorithms is compared using several different problem types, including random, flat, scale-free, planar, and timetabling graphs. Source code for each of these algorithms is also provided.

Chapter 6 considers several problems, both theoretical and practical, that can be expressed through graph colouring principles. Initial sections focus on special cases of the graph colouring problem, including map colouring (together with a history of the Four Colour Theorem), edge colouring, Latin squares, and Sudoku puzzles. The problems of colouring graphs where only limited information about connectivity is known, or where a graph is subject to change over time, are also considered, as are some natural extensions to graph colouring such as list colouring, equitable graph colouring, weighted graph colouring, and chromatic polynomials.

The final three chapters of this book examine three separate case studies in which graph colouring techniques can be used to find high-quality solutions to real-world problems. Chapter 7 looks at the problem of designing seating plans for large gatherings; Chap. 8 considers the creation of league schedules for sports competitions; Chap. 9 looks at timetabling in educational establishments. Each of these chapters is written so that, to a large extent, they can be read independently of the other chapters of this book.

Audience

This book is written for anyone with a background in mathematics, computer science, operational research, or management science. Initial sections are particularly appropriate for undergraduate learning and teaching; later sections are more suited for postgraduate and research levels.

This text has been written with the presumption that the reader has no previous experience of graph colouring or graph theory more generally. However, elementary knowledge of the notation and concepts surrounding sets, matrices, and enumerative combinatorics (particularly combinations and permutations) is assumed.

Supplementary Resources

All algorithms reviewed and tested in this book are available for free download at http://www.rhydlewis.eu/resources/gCol.zip. These implementations are written in C++ and can be compiled on Windows, macOS, and Linux. Full instructions on how to do this are provided in Appendix A. Readers are invited to experiment with these algorithms as they make their way through this book.

This book also shows how graph colouring problems can be generated and tackled using Sage and Python. Both of these programming languages are free to download. We also show how graph colouring problems can be solved via linear programming software, in this case using the commercial software FICO Xpress.

An online implementation of the seat-planning algorithm presented in Chap. 7 can be accessed at http://www.weddingseatplanner.com. C++ code for the algorithms described in Chaps. 8 and 9 can also be downloaded using the links given in the text.

I hope you find these implementations as useful as I did.

Cardiff, Wales, UK
June 2021

R. M. R. Lewis

Contents

1 Introduction to Graph Colouring 1
  1.1   Some Simple Practical Applications 2
         1.1.1   A Team Building Exercise 2
         1.1.2   Constructing Timetables 4
         1.1.3   Scheduling Taxis 5
         1.1.4   Compiler Register Allocation 6
  1.2   Why Colouring? 7
  1.3   Problem Description 9
  1.4   About This Book 12
         1.4.1   Algorithm Implementations 13
  1.5   A Note on Pseudocode and Notation 15
  1.6   Chapter Summary 16
  References 16

2 Problem Complexity 17
  2.1   Algorithm Complexity and Big O Notation 18
  2.2   Solving Graph Colouring via Exhaustive Search 20
  2.3   Problem Intractability 23
         2.3.1 $\mathcal{P}$ and $\mathcal{NP}$ 23
         2.3.2   Polynomial-Time Reductions 26
         2.3.3 $\mathcal{NP}$-Completeness 27
         2.3.4   Boolean Satisfiability Problems (SAT) 28
  2.4   Proofs of $\mathcal{NP}$-Completeness 30
  2.5   Graphs that are Easy to Colour Optimally 33
         2.5.1   Complete Graphs 34
         2.5.2   Cycle, Wheel, and Planar Graphs 35
         2.5.3   Grid Graphs 36
  2.6   Chapter Summary and Further Reading 37
  References 38

3 Bounds and Constructive Heuristics 39
  3.1   The Greedy Algorithm for Graph Colouring 41
  3.2   Bounds on the Chromatic Number 45

        3.2.1   Lower Bounds 45

        3.2.2   Upper Bounds 49

   3.3   The DSATUR Algorithm 54

   3.4   Colouring Using Maximal Independent Sets 58

        3.4.1   The RLF Algorithm 59

   3.5   Empirical Comparison. 62

        3.5.1   Experimental Considerations. 63

        3.5.2   Algorithm Complexities 65

        3.5.3   Results and Analysis 68

   3.6   Chapter Summary and Further Reading 71

   References 75

4 Advanced Techniques for Graph Colouring 77

   4.1   Exact Algorithms 77

        4.1.1   Backtracking Approaches 78

        4.1.2   Integer Programming 81

        4.1.3   Minimum Coverings and Column Generation 91

   4.2   Inexact Heuristics and Metaheuristics. 96

        4.2.1   Feasible-Only Solution Spaces 97

        4.2.2   Spaces of Complete, Improper k-Colourings 102

        4.2.3   Spaces of Partial, Proper k-Colourings 105

        4.2.4   Combining Solution Spaces 106

        4.2.5   Problems Related to Graph Colouring 106

   4.3   Reducing Problem Size. 107

        4.3.1   Removing Vertices and Splitting Graphs 107

        4.3.2   Extracting Independent Sets 108

   4.4   Chapter Summary. 110

   References 110

5 Algorithm Case Studies. 113

   5.1   The TABUCOL Algorithm 113

   5.2   The PARTIALCOL Algorithm 115

   5.3   The Hybrid Evolutionary Algorithm (HEA) 117

   5.4   The ANTCOL Algorithm. 118

   5.5   The Hill-Climbing (HC) Algorithm 121

   5.6   The Backtracking Algorithm 122

   5.7   Algorithm Comparison 124

        5.7.1   Random Graphs 125

        5.7.2   Flat Graphs 126

        5.7.3   Planar Graphs 129

        5.7.4   Scale-Free Graphs 134

        5.7.5   Exam Timetabling Graphs 137

        5.7.6   Social Network Graphs 141

        5.7.7   Comparison Discussion 143

	5.8	Further Improvements to the HEA	145
		5.8.1 Maintaining Diversity	146
		5.8.2 Recombination	149
		5.8.3 Local Search	151
	5.9	Chapter Summary and Further Reading	152
	References		153
6	**Applications and Extensions**		**155**
	6.1	Face Colouring	156
		6.1.1 Dual Graphs, Colouring Maps, and the Four Colour Theorem	158
		6.1.2 Four Colours Suffice	163
	6.2	Edge Colouring	166
	6.3	Precolouring	171
	6.4	Latin Squares and Sudoku Puzzles	172
		6.4.1 Relationship to Graph Colouring	173
		6.4.2 Solving Sudoku Puzzles	174
	6.5	Short Circuit Testing	179
	6.6	Graph Colouring with Incomplete Information	182
		6.6.1 Decentralised Graph Colouring	182
		6.6.2 Online Graph Colouring	185
		6.6.3 Dynamic Graph Colouring	187
	6.7	List Colouring	188
	6.8	Equitable Graph Colouring	189
	6.9	Weighted Graph Colouring	193
		6.9.1 Weighted Vertices	193
		6.9.2 Weighted Edges	195
		6.9.3 Multicolouring	196
	6.10	Chromatic Polynomials	196
	6.11	Chapter Summary	200
	References		200
7	**Designing Seating Plans**		**203**
	7.1	Problem Background	203
		7.1.1 Relation to Graph Problems	205
		7.1.2 Chapter Outline	206
	7.2	Problem Definition	206
		7.2.1 Objective Functions	207
		7.2.2 Problem Intractability	208
	7.3	Problem Interpretation and Tabu Search Algorithm	208
		7.3.1 Stage 1	209
		7.3.2 Stage 2	210
	7.4	Algorithm Performance	212
	7.5	Comparison to an IP Model	214

7.5.1   Results 217
7.6   Chapter Summary and Discussion 219
References 220

8 Designing Sports Leagues 221
8.1   Problem Background 221
8.1.1   Further Round-Robin Constraints 223
8.1.2   Chapter Outline 225
8.2   Representing Round-Robins as Graph Colouring Problems 225
8.3   Generating Valid Round-Robin Schedules 226
8.4   Extending the Graph Colouring Model.................. 227
8.5   Exploring the Space of Round-Robins 233
8.6   Case Study: Welsh Premiership Rugby 236
8.6.1   Solution Methods............................ 237
8.7   Chapter Summary and Discussion 244
References 245

9 Designing University Timetables 247
9.1   Problem Background 247
9.1.1   Designing and Comparing Algorithms.............. 249
9.1.2   Chapter Outline 250
9.2   Problem Definition and Preprocessing 251
9.2.1   Soft Constraints............................ 254
9.2.2   Problem Complexity 255
9.2.3   Evaluation and Benchmarking 255
9.3   Previous Approaches to This Problem 256
9.4   Algorithm Description: Stage One 257
9.4.1   Results 259
9.5   Algorithm Description: Stage Two..................... 261
9.5.1   SA Cooling Scheme.......................... 261
9.5.2   Neighbourhood Operators....................... 262
9.5.3   Dummy Rooms 264
9.5.4   Estimating Solution Space Connectivity 265
9.6   Experimental Results 266
9.6.1   Influence of the Neighbourhood Operators 266
9.6.2   Comparison to Published Results 269
9.6.3   Differing Time Limits 271
9.7   Chapter Summary and Discussion 271
References 274

Appendix A: Computing Resources 277

Appendix B: Table of Notation 297

Index 299

Introduction to Graph Colouring

1

In mathematics, a graph can be thought of as a set of objects in which some pairs of objects are connected by links. The interconnected objects are usually called *vertices* and the links connecting pairs of vertices are called *edges*. Graphs can be used to model a surprisingly large number of problem areas, including social networking, chemistry, scheduling, parcel delivery, satellite navigation, electrical engineering, and computer networking.

The graph colouring problem is one of the most famous problems in the field of graph theory and has a long and illustrious history. In a nutshell, it asks, given any graph, how might we go about assigning "colours" to the vertices so that (a) no vertices joined by an edge are given the same colour, and (b) the number of different colours used is minimised?

Figure 1.1 shows a picture of a graph with ten vertices (the circles), and 21 edges (the lines connecting the circles). It also shows an example colouring of this graph that uses five different colours. We can call this solution a "proper" colouring because all pairs of vertices joined by edges have been assigned to different colours, as required by the problem. Specifically, two vertices have been assigned to colour 1, three vertices to colour 2, two vertices to colour 3, two vertices to colour 4, and one vertex to colour 5.

This solution is not the only possible five-colouring for this graph. For example, swapping the colours of the bottom two vertices in the figure would give us a different proper five-colouring. It is also possible to colour the graph with anything between six and ten colours (where ten is the number of vertices in the graph), because assigning a vertex to an additional, newly created, colour still ensures that the colouring remains proper.

But what if we wanted to colour this particular graph using *fewer* than five colours? Is this possible? To answer this question, consider Fig. 1.2, where the dotted line indicates a selected portion of this graph. When we remove everything from outside this selection, we are left with a subgraph containing just five vertices. Importantly, we can see that every pair of vertices in this subgraph has an edge between them.

© The Author(s), under exclusive license to Springer Nature Switzerland AG 2021
R. M. R. Lewis, *Guide to Graph Colouring*, Texts in Computer Science,
https://doi.org/10.1007/978-3-030-81054-2_1

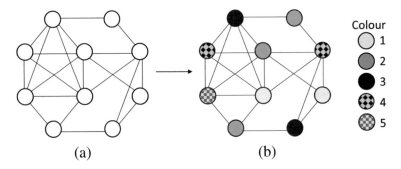

Fig. 1.1 A small graph (**a**), and corresponding five-colouring (**b**)

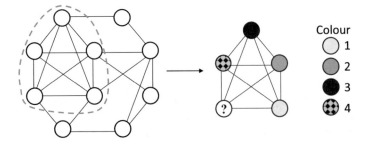

Fig. 1.2 If we extract the vertices in the dotted circle, we are left with a subgraph that clearly needs more than four colours

If we were to have only four colours available to us, as indicated in the figure we would be unable to properly colour this subgraph, since its five vertices all need to be assigned to a different colour in this instance. This allows us to conclude that the solution in Fig. 1.1 is *optimal*, since there is no solution available that uses fewer than five colours.

1.1 Some Simple Practical Applications

Let us now consider four simple practical applications of graph colouring to further illustrate the underlying concepts of the problem.

1.1.1 A Team Building Exercise

An instructive way of visualising the graph colouring problem is to imagine the vertices of a graph as a set of "items" that need to be divided into "groups". As an example, imagine we have a set of university students that we want to split into groups for a team-building exercise. Also, imagine we are interested in dividing the

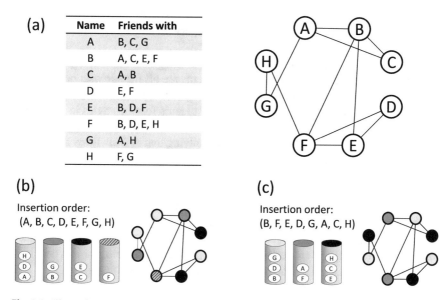

(a)

Name	Friends with
A	B, C, G
B	A, C, E, F
C	A, B
D	E, F
E	B, D, F
F	B, D, E, H
G	A, H
H	F, G

(b)

Insertion order:
(A, B, C, D, E, F, G, H)

(c)

Insertion order:
(B, F, E, D, G, A, C, H)

Fig. 1.3 Illustration of how proper five- and four-colourings can be constructed for the same graph

students so that no student is put in a group containing one or more of his friends, and so that the number of groups used is minimal. How might this be done?

Consider the example given in the table in Fig. 1.3a. Here, we have a list of eight students with names A through to H, together with information on who their friends are. From this information we can see that student A is friends with three students (B, C and G), student B is friends with four students (A, C, E, and F), and so on. Note that the information in this table is "symmetric" in that if student x lists student y as one of his friends, then student y also does the same with student x. This sort of relationship occurs in social network systems such as Facebook, where two people are only considered friends if both parties agree to be friends in advance. An illustration of this example in graph form is also given in the figure.

Let us now attempt to split the eight students of this problem into groups so that each student is put into a different group to that of his friends. A simple method to do this might be to take the students one by one in alphabetical order and assign them to the first group where none of their friends is currently placed. Walking through the process, we start by taking student A and assigning him to the first group. Next, we take student B and see that he is friends with someone in the first group (student A), and so we put him into the second group. Taking student C next, we notice that he is friends with someone in the first group (student A) and also the second group (student B), meaning that he must now be assigned to a third group. At this point, we have only considered three students, yet we have created three separate groups. What about the next student? Looking at the information, we can see that student D is only friends with E and F, allowing us to place him into the first group alongside student A. Following this, student E cannot be assigned to the first or second groups but can be assigned to the third. Continuing this process for all eight students gives us the solution shown in Fig. 1.3b. This solution uses four groups and also involves student F being assigned to a group by himself.

Can we do any better than this? By inspecting the graph in Fig. 1.3a, we can see that there are three separate cases where three students are all friends with one another. Specifically, these are students A, B, and C; students B, E, and F; and students D, E, and F. The edges between these triplets of students form triangles in the graph. Because of these mutual friendships, in each case these collections of three students will need to be assigned to different groups, implying that *at least* three groups will be needed in any valid solution. However, by visually inspecting the graph we can see that there is no occurrence of *four* students all being friends with one another. This hints that we may not necessarily need to use four groups in a solution.

A solution using three groups *is* possible in this case as Fig. 1.3c demonstrates. This solution has been achieved using the same assignment process as before but using a different ordering of the students, as indicated. Since we have already deduced that at least three groups are required for this particular problem, we can conclude that this solution is *optimal*—no proper solution using fewer colours exists.

The process we have used to form the solutions shown in Fig. 1.3b, c is known as the GREEDY algorithm for graph colouring. This is a fundamental part of the field of graph colouring and will be considered further in Chap. 3. Among other things, we will demonstrate that there will always be at least one ordering of the vertices that, when used with the GREEDY algorithm, will result in an optimal solution.

1.1.2 Constructing Timetables

A second important application of graph colouring arises in the production of timetables at colleges and universities. In these problems we are given a set of "events", such as lectures, exams, classroom sessions, together with a set of "timeslots" (e.g., Monday 09:00–10:00, Monday 10:00–11:00, and so on). Our task is to then assign the events to the timeslots in accordance with a set of constraints. One of the most important of these constraints is what is often known as the "event-clash" constraint. This specifies that if a person (or some other resource of which there is only one) is required to be present in a pair of events, then these events must not be assigned to the same timeslot since such an assignment will result in this person/resource having to be in two places at once.

Timetabling problems can be easily converted into an equivalent graph colouring problem by considering each event as a vertex, and then adding edges between any vertex pairs that are subject to an event-clash constraint. Each timeslot available in the timetable then corresponds to a colour, and the task is to find a colouring such that the number of colours is no larger than the number of available timeslots.

Figure 1.4 shows an example timetabling problem expressed as a graph colouring problem. Here we have nine events which we have managed to timetable into four timeslots. In this case, three events have been scheduled into timeslot 1, and two events have been scheduled into each of the remaining three. In practice, assuming

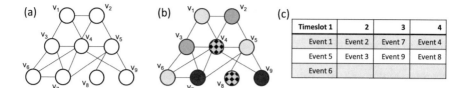

Fig. 1.4 A small timetabling problem (**a**), a feasible four-colouring (**b**), and its corresponding timetable solution using four timeslots (**c**)

that only one event can take place in a room at any one time, we would also need to ensure that three rooms are available during timeslot 1. If only two rooms are available in each timeslot, then an extra timeslot might need to be added to the timetable.

It should be noted that timetabling problems can often vary a great deal between educational institutions, and can also be subject to a wide range of additional constraints beyond the event-clash constraint mentioned above. Many of these will be examined further in Chap. 9.

1.1.3 Scheduling Taxis

A third example of how graph colouring can be used to model real-world problems arises in the scheduling of timed tasks. Imagine that a taxi firm has received n journey bookings, each of which has a start time, signifying when the taxi will leave the depot and a finish time telling us when the taxi is expected to return. How can we assign all of these bookings to vehicles so that the minimum number of vehicles is used?

Figure 1.5a shows an example problem where we have ten taxi bookings. For illustrative purposes, these have been ordered from top to bottom according to their start times. It can be seen, for example, that booking 1 overlaps with bookings 2, 3, and 4; hence any taxi carrying out booking 1 will not be able to serve bookings 2, 3, and 4. We can construct a graph from this information by using one vertex for each booking and then adding edges between any vertex pair corresponding to overlap-

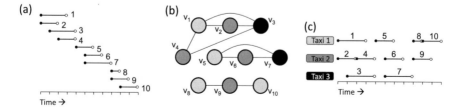

Fig. 1.5 A set of taxi journey requests over time (**a**), its corresponding interval graph and three-colouring (**b**), and (**c**) the corresponding assignment of journeys to taxis

ping bookings. A three-colouring of this example graph is shown in Fig. 1.5b, and the corresponding assignment of the bookings to three taxis (the minimum number possible) is shown in Fig. 1.5c.

A graph constructed from time-dependent tasks such as this is usually referred to as an *interval graph*. In Chap. 3, we will show that a simple inexpensive algorithm exists for interval graphs that will always produce a solution using that uses the fewest number of colours possible.

1.1.4   Compiler Register Allocation

Our final example in this section concerns the allocation of computer code variables to registers on a computer processor. When writing code in a particular programming language, whether it be C++, Fortran, Java, or Pascal, programmers are free to make use of as many variables as they see fit. When it comes to compiling this code, however, it is advantageous for the compiler to assign these variables to registers[1] on the processor since accessing and updating values in these locations are much faster than carrying out the same operations using the computer's RAM or cache.

Computer processors only have a limited number of registers. For example, most RISC processors feature 64 registers: 32 for integer values and 32 for floating-point values. However, not all variables in a computer program will be in use (or "live") at a particular time. We might therefore choose to assign multiple variables to the same register if they are not judged to interfere with one another.

Figure 1.6a shows an example piece of computer code making use of five variables, $v_1, \ldots, v_5$. It also shows the *live ranges* for each variable. So, for example, variable v_2 is live only in Steps (2) and (3), whereas v_3 is live from Steps (4) to (9). It can also be seen, for example, that the live ranges of v_1 and v_4 do not overlap. Hence we might use the same register for storing both of these variables at different periods during execution.

The problem of deciding how to assign the variables to registers can be modelled as a graph colouring problem by using one vertex for each live range and then adding edges between any pairs of vertices corresponding to overlapping live ranges. Such a graph is known as an *interference graph*, and the task is to now colour the graph using equal or fewer colours than the number of available registers. Figure 1.6b shows that, in this particular case, only three registers are needed: variables v_1 and v_4 can be assigned to register 1, v_2 and v_5 to register 2, and v_3 to register 3.

Note that in the example of Fig. 1.6, the resultant interference graph corresponds to an interval graph, rather like the taxi example from the previous subsection. Such graphs will arise in this setting when using straight-line code sequences or when using software pipelining. In most situations, however, the flow of a program is likely to be far more complex, involving if-else statements, loops, goto commands, and so on.

---

[1] Registers can be considered physical parts of a processor that are used for holding small pieces of data.

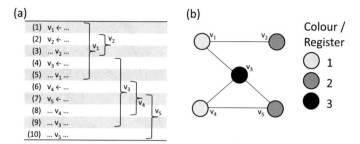

Fig. 1.6 a An example computer program together with the live ranges of each variable. Here, the statement "$v_i \leftarrow \ldots$" denotes the assignment of some value to variable v_i, whereas "$\ldots v_i \ldots$" is just some arbitrary operation using v_i. Part **b** shows an optimal colouring of the corresponding interference graph

In these cases, the more complicated process of *liveness analysis* will be needed for determining the live ranges of each variable. This could result in interference graphs of any arbitrary topology (see also [1]).

1.2   Why Colouring?

We have now seen four practical applications of the graph colouring problem. But why exactly is it concerned with "colouring" vertices? According to legend, the graph colouring problem was first noted in 1852 by a student of University College London, Francis Guthrie (1831–1899), who, while colouring a map of the counties of England, noticed that only four colours were needed to ensure that all neighbouring counties were allocated different colours.

To show how the colouring of maps relates to the colouring of vertices in a graph, consider the example map of the historical counties of Wales given in Fig. 1.7a. This particular map involves 16 "regions", including 14 counties, the sea on the left and England bordering on the right. Figure 1.7d shows that this map can indeed be coloured using just four colours (light grey, dark grey, black, and white). But how does the graph colouring problem itself inform this process?

As shown in Fig. 1.7b, we begin by placing a single vertex in the centre of each region of the map. Next, edges are drawn between any pair of vertices whose regions are seen to share a border. Thus, for example, the vertex appearing in England on the right will have edges drawn to the seven vertices in the seven neighbouring Welsh counties and also to the vertex appearing in the sea on the far left. If we take care in drawing these edges, we will always be able to draw a graph from a map so that no pair of edges need cross one another. Technically speaking, a graph that can be drawn with no crossing edges is known as a *planar graph*, of which Fig. 1.7c is an example.

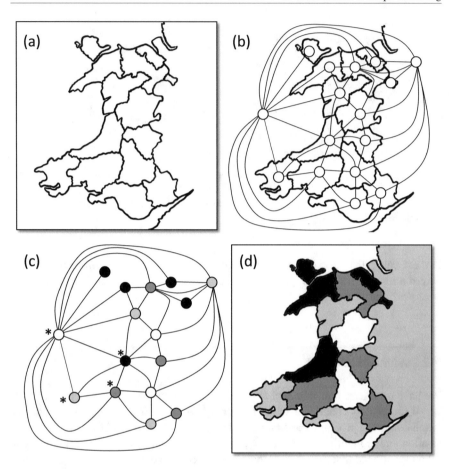

Fig. 1.7 Illustration of how graphs can be used to colour the regions of a map

Figure 1.7c illustrates how we might colour this planar graph using just four colours. The counties corresponding to these vertices can then be allocated the same colours in the actual map of Wales, as shown in Fig. 1.7d.

We might now ask whether we always need to use exactly four colours to successfully colour a map. In some cases, such as a map depicting a single island region surrounded by sea, fewer than four colours will be sufficient. On the other hand, for the map of Wales shown in Fig. 1.7, we can deduce that exactly four colours will be needed, (a) because a solution using four colours has already been constructed (as shown in the figure), and (b) because a solution using three or fewer colours is impossible. The latter point is because the planar graph in Fig. 1.7c contains a set of four vertices that each have an edge between them (indicated by asterisks in this figure). This tells us that different colours will be needed for each of these vertices.

The fact that, as Francis Guthrie suspected, four colours turn out to be sufficient to colour *any* map (or, equivalently, four colours are sufficient to colour any planar graph) is due to the celebrated Four Colour Theorem, which was ultimately proved in 1976 by Kenneth Appel and Wolfgang Haken of the University of Illinois—a full 124 years after it was first conjectured (see Sect. 6.1). However, it is important to stress at this point that the Four Colour Theorem does not apply to *all* graphs, but only to *planar* graphs. What can we say about the number of colours that are needed for colouring graphs that are not planar? Unfortunately, as we shall see, in these cases we do not have the luxury of a strong result like the Four Colour Theorem.

1.3 Problem Description

We are now in a position to define the graph colouring problem more formally. In graph theory, a graph G is usually defined by a pair of sets, V and E. The set V gives the names of all vertices in the graph, whereas E defines the set of edges.

Unless stated otherwise, in this book we will restrict our discussions to *simple* graphs. These are undirected graphs in which loops and multiple edges between vertices are forbidden. Consequently, each element of E is written as an unordered pair of vertices $\{u, v\}$ indicating the existence of an edge between vertices u and v (where $u \in V$, $v \in V$ and $u \neq v$). The number of vertices in G is denoted by n, while the number of edges is given by m.

To illustrate these ideas, the graph $G = (V, E)$ in Fig. 1.8 has a vertex set V containing $n = 10$ vertices as follows:

$$V = \{v_1, v_2, v_3, v_4, v_5, v_6, v_7, v_8, v_9, v_{10}\}.$$

The edge set E of this graph then contains $m = 21$ different edges as follows:

$$E = \{\{v_1, v_2\}, \{v_1, v_3\}, \{v_1, v_4\}, \{v_1, v_6\}, \{v_1, v_7\}, \{v_2, v_5\},$$
$$\{v_3, v_4\}, \{v_3, v_6\}, \{v_3, v_7\}, \{v_4, v_5\}, \{v_4, v_6\}, \{v_4, v_7\},$$
$$\{v_4, v_8\}, \{v_5, v_7\}, \{v_5, v_8\}, \{v_5, v_{10}\}, \{v_6, v_7\}, \{v_6, v_9\},$$
$$\{v_7, v_9\}, \{v_8, v_{10}\}, \{v_9, v_{10}\}\}.$$

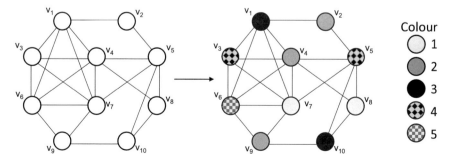

Fig. 1.8 A simple graph with $n = 10$ vertices and $m = 21$ edges, together with a corresponding five-colouring

Given a graph $G = (V, E)$, relationships between vertices and edges can be described using the following terms.

Definition 1.1 If $\{u, v\} \in E$, then the vertices u and v are said to be *adjacent*, else they are *nonadjacent*. Vertices u and v are also said to be the *endpoints* of the edge $\{u, v\} \in E$. An edge $\{u, v\} \in E$ is also said to be *incident* to the vertex u (and likewise for v).

Having gone over these basic definitions, we are now in a position to formally state the graph colouring problem.

Definition 1.2 Given a graph $G = (V, E)$, the *graph colouring problem* involves assigning each vertex $v \in V$ an integer $c(v) \in \{1, 2, \ldots, k\}$ such that

- $c(u) \neq c(v) \forall \{u, v\} \in E$; and
- k is minimal.

In this interpretation, instead of using actual colours such as grey, black, and white to colour the vertices, we use the labels 1, 2, 3, up to k. If we have a solution in which a vertex v is assigned to, say, colour 4, then this is then written by $c(v) = 4$. According to the first bullet above, pairs of vertices in G that are adjacent must be assigned to different colours. The second bullet then states that we are seeking to minimise the number of different colours being used.

Figure 1.8 also shows a five-colouring of our example graph. Using the above notation, this solution can be written as follows:

$$c(v_1) = 3, \ c(v_2) = 2, \ c(v_3) = 4, \ c(v_4) = 2, \ c(v_5) = 4,$$
$$c(v_6) = 5, \ c(v_7) = 1, \ c(v_8) = 1, \ c(v_9) = 2, \ c(v_{10}) = 3.$$

We now give some further definitions that help us to describe a graph colouring solution and its properties.

Definition 1.3 A colouring of a graph is called *complete* if all vertices $v \in V$ are assigned a colour $c(v) \in \{1, \ldots, k\}$; else the colouring is considered *partial*.

Definition 1.4 A *clash* describes a situation where a pair of adjacent vertices $u, v \in V$ are assigned the same colour (that is, $\{u, v\} \in E$ and $c(u) = c(v)$).

If a colouring has no clashes, then it is considered *proper*; else it is considered *improper*.

Definition 1.5 A colouring is *feasible* if and only if it is both complete and proper.

Definition 1.6 The *chromatic number* of a graph G, denoted by $\chi(G)$, is the minimum number of colours required in any feasible colouring of G. A feasible colouring of G using exactly $\chi(G)$ colours is considered *optimal*.

For example, the five-colouring shown in Fig. 1.8 is feasible because it is both complete (all vertices have been allocated colours) and proper (it contains no clashes). In this case, the chromatic number of this graph $\chi(G)$ is already known to be five, so the colouring can also be said to be optimal.

Some further useful definitions for the graph colouring problem involve colour classes and structures known as *cliques* and *independent sets*.

Definition 1.7 A *colour class* is a set containing all vertices in a graph that are assigned to a particular colour in a solution. That is, given a particular colour $i \in \{1, \ldots, k\}$, a colour class is defined by the set $\{v \in V : c(v) = i\}$.

Definition 1.8 An *independent set* is a subset of vertices $S \subseteq V$ that are mutually nonadjacent. That is, $\forall u, v \in S, \{u, v\} \notin E$.

Definition 1.9 A *clique* is a subset of vertices $S \subseteq V$ that are mutually adjacent: $\forall u, v \in S, \{u, v\} \in E$.

To illustrate these definitions, two example colour classes from Fig. 1.8 are $\{v_2, v_4, v_9\}$ and $\{v_7, v_8\}$. Example independent sets from this figure include $\{v_2, v_7, v_8\}$ and $\{v_3, v_5, v_9\}$. The largest clique in Fig. 1.8 is $\{v_1, v_3, v_4, v_6, v_7\}$, though numerous smaller cliques also exist, such as $\{v_6, v_7, v_9\}$ and $\{v_2, v_5\}$.

Given the above definitions, it is also useful to view graph colouring as a type of partitioning problem where a solution $\mathcal{S}$ is represented by a set of k colour classes

$\mathcal{S} = \{S_1, \ldots, S_k\}$. In order for $\mathcal{S}$ to be feasible it is then necessary that the following constraints be obeyed:

$$\bigcup_{i=1}^{k} S_i = V, \tag{1.1}$$

$$S_i \cap S_j = \emptyset \quad (1 \leq i \neq j \leq k), \tag{1.2}$$

$$\forall u, v \in S_i, \ \{u, v\} \notin E \ (1 \leq i \leq k), \tag{1.3}$$

with k being minimised. Here, Constraints (1.1) and (1.2) state that $\mathcal{S}$ should be a partition of the vertex set V (that is, all vertices should be assigned to exactly one colour class each). Constraint (1.3) then stipulates that no pair of adjacent vertices should be assigned to the same colour class (i.e., that all colour classes in the solution are independent sets). Referring to Fig. 1.8 once more, the depicted solution using this interpretation is now written $\mathcal{S} = \{\{v_1, v_{10}\}, \{v_2, v_4, v_9\}, \{v_3, v_5\}, \{v_6\}, \{v_7, v_8\}\}$.

1.4 About This Book

This book focuses on the problem of colouring *vertices* in graphs. Four different examples of this have already been discussed in this chapter. Sometimes the term "graph colouring" can also be used for the task of colouring the *edges* of a graph or the *faces* of a graph. However, as we will see in Chap. 6, edge and face colouring problems can easily be transformed into equivalent vertex-colouring problems using the concepts of line graphs and dual graphs, respectively. Consequently, unless explicitly stated otherwise, the term "graph colouring" in this book refers exclusively to vertex colouring.

This book is primarily concerned with algorithmic aspects of graph colouring. It focuses particularly on the characteristics of different heuristics for this problem and seeks answers to the following questions. Do these algorithms provide optimal solutions to some graphs? How do they perform on different topologies where the chromatic number is unknown? Why are some algorithms better for some types of graphs, but worse for others? What are the run time characteristics of these algorithms?

To help answer such questions, Chap. 2 of this book provides and in-depth treatment on algorithm complexity and $\mathcal{NP}$-completeness theory. This will help us to understand how algorithm efficiency can be gauged, and also see why graph colouring is such a difficult problem to solve to optimality. Chapters 3–5 then describe a number of different algorithms for graph colouring, ranging from simple constructive heuristics to complex metaheuristic- and mathematical programming-based techniques.

Another central aim of this book is to examine many of the real-world operational research problems that can be tackled using graph colouring techniques. As we will see, these include problems as diverse as the colouring of maps, the production of round-robin tournaments, solving Sudoku puzzles, assigning variables to computer registers, and checking for short circuits on circuit boards. These topics are considered in Chap. 6. Chapters 7–9 are also give in-depth treatments on the problems of designing seating plans, scheduling fixtures for sports leagues, and timetabling lectures at universities.

1.4.1 Algorithm Implementations

This book is accompanied by a suite of graph colouring algorithm implementations that can be downloaded from:

http://rhydlewis.eu/resources/gCol.zip

Each algorithm in this resource is described and analysed in this book. They are all are written in C++ and can be executed from the command line using common input and output protocols. A user manual and compilation instructions are provided in Appendix A. Readers are encouraged to make use of these algorithms on their graph colouring instances and are invited to modify the code in any way they see fit.

As we will see, when gauging the effectiveness of a graph colouring algorithm (or any algorithm for that matter) it is important to consider the amount of computational effort required to produce a solution of a given quality. Ideally, we should try to steer clear of measures such as wall-clock time or CPU time because these are largely influenced by the chosen hardware, operating systems, programming languages and compiler options. A more rigorous approach involves examining the number of *atomic operations* performed by an algorithm during execution. For classical computational problems such as searching or sorting the elements of a vector, these are usually considered to be the constant-time operations of comparing two elements and swapping two elements. For graph colouring algorithms, it is useful to follow this scheme by gauging computational effort via the number of *constraint checks* that are performed. Essentially, a constraint check occurs whenever an algorithm requests some information about a graph, such as whether two vertices are adjacent or not. We will define these operations presently, though it is first necessary to describe how graphs are to be stored in computer memory.

1.4.1.1 Representing Graphs

We saw in Sect. 1.3 that a graph G can be defined by a set V of n vertices and a set E of m edges. While the use of sets in this way is mathematically convenient, implementations of graph algorithms, including our own, usually make use of two different structures, namely *adjacency lists* and *adjacency matrices*. We now define these.

Definition 1.10 Given a graph $G = (V, E)$, an *adjacency list* is a vector of length n, where each element Adj_v is a list containing all vertices adjacent to vertex v, for all $v \in V$.

An example adjacency list for a ten-vertex graph is shown in Fig. 1.9. The length of an element, $|Adj_v|$, tells us the number of vertices that are adjacent to a vertex v. This is usually known as the *degree* of a vertex (see Definition 3.1). So, for example, vertex v_2 in this graph is seen to be adjacent to vertices v_1 and v_5 and therefore has a degree of 2. Note that the sum of all list lengths $\sum_{\forall v \in V} |Adj_v| = 2m$ since, if v appears in a vertex u's adjacency list, then u will also appear in v's adjacency list.

In algorithm implementations, adjacency lists are useful when we are interested in identifying all vertices that adjacent to a particular vertex v. On the other hand, they are less useful when we want to quickly answer the question "are vertices u and v adjacent?", as to do so would require searching through either Adj_u or Adj_v. For these situations, it is preferable to use an adjacency matrix.

Definition 1.11 Given a graph $G = (V, E)$, an *adjacency matrix* is a matrix $\mathbf{A}_{n \times n}$ for which $A_{ij} = 1$ if vertices v_i and v_j are adjacent, and $A_{ij} = 0$ otherwise.

An example adjacency matrix is also provided in Fig. 1.9. When considering graph colouring problems, note that edges are not directed, and graphs cannot contain loops. Consequently $\mathbf{A}$ is symmetric ($A_{ij} = A_{ji}$) and has zeros along its main diagonal ($A_{ii} = 0$).

When implemented, adjacency matrices require memory for storing n^2 pieces of information, regardless of the number of edges in the graph. Consequently, they can be quite bulky, particularly for sparse graphs.

Finally, as we will see in Chap. 5, our graph colouring algorithms will often attempt to improve solutions by changing the colours of certain vertices during a

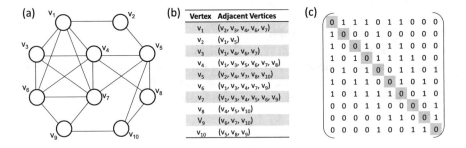

Fig. 1.9 A ten-vertex graph (**a**), its adjacency list (**b**), and its adjacency matrix (**c**)

run. In our implementations it is therefore also useful to make use of an additional matrix $\mathbf{C}_{n \times k}$ for representing graphs, where given a particular k-coloured solution $S = \{S_1, \ldots, S_k\}$, the element C_{vj} gives the number of vertices in colour class S_j that are adjacent to vertex v. Full descriptions of how this matrix is used are given in Sects. 5.1 and 5.2.

1.4.1.2   Measuring Computational Effort

Having specified how graphs are stored by our algorithm implementations, we are now in a position to define how constraint checks are counted during execution:

1. The task of checking whether two vertices u and v are adjacent is performed using the adjacency matrix $\mathbf{A}$. Accessing element A_{uv} counts as one constraint check.
2. The task of going through all vertices adjacent to a vertex v involves accessing all elements of the list Adj_v. This counts as $|Adj_v|$ constraint checks.
3. Determining the degree of a vertex v involves looking up the value $|Adj_v|$. This counts as one constraint check.
4. Determining the number of vertices in colour class $S_i \in S$ that is adjacent to a particular vertex v involves accessing element C_{vi}. This counts as one constraint check.

1.5   A Note on Pseudocode and Notation

While many of the algorithms featured in this book are described within the main text, others are more conveniently defined using pseudocode. The benefit of pseudocode is that it enables readers to concentrate on the algorithmic process without worrying about the syntactic details of any particular programming language. Our pseudocode style is based on that of the well-known textbook *Introduction to Algorithms* by Cormen et al. [2]. This particular pseudocode style makes use of all the usual programming constructs such as while-loops, for-loops, if-else statements, break statements, and so on, with indentation being used to indicate their scope. To avoid confusion, different symbols are also used for assignment and equality operators. For assignment, a left arrow ($\leftarrow$) is used. So, for example, the statement $x \leftarrow 10$ should be read as "x becomes equal to 10", or "let x be equal to 10". On the other hand, an equals symbol is used only for equality *testing*; hence a statement such as $x = 10$ will only evaluate to true or false (x is either equal to 10, or it is not).

All other notation used within this book is defined when the necessary concepts arise. Throughout the text, the notation $G = (V, E)$ is used to denote a graph G comprising a "vertex set" V and an "edge set" E. The number of vertices and edges in a graph is denoted by n and m, respectively. The colour of a particular vertex $v \in V$ is written $c(v)$, while a candidate solution to a graph colouring problem is usually defined as a partition of the vertices into k subsets $S = \{S_1, S_2, \ldots, S_k\}$.

A table of our notation is included in Appendix B. Further details can be found in the various definitions within this chapter and Chap. 3.

1.6 Chapter Summary

In this introductory chapter, we have defined the graph colouring problem and provided several examples of its practical applications. In the next chapter, we will consider the reasons as to why graph colouring should be considered an "intractable" problem. This characteristic implies that we are very unlikely to be able to create an algorithm that can find an optimal solution to an arbitrary graph in reasonable time frames. Heuristic algorithms for graph colouring will be considered in Chap. 3 onwards.

References

1. Chaitin G (2004) Register allocation and spilling via graph coloring. SIGPLAN Not. 39(4):66–74
2. Cormen T, Leiserson C, Rivest R, Stein C (2009) Introduction to algorithms, 3rd edn. The MIT Press

Problem Complexity

2

The previous chapter defined the graph colouring problem and gave some examples of practical applications. A question that we should now ask is "What algorithm can be used to solve this problem?" Here we use the word "solve" in the strong sense, in that an algorithm solves a problem only if it can take *any* problem instance and always return an optimal solution. For the graph colouring problem, this involves taking any graph G and returning a feasible solution using exactly $\chi(G)$ colours. Algorithms that solve a problem in this way are known as *exact* algorithms.

In this chapter, we will see that an exact algorithm with acceptable run times almost certainly does not exist for the graph colouring problem. Graph colouring can, therefore, be considered a type of "intractable" problem that will usually need to be tackled using inexact algorithms. To reach this conclusion, this chapter begins by first providing an overview of how algorithm time requirements are measured (Sect. 2.1). In Sect. 2.2, we then consider an intuitive algorithm for graph colouring that can find the optimal solution by going through every possible assignment of colours to vertices. While this algorithm is indeed exact, we will see that it is far too slow to be of any practical use.

In Sect. 2.3 we define the notion of problem intractability more rigorously. In particular, we consider the contrasting classes of polynomially solvable problems and $\mathcal{NP}$-complete problems. Several examples are then introduced to help describe the concepts of $\mathcal{P}$ and $\mathcal{NP}$, polynomial-time reductions, and Boolean satisfiability. Section 2.4 then uses these concepts to prove that graph colouring is $\mathcal{NP}$-complete.

Despite being $\mathcal{NP}$-complete in the general case, there are still certain conditions that, when satisfied, can make graph colouring an easy problem to solve. Section 2.5 surveys some of these. Section 2.6 then summarises this chapter and makes suggestions for further reading.

R. M. R. Lewis, *Guide to Graph Colouring*, Texts in Computer Science,
https://doi.org/10.1007/978-3-030-81054-2_2

2.1 Algorithm Complexity and Big O Notation

When solving a computational problem such as graph colouring, we are usually interested in finding the most efficient algorithm for the task. The word "efficient" in our case means the fastest; hence, we are interested in assessing the amount of time that an algorithm will take to solve the given problem. In practice, the run time of an algorithm can depend on many things, including:

- The size of the problem being tackled;
- The programming language and compiler used for the implementation; and
- The available memory and speed of the hardware being used.

From an analytical point of view, it is usually enough to only consider the first of these factors. The *complexity* of an algorithm will then usually be measured in terms of the number of "basic operations" that it needs to carry out to solve the given problem. These basic operations are constant-time actions that are deemed necessary for the problem at hand and can include things such as memory lookups, comparisons and swaps. For graph colouring, these basic operations are the constraint checks described in Sect. 1.4.1.

· When analysing an algorithm's complexity, it is usual to consider its *worst-case* run times. That is, we are interested in determining the maximum number of basic operations that will need to be performed for input of a particular size. Using the worst case might seem rather pessimistic, but is useful for the following reasons.

- It gives us an upper bound on the amount of time that the algorithm will require.
- The worst case will often occur. For example, consider the problem of searching for a number x in an unordered list of n integers. If x is not present in the list, we will still need to go through the entire list (and perform n comparisons) to confirm this.
- For many problems, worst-case run times are equal to the average case and the best case. For example, in the task of calculating the maximum value in an unordered list of n numbers, $n - 1$ comparisons are required in all cases.

Generally, we are not interested in the *exact* number of operations that are performed by an algorithm; instead, we are concerned with the algorithm's *order of growth* with respect to problem size. For example, if the problem size is doubled, does the number of operations performed by the algorithm stay the same? Does it double? Or is it something else?

The worst-case complexity of an algorithm is therefore usually expressed using what we call "big O" notation. This is formally defined as follows.

> **Definition 2.1** A function $f(n) = \mathcal{O}(g(n))$ if there exist positive constants c
> and n_0 such that
> $$0 \leq f(n) \leq c \cdot g(n) \; \forall n \geq n_0.$$
> In this case we say that the function $f(n)$ is *of order* $g(n)$.

Another way of expressing this definition is to say that "$f(n)$ tends to $g(n)$ as n approaches infinity". In practice, this allows us to drop any multiplicative constants and lower order terms from a function. In analysing algorithm complexity, we can therefore say things like the following.

* If, for an input of size n, an algorithm needs to perform a maximum of $6n^4 + 2n^2 + 53$ operations, then the complexity of this algorithm is simply $\mathcal{O}(n^4)$.
* If an algorithm performs a maximum of $4n^2 - 2n + 2$ operations, then its complexity is $\mathcal{O}(n^2)$.
* If an algorithm performs a maximum of $100n$ operations, then its complexity is $\mathcal{O}(n)$.

Big O notation, therefore, gives us a convenient way of classifying algorithms according to their worst-case run times. In this regard, it also allows us to easily compare the efficiency of different algorithms for the same problem.

It is usual to classify algorithms into two types. *Polynomial time* algorithms are those whose complexities are $\mathcal{O}(p(n))$ for some polynomial function p. On the other hand, any algorithm whose complexity cannot be bounded as such is known as an *exponential-time* algorithm. Examples of common polynomial growth rates for algorithms include $\mathcal{O}(1)$ (constant-time algorithms); $\mathcal{O}(\lg n)$ (log-time algorithms); $\mathcal{O}(n)$ (linear-time algorithms); $\mathcal{O}(n \lg n)$ ("n-log-n" algorithms); and $\mathcal{O}(n^2)$ (quadratic-time algorithms). Common exponential growth rates include $\mathcal{O}(2^n)$, $\mathcal{O}(3^n)$, $\mathcal{O}(n!)$ and $\mathcal{O}(n^n)$.

Figure 2.1 illustrates the time requirements that occur for algorithms with differing growth rates over various input sizes n. From this, we see that it is desirable to use algorithms that feature low growth rates and, in particular, to try and avoid exponential-time algorithms whenever possible.

When an algorithm of complexity $\mathcal{O}(g(n))$ is known to solve all instances of a particular problem, we say that this problem is "solvable in $\mathcal{O}(g(n))$ time". However, we will see in this chapter that there are many problems for which no polynomial-time algorithms exist. Unfortunately for us, graph colouring happens to be one such problem.

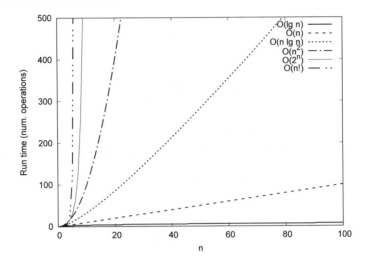

Fig. 2.1 Comparison of various growth rates according to increases in problem size n

2.2 Solving Graph Colouring via Exhaustive Search

Let us now consider the growth rate of one very simple (though ultimately foolhardy) algorithm for solving the graph colouring problem. The idea behind this method will be to define a solution space of candidate solutions, where each candidate solution specifies a different assignment of colours to vertices. Our task will then be to go through all members of this space and return the best among them; that is, the algorithm will return the candidate solution that is both feasible and seen to be using the fewest colours. Such a method is indeed guaranteed to return an optimal solution and is therefore exact. But will it be useful in practice?

If we were to follow such an approach, we would first need to decide the maximum number of colours that any solution might need. In practice, this could be estimated as some value between one and n, where n is the number of vertices in our graph. For now though, we will use the maximum number of colours n, since no feasible solution will ever require more colours than vertices.

Consider now the number of candidate solutions within this solution space. Since there would be n choices of colour for each of the n vertices, this gives a solution space containing a total of n^n candidate solutions, each that will need to be evaluated. Our algorithm will therefore feature an exponential complexity of $\mathcal{O}(n^n)$. This growth rate increases rapidly with regard to n, meaning that the number of candidate solutions to be checked will quickly become too large for even the most powerful computer to tackle. To illustrate, a graph with $n = 50$ vertices would lead to over $50^{50} \approx 8.8 \times 10^{84}$ different assignments: a truly astronomical number. This would make the task of creating and checking all of these assignments, even for this modestly sized problem, far beyond the computing power of all of the world's computers combined. (For comparison's sake, the number of atoms in the observable universe is thought to be around 10^{82}.) Also, even if we were to limit ourselves to candidate solutions using a maximum of $k < n$ colours labelled $1, \ldots, k$, this would still lead to a solution space containing $\mathcal{O}(k^n)$ candidate solutions. This function is still subject to an exponential growth rate for any $k > 1$.

A slightly better, though still ultimately doomed, approach might consider the symmetry that exists in the assignment method just described. Note that when we allocate "colours" to vertices in a graph, these are essentially arbitrary labels, and what we are more interested in is the *number of different colours* being used as opposed to what these labels actually are. To these ends, a solution such as

$$c(v_1) = 3, \ c(v_2) = 2, \ c(v_3) = 4, \ c(v_4) = 2, \ c(v_5) = 4,$$
$$c(v_6) = 5, \ c(v_7) = 1, \ c(v_8) = 1, \ c(v_9) = 2, \ c(v_{10}) = 3$$

can be considered identical to the solution

$$c(v_1) = 1, \ c(v_2) = 2, \ c(v_3) = 4, \ c(v_4) = 2, \ c(v_5) = 4,$$
$$c(v_6) = 5, \ c(v_7) = 3, \ c(v_8) = 3, \ c(v_9) = 2, \ c(v_{10}) = 1,$$

because the makeup of the five colour classes in both cases are equivalent, with only the labels of the colours being different (in this example, colours 1 and 3 have been swapped). Stated more precisely, this means that if we are given a candidate solution using k different colours with labels $1, 2, \ldots, k$, then the number of different assignments that would actually specify the same solution would be $k! = k \times (k - 1) \times (k - 2) \times \cdots \times 2 \times 1$, representing all possible assignments of the k labels to the colour classes. The space of all possible assignments of vertices to colours is therefore far larger (exponentially so) than it needs to be.

Perhaps a better approach than considering all possible assignments is to therefore turn to the alternative but equivalent statement of the graph colouring problem from Sect. 1.3. Here we will seek to partition the vertices into a set of k colour classes $S = \{S_1, \ldots, S_k\}$, where each set $S_i \in S$ is an independent set, and where k is minimised.

The number of ways of partitioning a set with n items into nonempty subsets is given by the nth Bell number, denoted by B_n. For example, the third Bell number $B_3 = 5$ because a set with three elements, v_1, v_2, and v_3, can be partitioned in five separate ways:

$$\{\{v_1, v_2, v_3\}\},$$
$$\{\{v_1\}, \{v_2\}, \{v_3\}\},$$
$$\{\{v_1\}, \{v_2, v_3\}\},$$
$$\{\{v_1, v_3\}, \{v_2\}\}, \text{ and}$$
$$\{\{v_1, v_2\}, \{v_3\}\}.$$

In contrast to the previous case where n^n assignments need to be considered, the use of sets here means that the *labelling* of the colour classes is not now relevant, meaning that the issues surrounding solution symmetry have been resolved. A suitable approach for colouring a graph with n vertices might now involve enumerating all B_n possible partitions, identifying those that only consist of independent sets, and then finding the solution amongst these that uses the smallest number of independent sets. Unfortunately, however, the growth rate of Bell numbers is still exponential at $\mathcal{O}(n^n)$, so such a method will still be infeasible for non-trivial values of n.

Alternatively, we might choose to limit the number of available colours to some value $k < n$ and then seek to partition the vertices into k colour classes. The number of ways of partitioning n items into exactly k nonempty subsets are given by Stirling numbers of the second kind. These are denoted by $\left\{{n \atop k}\right\}$, and can be calculated by the formula:

$$\left\{{n \atop k}\right\} = \frac{1}{k!} \sum_{i=0}^{k} (-1)^i \binom{k}{i} (k-i)^n. \tag{2.1}$$

So, for instance, the number of ways of partitioning three items into exactly two nonempty subsets is $\left\{{3 \atop 2}\right\} = 3$, because we have three different options:

$$\{\{v_1\}, \{v_2, v_3\}\},$$
$$\{\{v_1, v_3\}, \{v_2\}\}, \text{ and}$$
$$\{\{v_1, v_2\}, \{v_3\}\}.$$

Note that summing Stirling numbers of the second kind for all values of k from 1 to n leads to the nth Bell number:

$$B_n = \sum_{k=1}^{n} \left\{{n \atop k}\right\}. \tag{2.2}$$

We might now choose to employ an enumeration algorithm that starts by considering $k = 1$ colour. At each step, the algorithm then simply needs to check all $\left\{{n \atop k}\right\}$ possible partitions to see if any correspond to a feasible solution (k-colouring). If such a solution is found, the algorithm can halt immediately with the knowledge that an optimal solution has been found. Otherwise, k can be incremented by 1 and the process should be repeated. Ultimately such an algorithm will need to consider a maximum of

$$\sum_{k=1}^{\chi(G)} \left\{{n \atop k}\right\} \tag{2.3}$$

candidate solutions.

But is such an approach useful in any practical sense? Unfortunately not. Even though our original solution space size of n^n has been reduced, Stirling numbers of the second kind still exhibit an exponential growth rate of $\mathcal{O}(k^n)$. As an example, if we were seeking to produce and examine all partitions of 50 items into 10 subsets, which is again quite a modestly sized graph colouring problem, this would lead to $\left\{{50 \atop 10}\right\} \approx 2.6 \times 10^{43}$ candidate solutions. Such a figure is still well beyond the reach of any contemporary computing resources.

2.3   Problem Intractability

The discussions in the previous subsection have demonstrated that a graph colouring algorithm based on enumerating and checking all possible candidate solutions is not sensible because, on anything except trivial problem instances, its execution will simply take too long. However, the exponential growth rate of the solution space is not the sole reason why the graph colouring problem is so troublesome since many "easy to solve" problems also feature similarly large solution spaces. As an example, consider the computational problem of sorting a collection of integers into ascending order. Given a set of n unique integers, there are a total of $n!$ different ways of arranging these, and only one of these "candidate solutions" will give us the required answer. However, it would be foolish and unnecessary to employ an algorithm that went about checking all of the $n!$ possibilities, because a multitude of polynomial-time algorithms are available for the sorting problem, including the $\mathcal{O}(n \lg n)$ Merge Sort and Heap Sort algorithms. In this sense, we can say that the sorting problem is "solvable in $\mathcal{O}(n \lg n)$ time".

To isolate just what it is that makes a problem "intractable", it is, therefore, necessary to identify features beyond the number of possible candidate solutions. These features will be examined in the remainder of this subsection. To aid our discussions, we will consider five additional computational problems on graphs. These are defined as follows.

Definition 2.2 Consider a graph $G = (V, E)$.

1. Given two vertices $u, v \in V$, the *minimum distance problem* involves calculating the minimum number of edges in any path that connects u to v.
2. The *maximum clique problem* involves calculating the size of the largest clique in G. (See also Definition 1.9.)
3. The *Hamiltonian cycle problem* involves calculating a cycle in G that visits every vertex exactly once and that has no repeated edges.
4. The *maximum independent set problem* involves calculating the size of the largest independent set in G. (See also Definition 1.8.)
5. Assuming that integer weights are allocated to each edge in G, the *travelling salesman problem* involves calculating a Hamiltonian cycle in G whose edge-weight sum is minimal.

Examples of these five problems are provided in Fig. 2.2.

2.3.1 $\mathcal{P}$ and $\mathcal{NP}$

For the time being, we will restrict our discussions to so-called "decision problems". These are computational problems that have only two outcomes, denoted by "yes" and "no" (or, if we prefer, "true" and "false"). Although this might sound rather

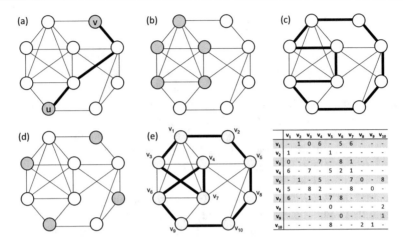

Fig. 2.2 Illustration of the problems given in Definition 2.2. Part **a** demonstrates the minimum distance problem, using bold edges to indicate a shortest path between vertices u and v. In this case, the minimum distance between u and v is three (edges). Part **b** illustrates the maximum clique problem. The shaded vertices indicate the largest clique in this graph. Part **c** demonstrates the Hamiltonian cycle problem, using bold edges to indicate that a Hamiltonian does indeed exist in this graph. Part **d** illustrates the maximum independent set problem. The largest independent set in this graph is shown by the shaded vertices. Finally, Part **e** gives an example of the travelling salesman problem. Edge weights are shown in the table. The bold edges in the graph give an optimal solution for this graph, which has an edge-weight total of nine

restrictive, most computational problems can be expressed as some form of decision problem, including all of the example problems listed in Definition 2.2. We now define these corresponding decision variants, using upper case letters for their names.

Definition 2.3 Consider a graph $G = (V, E)$. The following are all decision problems.

1. *k-DISTANCE*: Given two vertices $u, v \in V$ and an integer k, is there a path from u to v using k or fewer edges?
2. *k-CLIQUE*: Given an integer k, is there a clique in G with k or more vertices?
3. *HAM-CYCLE*: Does G contain a Hamiltonian cycle?
4. *k-I-SET*: Given an integer k, is there an independent set in G with k or more vertices?
5. *TSP*: Assuming that integer weights are allocated to all edges in G, and given an integer k, is there Hamiltonian cycle in G whose total edge weight is less than or equal to k?

Graph colouring can itself also be expressed as a decision problem as follows.

Definition 2.4 Consider a graph $G = (V, E)$. The decision problem k-*COL*
asks: Given an integer k, can a feasible colouring of G be achieved using k or
fewer colours?

We can now introduce the following definition concerning decision problems.

Definition 2.5 If a decision problem can be solved by a polynomial-time algorithm, then it is said to belong to the class $\mathcal{P}$. If a candidate solution to a particular decision problem can be *verified* in polynomial time, then it belongs to
the class $\mathcal{NP}$.[a]

[a]The initials $\mathcal{NP}$ stand for "nondeterministic polynomial time". This name originates from
early research that focussed on nondeterministic models of computation that do not need to
be considered here.

To exemplify this notion of polynomial-time verification, suppose that a friend
presented you with a colouring of a graph G and claimed that it was both feasible and
using k colours. It would be easy to verify the validity of their claim in polynomial
time by simply looking at each edge in G, checking that the colours of its endpoints
were different, and then tallying up the total number of colours used. If their claim
turned out to be true, this would confirm that, for this particular graph, k-COL is
answerable with "yes". In a similar fashion, if your friend was to give you a set of k
vertices and claim that they formed a clique in G, it would also be simple to verify
this claim (in polynomial time) by going through each pair of vertices u, v in the
set and checking for the presence of the corresponding edge $\{u, v\}$ in G. Again, if
their claim turned out to be correct, this would provide us with a "yes" answer for
this particular instance of k-CLIQUE. Similar arguments can be made for all of the
decision problems listed in Definition 2.3. Hence, we can confirm that they all belong
to the set $\mathcal{NP}$.

Amongst our example decision problems, we can also show that k-DISTANCE
is a member of the set $\mathcal{P}$. This is because the shortest path between any two vertices
in a graph can be calculated in polynomial time. An example of such an algorithm
is breadth-first search, which operates in $\mathcal{O}(n + m)$ time, where n and m are the
number of vertices and edges in the graph, respectively [1]. Importantly, note that
when a polynomial-time algorithm is available for solving a decision problem, then
this algorithm can also be used as the polynomial-time verification procedure. This
implies that if a problem belongs to $\mathcal{P}$ then it also belongs to $\mathcal{NP}$, meaning that
$\mathcal{P} \subseteq \mathcal{NP}$. Currently, it is not known whether $\mathcal{P} = \mathcal{NP}$, though most experts believe
that $\mathcal{NP}$ contains problems that are not in $\mathcal{P}$, implying $\mathcal{P} \neq \mathcal{NP}$. For the remainder
of this chapter, we will make this assumption. The uncertainty surrounding this issue
will be considered further in Sect. 2.6.

2.3.2 Polynomial-Time Reductions

If $\mathcal{P} \neq \mathcal{NP}$, then the differences between these two classes of problem are important. Indeed, all decision problems in $\mathcal{P}$ can be solved in polynomial time, whereas those in the set $\mathcal{NP} - \mathcal{P}$ cannot. But how can we show that a problem belongs to the set $\mathcal{NP} - \mathcal{P}$? One convenient way is to use what are known as polynomial-time reductions.

> **Definition 2.6** A *polynomial-time reduction* from a decision problem Π_1 to another decision problem Π_2 is a polynomial-time algorithm that transforms instances of problem Π_1 into instances of problem Π_2 such that Π_2 is answerable with "yes" if and only if Π_1 is answerable with "yes". If such a reduction exists, we write $\Pi_1 \propto \Pi_2$.

A polynomial-time reduction provides us with a method of transforming all instances of a problem Π_1 into a corresponding instance of another problem Π_2. This means that if we know that $\Pi_1 \propto \Pi_2$ and, in addition, we already have an algorithm that solves problem Π_2, then all instances of Π_1 can be solved by simply transforming them to a corresponding instance of Π_2. The statement $\Pi_1 \propto \Pi_2$ can therefore be interpreted in the following ways.

- Problem Π_1 is at most as hard as Π_2.
- Problem Π_1 is a special case of Π_2.
- Problem Π_2 is at least as hard as Π_1.

Figure 2.3 shows two simple examples of polynomial-time reductions using problems from Definition 2.3. The first of these illustrates how any instance of HAM-CYCLE can be converted into an instance of the TSP, therefore showing that HAM-CYCLE $\propto$ TSP. Consequently, a graph G will have a Hamiltonian cycle (and be answerable with "yes") if and only if G' has a TSP solution with a total edge weight of zero.

What does this relationship tell us? Suppose that an algorithm A can solve any instance of the TSP. This implies that we could also use A to solve any instance of HAM-CYCLE by simply applying the illustrated polynomial-time reduction first. Moreover, if algorithm A was known to operate in polynomial time (implying that TSP $\in \mathcal{P}$), then this would also mean that HAM-CYCLE $\in \mathcal{P}$, since a polynomial-time algorithm for HAM-CYCLE would involve simply applying the polynomial-time reduction and then executing A. (In fact, we will see in the next subsection that neither of these problems is actually in $\mathcal{P}$.)

The second example of Fig. 2.3 shows how instances of k-CLIQUE can be converted into instances of k-I-SET by taking the complement of the graph. Consequently, k-CLIQUE $\propto k$-I-SET. This process can also be carried out in reverse here, meaning that k-I-SET $\propto k$-CLIQUE. In this case, we can therefore say that k-CLIQUE and k-I-SET are *polynomially equivalent*.

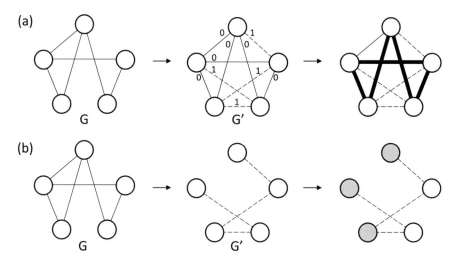

Fig. 2.3 Two examples of polynomial-time reductions on a graph $G = (V, E)$. Part **a** shows how HAM-CYCLE can be transformed into TSP. This involves creating a new graph $G' = (V, E')$ with edge weights of $w(\{u, v\}) = 0$ if $\{u, v\} \in E$ and $w(\{u, v\}) = 1$ if $\{u, v\} \notin E$. A TSP solution for G' with a total edge weight of zero corresponds to a Hamiltonian cycle, as indicated on the right. Part **b** shows how k-CLIQUE can be transformed into a corresponding instance of k-I-SET. This involves constructing the complement graph G' of G (that is, two vertices are adjacent in G' if and only if they are not adjacent in G). An independent set of k vertices in G' (as shown on the right) corresponds to a clique of size k in G

2.3.3 $\mathcal{NP}$-Completeness

We now turn our attention towards the set of so-called $\mathcal{NP}$-complete decision problems. When a problem is $\mathcal{NP}$-complete it is a member of $\mathcal{NP}$ and so candidate solutions can be verified in polynomial time. However, there is no known way of *solving* the problem in polynomial time. Instead, the time required to solve an $\mathcal{NP}$-complete problem grows exponentially with regards to problem size. In fact, Problems 2 to 5 from Definition 2.3 all turn out to be $\mathcal{NP}$-complete.

A formal definition of $\mathcal{NP}$-completeness can be made using polynomial-time reductions.

> **Definition 2.7** A decision problem Π is $\mathcal{NP}$-*complete* if $\Pi \in \mathcal{NP}$ and, for all other decision problems $\Pi' \in \mathcal{NP}$, $\Pi' \propto \Pi$.

In other words, a decision problem Π is $\mathcal{NP}$-complete only when every other decision problem in $\mathcal{NP}$ can be reduced to Π in polynomial time. If this is true, then Π can be considered "at least as hard" as all other problems in $\mathcal{NP}$. In addition, an algorithm able to solve Π could also be used to solve *all* problems in $\mathcal{NP}$.

x_1	x_2	$x_1 \wedge x_2$
T	T	T
T	F	F
F	T	F
F	F	F

AND operator

x_1	x_2	$x_1 \vee x_2$
T	T	T
T	F	T
F	T	T
F	F	F

OR operator

x_1	$\neg x_1$
T	F
F	T

NOT operator

Fig. 2.4 Truth tables for the AND, OR, and NOT operators

Definition 2.7 seems to suggest that in order for a problem to be proved $\mathcal{NP}$-complete, we need to show that every problem in $\mathcal{NP}$ is polynomially reducible to it. This formidable task can be avoided, however, by making use of the following theorem.

> **Theorem 2.1** Let $\Pi_1 \in \mathcal{NP}$, $\Pi_2 \in \mathcal{NP}$, and let Π_1 be $\mathcal{NP}$-complete. Then Π_2 is also $\mathcal{NP}$-complete whenever $\Pi_1 \propto \Pi_2$.

Proof Since Π_1 is $\mathcal{NP}$-complete, by Definition 2.7 we have $\Pi' \propto \Pi_1$ for all $\Pi' \in \mathcal{NP}$. Since $\Pi_1 \propto \Pi_2$, it is also true that $\Pi' \propto \Pi_2$ for all $\Pi' \in \mathcal{NP}$. Hence Π_2 is $\mathcal{NP}$-complete.   □

Theorem 2.1 implies that in order to show that a decision problem Π_2 is $\mathcal{NP}$-complete, we simply need to identify a single $\mathcal{NP}$-complete problem Π_1 for which $\Pi_1 \propto \Pi_2$. In other words, if we are given a decision problem Π_2 that we suspect might be $\mathcal{NP}$-complete, a convenient way of proving this is to identify another problem Π_1 that is already known to be $\mathcal{NP}$-complete, and then design a polynomial-time reduction method that converts all instances of Π_1 into corresponding instances of Π_2. Of course, to do this we must already have at least one problem that is known to be $\mathcal{NP}$-complete. One such problem is the Boolean satisfiability (SAT) problem, which we now consider.

2.3.4 Boolean Satisfiability Problems (SAT)

The Boolean satisfiability problem (SAT) is the problem of determining whether or not there exists an assignment of true and false values to the variables of a Boolean formula such that the overall formula evaluates to true. If this is the case, the formula is called *satisfiable*, otherwise it is *unsatisfiable*.

In SAT problems we can denote the Boolean variables by $x_1, x_2, \ldots, x_n$. The Boolean operators of interest are then AND, denoted by $\wedge$; OR, denoted by $\vee$; and the unary NOT operator, denoted by $\neg$. The truth tables for these operators are shown in Fig. 2.4.

To exemplify these ideas, consider the following small Boolean formula

$$x_1 \wedge \neg x_2.$$

This formula is satisfiable because an assignment of, say, true to x_1 and false to x_2 leads to

$$T \wedge \neg F = T \wedge T = T$$

as required. On the other hand, the formula

$$x_1 \wedge \neg x_1$$

is unsatisfiable, because all possible assignments to the variables eventually evaluate to false. Specifically,

$$T \wedge \neg T = T \wedge F = F$$

and

$$F \wedge \neg F = F \wedge T = F.$$

SAT was the first decision problem to be proven $\mathcal{NP}$-complete. This is due to the seminal work of Cook [2], who showed how any decision problem in $\mathcal{NP}$ can be converted in polynomial time to a corresponding instance of SAT such that the instance of SAT will be satisfiable if and only if the original decision problem is answerable with "yes". This allows us to consider SAT to be the "hardest" problem in $\mathcal{NP}$ since, if it were solvable in polynomial time, we would be able to solve *all* problems in $\mathcal{NP}$ in polynomial time.[1]

When studying SAT, it is often useful to state all Boolean formulas in 3-conjunctive normal form.

Definition 2.8 In a Boolean formula, a *literal* is an occurrence of a variable x_i, or its negation $\neg x_i$. A Boolean formula is considered to be in *3-conjunctive normal form* (3-CNF) if: (a) it is expressed as a series of AND clauses, and (b) each clause is the OR of exactly three distinct literals.

It is known that any Boolean formula can be converted into 3-CNF in polynomial time such that the original formula is satisfiable if and only if the corresponding 3-CNF formula is satisfiable. For example, the Boolean formula

$$(x_1 \wedge x_2 \wedge \neg x_3) \vee (\neg x_1 \wedge x_4) \vee x_5$$

---

[1]This result is sometimes referred to as the Cook–Levin theorem. Cook's proof involves using a mathematical model of computing known as a deterministic Turing machine, the details of which are given in the original manuscript.

can be written in 3-CNF as

$$(x_4 \vee x_1 \vee x_5) \wedge (\neg x_1 \vee x_2 \vee x_5) \wedge (x_4 \vee x_2 \vee x_5) \wedge (\neg x_1 \vee \neg x_3 \vee x_5) \wedge (x_4 \vee \neg x_3 \vee x_5),$$

which features five clauses. Note that when written in 3-CNF, a Boolean formula will be satisfiable only if all individual clauses evaluate to true. Also, individual clauses will evaluate to true only when at least one of its literals is true. In this particular example, the formula and its original are satisfiable by assigning all variables $x_1, \ldots, x_6$ to true.

As another example, consider the Boolean formula

$$x_1 \vee x_2 \vee x_3 \vee x_4 \vee x_5.$$

This can be written in 3-CNF as follows

$$(x_1 \vee x_2 \vee x_6) \wedge (\neg x_6 \vee x_3 \vee x_7) \wedge (\neg x_7 \vee x_4 \vee x_5)$$

though we note that this conversion involves introducing two additional dummy variables x_6 and x_7. These particular formulas are also satisfiable by setting all variables to true.

Now let 3-CNF-SAT be the decision problem of determining whether a Boolean formula in 3-CNF is satisfiable. Since, as we have noted, any Boolean formula can be converted into an equivalent 3-CNF formula in polynomial time, this tells us that SAT $\propto$ 3-CNF-SAT. Moreover, since SAT is $\mathcal{NP}$-complete, this also means that 3-CNF-SAT is $\mathcal{NP}$-complete.

2.4   Proofs of $\mathcal{NP}$-Completeness

We are now is a position to show that k-COL—the decision variant of the graph colouring problem—is $\mathcal{NP}$-complete. We begin by showing that this is true even when limited to just three colours. This result is then generalised to all values of $k \geq 3$.

> **Theorem 2.2** (Karp [3]) *3-COL is $\mathcal{NP}$-complete.*

Proof Given a candidate solution to 3-COL, it is easy to verify in polynomial time whether it represents a feasible colouring using three colours: we simply consider each edge in turn, check that its endpoints have different colours, and tally up the total number of colours used. Such a process can be achieved in $\mathcal{O}(m)$ time, where m is the number of edges in the graph. Hence 3-COL $\in \mathcal{NP}$.

We now show that 3-CNF-SAT $\propto$ 3-COL. That is, we show how any Boolean formula ϕ in 3-CNF can be converted into a corresponding graph G. A three-colouring of G is then possible if and only if ϕ is satisfiable.

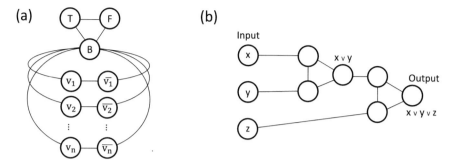

Fig. 2.5 a A partially constructed graph for showing 3-CNF-SAT $\propto$ 3-COL; **b** The OR-gadget graph

Let ϕ be an instance of 3-CNF-SAT, and let $C_1, C_2, \ldots, C_l$ be the clauses of ϕ defined over the variables $x_1, x_2, \ldots, x_n$. The constructed graph G should be such that a feasible three-colouring of G will give us the necessary truth values of $x_1, x_2, \ldots, x_n$ such that ϕ evaluates to true. To do this, let G start as a clique of three vertices. We label these vertices as T, F, and B, standing for true, false, and base respectively. Note that this clique requires three colours to be coloured feasibly. Without loss of generality assume that vertex T is assigned to the colour white, B to grey, and F to black. Next, we add two vertices labelled v_i and $\bar{v}_i$ for each variable x_i in ϕ. We also add edges $\{B, v_i\}$, $\{B, \bar{v}_i\}$, and $\{v_i, \bar{v}_i\}$ for each x_i.

An example of this process is shown in Fig. 2.5a. Note that in any feasible three-colouring of this current graph, the vertex labelled B will be unique in colour since it is adjacent to all other vertices. Moreover, each vertex v_i will have one of two colours, white or black. If v_i is white, then $\bar{v}_i$ will be black; if v_i is black, then $\bar{v}_i$ will be white. The colours of these vertices can therefore be interpreted as representing a truth assignment to all variables x_i in ϕ: if v_i is white, x_i is set to true; else x_i is set to false.

Further additions now need to be made to G to capture the satisfiability of each clause C_i in ϕ. To do this we need to introduce a small graph known as an OR-gadget for each C_i. An illustration of the OR-gadget for a clause $C_i = (x \vee y \vee z)$ is shown in Fig. 2.5b. Note that the OR-gadget graph has the following properties, which can be confirmed by inspection.

1. If, in a three-colouring of an OR-gadget, the input vertices (x, y, and z) are all coloured black (false), then the output vertex must also be coloured black.
2. If, in a three-colouring of an OR-gadget, any of the input vertices are coloured white (true), then there exists a three-colouring where the output vertex is also coloured white.

To finish the construction of G we now add an OR-gadget graph for each clause C_i in ϕ, using the appropriate vertices from $\{v_1, \bar{v}_1, \ldots, v_n, \bar{v}_n\}$ as the inputs. We then

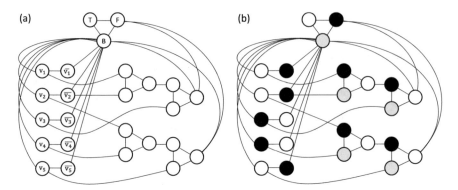

Fig. 2.6 Graph constructed from the 3-CNF Boolean formula $\phi = (x_1 \vee \neg x_2 \vee x_3) \wedge (x_2 \vee x_4 \vee \neg x_5)$. Part **b** shows a feasible three-colouring of this graph, using white for true and black for false. This colouring gives $x_1 = T$, $x_2 = T$, $x_3 = F$, $x_4 = F$, and $x_5 = T$ and therefore gives $(T \vee \neg T \vee F) \wedge (T \vee F \vee \neg T) = (T \vee F \vee F) \wedge (T \vee F \vee \neg F) = T \wedge T = T$. Hence ϕ is satisfiable as expected

connect the output vertices of these gadgets to the B and F vertices. A full example of this construction is shown in Fig. 2.6.

We now need to show that ϕ is satisfiable if and only if a feasible three-colouring of G is possible. First, suppose that ϕ is satisfiable. This means that in each clause $C_i = (x \vee y \vee z)$, at least one of x, y, or z must be true; consequently, at least one of the input vertices of the corresponding OR-gadget in G should be white. According to Property 2 above, this implies that the output vertex of this gadget can also be white. Since this output vertex is adjacent to the F and B vertices (coloured black and grey respectively), this three-colouring will be feasible.

Now suppose that G is three-colourable. If v_i is coloured white, then x_i is set to true, else x_i is set to false. This gives a legal truth assignment. Now consider the clause $C_i = (x \vee y \vee z)$. It cannot be that all input vertices to the corresponding OR-gadget are black (false) because, according to Property 1, this would force the output vertex to also be black. However, since this output vertex is adjacent to the B and F vertices, this would lead to an improper colouring, contradicting the assumption that a feasible three-colouring of G exists. □

Having shown that 3-COL is $\mathcal{NP}$-complete, it is now just a little more work to generalise this result to k-COL.

Theorem 2.3 (Karp [3]) *k-COL is $\mathcal{NP}$-complete for $k \geq 3$.*

Proof Checking whether a candidate solution represents a feasible k-colouring can be carried out in $\mathcal{O}(m)$ time, as described in Theorem 2.2. Hence k-COL $\in \mathcal{NP}$.

We now show that k-COL $\propto (k+1)$-COL. Because 3-COL is $\mathcal{NP}$-complete (as shown in Theorem 2.2), we also assume that $k \geq 3$.

Given a graph $G = (V, E)$, we can construct a new graph $G' = (V', E')$ such that G is k-colourable if and only if G' is $(k+1)$-colourable. To do this, G' is simply a copy of G with one additional vertex u that is made adjacent to all other vertices. That is $V' = V \cup \{u\}$ and $E' = E \cup \{\{u, v\} : v \in V\}$.

Now observe that if G is k-colourable, then G' is $(k + 1)$-colourable by using the additional colour for u. Conversely, if G' is $(k+1)$-coloured, removing u will reduce the number of colours by one, giving a k-colouring. Hence k-COL $\propto (k+1)$-COL.   □

To summarise these proofs, what we have done is first shown that any instance ϕ of 3-CNF-SAT can be converted in polynomial time to a graph G that is three-colourable if and only if ϕ is satisfiable. Consequently, we can say that 3-COL is *at least as hard* as 3-CNF-SAT. We have then shown how this result can also be extended to other values of $k \geq 3$. Again, this can be interpreted as saying that k-COL is at least as hard as 3-COL. These results imply that in the unlikely event that we were to discover an algorithm for solving k-COL in polynomial time, this would also allow us to solve both 3-CNF-SAT and SAT in polynomial time.

Similar proofs to that of Theorem 2.2 have also been used to show that 3-CNF $\propto$ HAM-CYCLE and that 3-CNF $\propto k$-CLIQUE [3]. The polynomial-time reduction methods described in Fig. 2.3 also demonstrate that HAM-CYCLE $\propto$ TSP and that k-CLIQUE $\propto k$-I-SET. These results allow us to conclude that Problems 2 to 4 from Definition 2.3 are all $\mathcal{NP}$-complete. A summary of how these results have been derived is shown in Fig. 2.7.

2.5 Graphs that are Easy to Colour Optimally

Although graph colouring should be considered an "intractable" problem, there are still various situations where a graph will be quite easy to colour optimally in polynomial time. For example, in the previous section, we saw that the decision problem

Fig. 2.7 Summary of the polynomial-time reductions discussed in this chapter

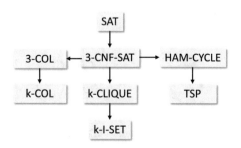

k-COL is $\mathcal{NP}$-complete for $k \geq 3$. What can we say about situations where $k = 1$ or $k = 2$?

For $k = 1$, consider a graph comprising a vertex set $V = \{v_1, \ldots, v_n\}$ and edge set $E = \emptyset$. Such graphs are commonly known as the *empty graph on n vertices*. It is easy see that a graph will be one-colourable if and only if it is an empty graph. Also, checking whether $E = \emptyset$ can be done in constant time; consequently, the decision problem 1-COL $\in \mathcal{P}$.

It is also straightforward to show that 2-COL $\in \mathcal{P}$ by observing that a graph will be two-colourable if and only if it is bipartite. Bipartite graphs, denoted by $G = (V_1, V_2, E)$, are graphs whose vertices can be partitioned into two sets V_1 and V_2 such that edges only exist between vertices in V_1 and vertices in V_2. As a result, V_1 and V_2 are independent sets and can each be allocated their own colour, giving an optimal solution. Figure 2.8 shows three examples of bipartite graphs and their optimal colourings: an arbitrary bipartite graph; a tree (that is, a graph containing no cycles); and a star graph. Algorithms such as the $\mathcal{O}(n + m)$ breadth-first search method can be used to determine whether a graph is bipartite. Polynomial-time graph colouring heuristics such as DSATUR and RLF are also suitable for this purpose, as we will see in Chap. 3.

In addition to situations where $k \leq 2$, various graph topologies can also be optimally coloured in polynomial time. The following subsections now describe some of these. In Chap. 3, we will also see some heuristic algorithms for graph colouring that, in addition to producing good results on arbitrary graphs, also turn out to be exact for some of these examples.

2.5.1   Complete Graphs

Complete graphs with n vertices, denoted by K_n, are graphs that feature an edge between every pair of vertices, giving a set E of $m = \binom{n}{2} = \frac{n(n-1)}{2}$ edges. It is obvious that because all vertices in the complete graph are mutually adjacent, all vertices must be assigned to their own individual colour. Hence the chromatic number of a complete graph $\chi(K_n) = n$. Example optimal solutions for K_1 to K_5 are shown in Fig. 2.9.

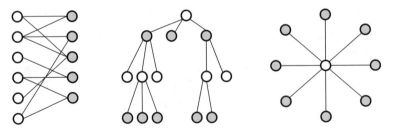

Fig. 2.8 Optimal colourings of (from left to right) an arbitrary bipartite graph, a tree, and a star graph

2.5.2 Cycle, Wheel, and Planar Graphs

Cycle graphs, denoted by C_n (where $n \geq 3$), comprise a set of vertices $V = \{v_1, \ldots, v_n\}$ and a set of edges $E = \{\{v_1, v_2\}, \{v_2, v_3\}, \ldots, \{v_{n-1}, v_n\}, \{v_n, v_1\}\}$. The cycle graphs C_3, C_4, C_8, and C_9 are shown in Fig. 2.10a.

Only two colours are needed to colour C_n when n is even (an even cycle); hence even cycles are a type of bipartite graph. However, three colours are needed when n is odd (an odd cycle). This is illustrated in the figure, where $\chi(C_4) = \chi(C_8) = 2$ but $\chi(C_3) = \chi(C_9) = 3$.

To explain this result, first consider the even cycle case. To construct a two-colouring we simply choose an arbitrary vertex and colour it white. We then proceed around the graph in a clockwise direction colouring the second vertex grey, the third vertex white, the fourth grey, and so on. When we reach the nth vertex, this can be coloured grey because the two vertices adjacent to it, namely the first and $(n-1)$th vertex will both be coloured white. Hence only two colours are required.

On the other hand, when n is odd (and $n \geq 3$), three colours will be required. Following the same pattern as the even case, an initial vertex is chosen and coloured white, with other vertices in a clockwise direction being assigned grey, white, grey, white, as before. However, when the nth vertex is reached, this will be adjacent to

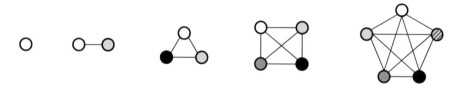

Fig. 2.9 Optimal colourings of the complete graphs (from left to right) K_1, K_2, K_3, K_4 and K_5

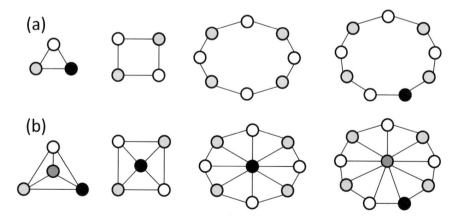

Fig. 2.10 Optimal colourings of **a** cycle graphs C_3, C_4, C_8, and C_9, and **b** wheel graphs W_4, W_5, W_9, and W_{10}

both the $(n-1)$th vertex (coloured grey), and the first vertex (coloured white). Hence a third colour will be required.

Wheel graphs with n vertices, denoted by W_n, are obtained from the cycle graph C_{n-1} by adding a single extra vertex v_n together with the additional edges $\{v_1, v_n\}, \{v_2, v_n\}, \ldots, \{v_{n-1}, v_n\}$. Example wheel graphs are shown in Fig. 2.10b. It is clear that similar results to cycle graphs can be stated for wheel graphs. Specifically, when n is odd, three colours will be required to colour W_n because the graph will be composed of the even cycle C_{n-1}, requiring two colours, and the additional vertex v_n which, being adjacent to all vertices in C_{n-1}, will require a third colour. Similarly, when n is even, $\chi(W_n) = 4$ because the graph will be composed of the odd cycle C_{n-1}, requiring three colours, together with vertex v_n, which will require a fourth colour.

From the illustrations in Fig. 2.10 we see that cycle graphs and wheel graphs (of any size) are both particular cases of planar graphs, in that they can be drawn on a two-dimensional plane without any of the edges crossing. This fits with the Four Colour Theorem, which states that if a graph is planar then it can be feasibly coloured using four or fewer colours (see Sects. 1.2 and 6.1). However, the Four Colour Theorem does not imply that if a graph is four-colourable then it must also be planar, as the next example illustrates.

2.5.3 Grid Graphs

Grid graphs can be formed by placing all vertices in a lattice formation on a two-dimensional plane. In a *sparse* grid graph, each vertex is adjacent to four vertices: the vertex above it, the vertex below it, the vertex to the right, and the vertex to the left (see Fig. 2.11a). For a *dense* grid graph, a similar pattern is used, but vertices are also adjacent to vertices on their surrounding diagonals (Fig. 2.11b).

A practical application of such graphs occurs in the arrangement of seats in exam venues. Imagine a large examination venue where the desks have been placed in a grid formation. In such cases, we might want to avoid instances of students copying from each other by making sure that each student is always seated next to students taking different exams. What is the minimum number of exams that can take place in the venue? This problem can be posed as a graph colouring problem by representing each desk as a vertex, with edges representing pairs of desks that are close enough for students to copy from.

If it is assumed that students can only copy from students seated in front, behind, to their left, or to their right, then we get the sparse grid graph shown in Fig. 2.11a. This graph is a type of bipartite graph since it can be coloured using just two colours according to the pattern shown. Hence a minimum of two exams can take place in this venue at any one time.

In circumstances where students can copy from students sitting on any of the eight desks surrounding them, we get the dense grid graph shown in Fig. 2.11b. As illustrated, this grid can be coloured using four colours according to the pattern shown. In this graph, each vertex, together with the vertex above, the vertex on the

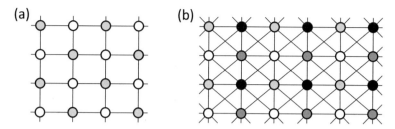

Fig. 2.11 Optimal colourings of **a** a sparse grid graph and **b** a dense grid graph

right, and the vertex on the upper right diagonal, forms a clique of size four. Hence we can conclude that a feasible colouring using fewer than four colours is not possible.

The dense grid graph also provides a simple example of a graph that is nonplanar but is still four-colourable. Although cliques of size 4 are themselves planar, the way in which the various cliques interlock in this example means that some edges will always need to cross one another. This illustrates that a graph does not need to be planar for it to be colourable with four or fewer colours.

2.6   Chapter Summary and Further Reading

This chapter has looked at the issues surrounding the intractable nature of various computational problems, including graph colouring. Much of this chapter's content is due to the work of Cook [2], who first identified the existence of $\mathcal{NP}$-complete problems, and Karp [3], who used polynomial-time reductions to prove the $\mathcal{NP}$-completeness of a whole host of combinatorial problems, including the examples considered here.

When talking about intractable problems, it is common to hear the term "$\mathcal{NP}$-hard" being used instead of $\mathcal{NP}$-complete. $\mathcal{NP}$-hard problems can be considered "at least as hard" as $\mathcal{NP}$-complete problems because they are not required to be in the class $\mathcal{NP}$ and therefore do not have to be stated as decision problems. $\mathcal{NP}$-hard problems will often be stated as an optimisation version of a corresponding $\mathcal{NP}$-complete problem. So with graph colouring for instance, rather than asking "is there a feasible colouring of G that uses k or fewer colours?" (to which the answer is "yes" or "no"), the problem will now be stated: "What is the minimum number of colours k needed to feasibly colour G, and how can we assign colours to the vertices of G to achieve this? Note that the non-decision version of the graph colouring algorithm as stated in Definition 1.2 is itself $\mathcal{NP}$-hard, as are Problems 2 to 5 from Definition 2.2.

When a problem is known to be $\mathcal{NP}$-complete (or $\mathcal{NP}$-hard) then there will be no algorithm that solves it in polynomial time. As we have discussed, this statement depends on the conjecture that $\mathcal{P} \neq \mathcal{NP}$. Though this inequality is generally thought to be true, the issue of formally proving (or disproving) it is actually one of the most famous unsolved problems in the mathematical sciences. Indeed, the Clay

Mathematics Institute has included the $P \neq NP$ question as one of their millennium prizes, offering one million US dollars to anyone who can offer a formal proof one way or the other (https://www.claymath.org/millennium-problems).

One way of disproving $P \neq NP$ would be to create a polynomial-time algorithm for an arbitrary NP-complete problem. If this could be done, then, as a consequence of Definition 2.7, *all* problems in NP could also be solved in polynomial time. Such a discovery would have profound scientific consequences, not least because it would lead to the availability of polynomial-time algorithms for large numbers of computational problems previously thought to be intractable, including graph colouring itself. Although now more than forty years old, an excellent textbook on these issues is due to Garey and Johnson [4]. This book also includes an appendix listing over 300 different NP-hard problems.

Although we should expect any exact algorithm for graph colouring to feature an exponential growth rate, research is still ongoing as to how to reduce these rates as far as possible. One exact algorithm for three-colouring is due to Beigel and Eppstein [5] and operates in $\mathcal{O}(1.3289^n)$ time. The basic idea is to carry out a series of trials in which each vertex is restricted to just two of the three possible colours. Each trial can then be solved in polynomial time by treating it as a 2-CNF-SAT problem. An algorithm for determining the chromatic number of a graph has also been proposed by Eppstein [6]. This operates by using dynamic programming to calculate the chromatic number of a graph together with all of its induced subgraphs and has a complexity of $O(2.4151^n)$.

Other ways of achieving exact algorithms for graph colouring are through integer programming and backtracking techniques. These will be considered further in Chaps. 4 and 5.

References

1. Cormen T, Leiserson C, Rivest R, Stein C (2009) Introduction to algorithms, 3rd edn. The MIT Press
2. Cook S (1971) The complexity of theorem-proving procedures. In: Proceedings of the third annual ACM symposium on theory of computing, STOC'71, New York, NY, USA. ACM, pp 151–158. https://doi.org/10.1145/800157.805047
3. Karp M (1972) Complexity of computer computations. In: Reducibility among combinatorial problems. Plenum, New York, pp 85–103
4. Garey M, Johnson D (1979) Computers and intractability: a guide to NP-completeness, 1st edn. W. H. Freeman and Company, San Francisco
5. Beigel R, Eppstein D (2005) 3-colouring in time $\mathcal{O}(1.3289^n)$. J Algorithms 54:168–204
6. Eppstein D (2003) Small maximal independent sets and faster exact graph coloring. J Graph Algorithms Appl 7(2):131–140

Bounds and Constructive Heuristics

3

In the previous chapter, we arrived at the inconvenient conclusion that no polynomial-time algorithm exists for solving the graph colouring problem.[1] The fact that graph colouring is "intractable" in this way implies that there is a limited amount that we can say about the chromatic number of an arbitrary graph. One simple rule is that, given a graph G with n vertices and m edges, if $m > \lfloor n^2/4 \rfloor$ then $\chi(G) \geq 3$, since any graph meeting this criterion must contain a triangle and therefore cannot be bipartite [1]. However, as we saw in the previous chapter, even the problem of deciding whether $\chi(G) = 3$ is $\mathcal{NP}$-complete for arbitrary graphs.

In this chapter, we will review several upper and lower bounds on the chromatic number. We will also examine five different constructive heuristics for the graph colouring problem. In general, a heuristic can be considered a type of algorithm that is designed to produce approximate solutions for a problem in short amounts of time. They, therefore, offer a practical alternative to exact algorithms, sacrificing optimality for speed. The heuristics described in this chapter—which include the well-known GREEDY, DSATUR and RLF methods—operate by assigning each vertex to a colour one at a time, using rules that attempt to keep the number of colours being used as small as possible. For certain topologies, these heuristics are exact, though in many other cases they produce suboptimal solutions. Towards the end of the chapter, we carry out an empirical comparison of these heuristics to provide information on their relative strengths and weaknesses.

At this point it is necessary to introduce some further graph terminology. Recall that a graph $G = (V, E)$ is defined by a vertex set V and an edge set E, where $|V| = n$ and $|E| = m$.

---

[1]Unless $\mathcal{P} = \mathcal{NP}$. See Sect. 2.6.

© The Author(s), under exclusive license to Springer Nature Switzerland AG 2021
R. M. R. Lewis, *Guide to Graph Colouring*, Texts in Computer Science,
https://doi.org/10.1007/978-3-030-81054-2_3

Definition 3.1 Given the graph $G = (V, E)$, the *neighbourhood* of a vertex v, written $\Gamma(v)$, is the set of vertices adjacent to v. That is, $\Gamma(v) = \{u \in V : \{v, u\} \in E\}$. The *degree* of a vertex v is the cardinality of its neighbourhood set, $|\Gamma(v)|$, and is usually written $\deg(v)$.

Definition 3.2 A graph $G' = (V', E')$ is a *subgraph* of G, denoted by $G' \subseteq G$, if $V' \subseteq V$ and $E' \subseteq E$. If G' contains all edges of G that join two vertices in V' then G' is said to be the graph *induced* by V'.

Definition 3.3 Let $W \subseteq V$, then $G - W$ is the subgraph obtained by deleting the vertices in W from G, together with the edges incident to them.

Definition 3.4 A *path* is a sequence of edges that connect a sequence of distinct vertices. A path between two vertices u and v is called a *uv-path*. If a uv-path exists between all pairs of vertices $u, v \in V$, then G is said to be *connected*; otherwise it is *disconnected*.

Definition 3.5 The *length* of a uv-path $P = (u = v_1, v_2, \ldots, v_l = v)$, is the number of edges it contains, equal to $l - 1$. The *distance* between two vertices u and v is the minimal path length between u and v.

Definition 3.6 A *cycle* is a uv-path for which $u = v$. All other vertices in the cycle must be distinct. A graph containing no cycles is said to be *acyclic*.

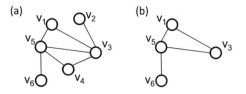

Fig. 3.1 **a** An example graph G, and **b** a subgraph G' of G

GREEDY $(G = (V, E))$
---
(1)   Let P be a permutation of the vertices in V
(2)   **for each** v **in** P **do**
(3) $c(v) \leftarrow j$, where j is the lowest colour label not assigned to any neighbours of v

Fig. 3.2   The GREEDY algorithm for graph colouring

Definition 3.7 The *density* of a graph $G = (V, E)$ is the ratio of the number of edges to the number of pairs of vertices. For a simple graph with no loops, this is calculated by $m / \binom{n}{2}$. Graphs with low densities are often referred to as *sparse* graphs; those with high densities are known as *dense* graphs.

To illustrate these definitions, Fig. 3.1a shows a graph G where, for example, the neighbourhood of v_1 is $\Gamma(v_1) = \{v_3, v_5\}$, giving $\deg(v_1) = 2$. The density of G is $7/15 = 0.467$. The subgraph G' in Fig. 3.1b has been created via the operation $G - \{v_2, v_4\}$. G' is also the subgraph induced from G using the set of vertices $\{v_1, v_3, v_5, v_6\}$. In this particular case both G and G' are connected. Paths in G from, for example, v_1 to v_6 include $(v_1, v_3, v_4, v_5, v_6)$ (of length 4) and (v_1, v_5, v_6) (of length 2). Since the latter path is also the shortest path between v_1 to v_6, the distance between these vertices is also 2. Cycles also exist in both G and G', such as (v_1, v_3, v_5, v_1).

3.1   The Greedy Algorithm for Graph Colouring

Recall the example from Sect. 1.1.1 where we sought to partition some students into a minimal number of groups for a team-building exercise. The process we used to try to achieve this is known as the GREEDY algorithm for graph colouring. Pseudocode for this very simple constructive heuristic is shown in Fig. 3.2.

As shown, GREEDY operates by taking vertices one at a time according to some predefined ordering P. Using colour labels $1, 2, 3, \ldots$, each vertex is then simply assigned to the lowest colour not being used by any of its neighbours.

An example run of GREEDY using the vertex permutation $P = v_1, v_2, \ldots, v_n$ is shown in Fig. 3.3. In the first iteration we see that v_1 is assigned to colour 1 (that is,

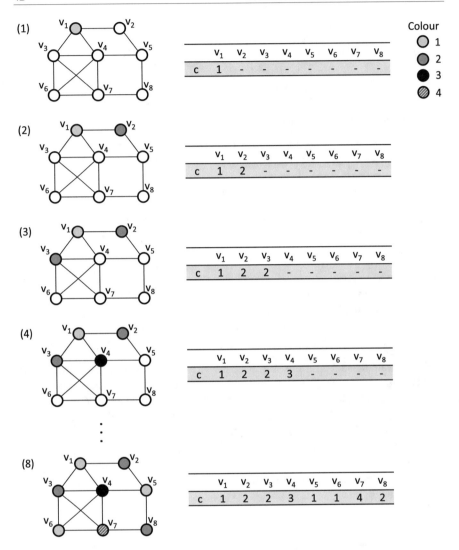

Fig. 3.3 Example application of the GREEDY algorithm using the permutation $P = v_1, v_2, \ldots, v_8$. Here, uncoloured vertices are shown in white. As a partition, the final solution is written $\mathcal{S} = \{\{v_1, v_5, v_6\}, \{v_2, v_3, v_8\}, \{v_4\}, \{v_7\}\}$, where each $S \in \mathcal{S}$ is an independent set

$c(v_1)$ is set to 1). In the second iteration, v_2 is adjacent to a vertex with colour 1, so it is assigned to colour 2. The same thing happens in the third iteration: v_3 is seen to be adjacent to colour 1, so it is also assigned to colour 2. In the fourth iteration, v_4 is seen to be adjacent to colours 1 and 2 and is therefore assigned to colour 3. The process continues in this fashion until all vertices have been coloured. This gives the four-colouring shown in the figure.

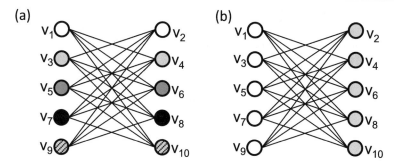

Fig. 3.4 Two different colourings of a bipartite graph achieved by the GREEDY algorithm. The graph in **a** is using $n/2$ colours. Part **b** shows an optimal two-colouring

The solution produced by GREEDY in Fig. 3.3 actually happens to be optimal; indeed, in Theorem 3.2 we will show that there always exists a permutation of a graph's vertices that will allow GREEDY to form an optimal solution. On the other hand, some permutations can produce very poor solutions. For example, consider the bipartite graph $G = (V_1, V_2, E)$ in which n is even and where the vertex sets and edge set are defined $V_1 = \{v_1, v_3, \ldots, v_{n-1}\}$, $V_2 = \{v_2, v_4, \ldots, v_n\}$, and $E = \{\{v_i, v_j\} : v_i \in V_1 \wedge v_j \in V_2 \wedge i + 1 \neq j\}$. Figure 3.4 shows an example of such a graph using $n = 10$. For $n \geq 4$, such graphs will have a chromatic number $\chi(G) = 2$; however, observe that an application of GREEDY using the permutation $P = v_1, v_2, v_3, \ldots, v_n$ will actually lead to a solution using $n/2$ colours, as Fig. 3.4a illustrates. On the other hand, the permutation $P = v_1, v_3, \ldots, v_{n-1}, v_2, v_4, \ldots, v_n$ will give the optimal solution shown in Fig. 3.4b. Clearly then, the order that the vertices are fed into the GREEDY algorithm can be very important with regards to the quality of the solution that results.

One useful feature of the GREEDY algorithm involves using existing feasible colourings of a graph to help generate new permutations of the vertices that can then be fed back into the algorithm. Consider the situation where we have a feasible colouring S of a graph G. Consider further a permutation P of G's vertices that is generated such that the vertices occurring in each colour class of S are placed into adjacent locations in P. If we now use this permutation with GREEDY, the result will be a new solution S' that uses no more colours than S, but possibly fewer. This is stated more concisely as follows.

Theorem 3.1 *Let S be a feasible colouring of a graph G. If each colour class $S_i \in S$ (for $1 \leq i \leq |S|$) is considered in turn, and all vertices of S_i are fed one by one into the Greedy algorithm, the resultant solution S' will also be feasible, with $|S'| \leq |S|$.*

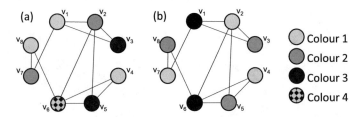

Fig. 3.5 Feasible four- and three-colourings of a graph

Proof Because $\mathcal{S} = \{S_1, \ldots, S_{|\mathcal{S}|}\}$ is a feasible solution, each $S_i \in \mathcal{S}$ is an independent set. Obviously, any subset $T \subseteq S_i$ is also an independent set. Now consider an application of GREEDY using $\mathcal{S}$ to build a new candidate solution $\mathcal{S}'$. In applying this algorithm, each set $S_1, \ldots, S_{|\mathcal{S}|}$ is considered in turn, and all vertices $v \in S_i$ are assigned one by one to some set $S'_j \in \mathcal{S}'$ according to the rules of GREEDY (that is, v is first considered for inclusion in S'_1, then S'_2, and so on). Considering each vertex $v \in S_i$, two situations and resultant actions will occur in the following order of priority:

Case 1: An independent set $S'_{j<i} \in \mathcal{S}'$ exists such that $S'_j \cup \{v\}$ is also an independent set. In this case v will be assigned to the jth colour class in $\mathcal{S}'$.

Case 2: An independent set $S'_{j=i} \in \mathcal{S}'$ exists such that $S'_j \cup \{v\}$ is also an independent set.

In both cases, it is clear that v will always be assigned to a set in $\mathcal{S}'$ with an index that is less than or equal to that of its original set in $\mathcal{S}$. Of course, if a situation arises by which *all* items in a particular set S_i are assigned according to Case 1, then at the termination of GREEDY, $\mathcal{S}'$ will contain fewer colours than $\mathcal{S}$.

Now assume that it is necessary to assign a vertex $v \in S_i$ to a set $S'_{j>i}$. For this to occur, it is first necessary that the proposed actions of Cases 1 and 2 (i.e., adding v to a set $S'_{j\leq i}$) cause a clash. However, $S'_i \subset S_i$ and is, therefore, an independent set. By definition, $S'_i \cup \{v\} \subseteq S_i$ is also an independent set, contradicting the assumption. $\square$

To show these concepts in action, the colouring shown in Fig. 3.5a has been generated by the GREEDY algorithm using the permutation $P = v_1, v_2, v_3, v_4, v_5, v_6, v_7, v_8$. This gives the four-colouring $\mathcal{S} = \{\{v_1, v_4, v_8\}, \{v_2, v_7\}, \{v_3, v_5\}, \{v_6\}\}$. This solution might then be used to form a new permutation $P = v_1, v_4, v_8, v_2, v_7, v_3, v_5, v_6$ which can then be fed back into GREEDY. However, our use of sets in defining a solution $\mathcal{S}$ means that we are free to use any ordering of the colour classes in $\mathcal{S}$ to form P, and indeed any ordering of the vertices within each colour class. One alternative permutation of the vertices formed from $\mathcal{S}$ in this way is $P = v_2, v_7, v_5, v_3, v_6, v_4, v_8, v_1$. This permutation has been used with GREEDY to give the solution shown in Fig. 3.5b, which we see is using fewer colours than the solution from which it was formed.

These concepts demonstrate the following theorem:

Theorem 3.2 *Let G be graph with an optimal graph colouring solution S. Then there are at least*

$$\chi(G)! \prod_{i=1}^{\chi(G)} |S_i|! \tag{3.1}$$

permutations of the vertices which, when fed into GREEDY, *will result in an optimal solution.*

Proof This arises immediately from Theorem 3.1. Since S is optimal, an appropriate permutation can be generated from S in the manner described. Moreover, because the colour classes and vertices within each colour class can themselves be permuted, the above formula holds. □

Note that if $\chi(G) = 1$ or $\chi(G) = n$ then, trivially, the number of permutations decoding into an optimal solution will be $n!$. That is, for empty and complete graphs *every* permutation of the vertices will decode to an optimal colouring.

3.2 Bounds on the Chromatic Number

In this section, we now review some of the upper and lower bounds that can be stated about the chromatic number of a graph. Some of these bounds make use of the GREEDY algorithm in their proofs, helping us to further understand the behaviour of the algorithm. While these bounds can be quite useful, or even exact for some topologies, in other cases they are either too difficult to calculate or give us bounds that are too loose to be of much practical use. This latter point will also be demonstrated empirically in Sect. 3.5.

3.2.1 Lower Bounds

To start, we observe that if a graph G contains as a subgraph the complete graph K_k, then a feasible colouring of G will require at least k colours. Stating this in another way, let $\omega(G)$ denote the number of vertices contained in the largest clique in G (this is sometimes known as G's *clique number*). Since $\omega(G)$ different colours will be needed to colour this clique, we deduce that $\chi(G) \geq \omega(G)$.

From another perspective, we can also consider the independent sets of a graph. Let $\alpha(G)$ denote the *independence number* of a graph G, defined as the number of vertices contained in its largest independent set. In this case, $\chi(G)$ must be at least

$\lceil n/\alpha(G) \rceil$ since, to be less than this value would require an independent set that is larger than $\alpha(G)$.

These two bounds can be combined into the following:

$$\chi(G) \geq \max\{\omega(G), \lceil n/\alpha(G) \rceil\} \tag{3.2}$$

The accuracy of the bounds given in Eq. (3.2) will vary on a case to case basis. Their major drawback is that the tasks of calculating $\omega(G)$ and $\alpha(G)$ are themselves $\mathcal{NP}$-hard problems, namely, the maximum clique problem and the maximum independent set problem (see Definition 2.2). However, this does not mean that these bounds are useless: in some practical applications, the sizes of the largest cliques and/or independent sets might be quite obvious from the graph's topology, or even specified as part of the problem itself (see, for example, the sport scheduling models used in Chap. 8). In other cases, we might also be able to approximate $\omega(G)$ and/or $\alpha(G)$ using heuristics or by applying probabilistic arguments.

To illustrate how we might estimate the size of a maximum clique in probabilistic terms, consider a graph G that is generated such that each pair of vertices is joined by an edge with probability p. Assuming independence, the probability that a subset of $x \leq n$ vertices forms a clique K_x is calculated by $p^{\binom{x}{2}}$, since there are $\binom{x}{2}$ edges that are required to be present among the x vertices. The probability that the x vertices do not form a clique is therefore simply $1 - p^{\binom{x}{2}}$. Since there are $\binom{n}{x}$ different subsets of x vertices in G, the probability that none of these form cliques is calculated as $(1 - p^{\binom{x}{2}})^{\binom{n}{x}}$. Hence the probability that there exists at least one clique of size x in G is defined as

$$P(\exists K_x \subseteq G) = 1 - (1 - p^{\binom{x}{2}})^{\binom{n}{x}} \tag{3.3}$$

for $2 \leq x \leq n$.

In practice, we might use this formula to estimate a lower bound with a certain confidence. For example, we might say "with greater than 99% confidence we can say that G contains a clique of size y", where y represents the largest x value for which Eq. 3.3 is greater than 0.99. We might also collect similar information on the size of the largest maximum independent set in G by simply replacing p with $(1 - p)$ in the above formula. We must be careful in calculating the latter, however, because dividing n by an underestimation of $\alpha(G)$ could lead to an invalid bound that exceeds $\chi(G)$. We should also be mindful that, for larger graphs, the numbers involved in calculating Eq. (3.3) might be very large indeed, perhaps requiring rounding and introducing inaccuracies.

Even if we can estimate or determine values such as $\omega(G)$, we must still bear in mind that they can still give very weak lower bounds in many cases. Consider, for example, the graph shown in Fig. 3.6, known as the Grötzch graph. This graph is "triangle-free" in that it contains no cliques of size 3 or above; hence $\omega(G) = 2$. However, as illustrated in the figure, the chromatic number of the Grötzch graph is four: double the lower bound determined by $\omega(G)$. In fact, the Grötzch graph is the smallest graph in a set of graphs known as the Mycielskians, named after their discoverer Jan Mycielski [2]. Mycielskian graphs demonstrate the potential inaccuracies involved in using $\omega(G)$ as a lower bound by showing that, for any $q \geq 1$, there exists a graph G with $\omega(G) = 2$ but for which $\chi(G) > q$. Hence we

Fig. 3.6 Optimal four-colouring of the Grötzch graph

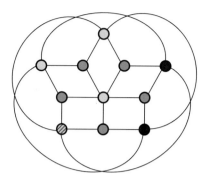

can encounter graphs for which $\omega(G)$ gives us a lower bound of 2, but for which the chromatic number is arbitrarily large.

3.2.1.1 Bounds on Interval and Chordal Graphs

While topologies such as the Mycielskian graphs demonstrate the potential for $\omega(G)$ to produce very poor lower bounds, in other cases this bound turns out to be both exact and easy to calculate. One practical application where this occurs is with *interval graphs*.

Given a set of intervals defined on the real line, an interval graph is defined as a graph in which adjacent vertices correspond to overlapping intervals. More formally:

Definition 3.8 Let $\mathcal{I} = \{I_1, \ldots, I_n\}$ be a set of intervals defined on the real line such that each interval $I_i = \{x \in \mathbb{R} : a_i \leq x < b_i\}$, where a_i and b_i define the start and end values of interval I_i. The *interval graph* of $\mathcal{I}$ is the graph $G = (V, E)$ for which $V = \{v_1, \ldots, v_n\}$ and where $E = \{\{v_i, v_j\} : I_i \cap I_j \neq \emptyset\}$.

An example interval graph has already been seen in Sect. 1.1.3 where we sought to assign taxi journeys with known start and end times to a minimal number of vehicles. This set of intervals, together with its interval graph is reproduced in Fig. 3.7. In this case, the real line is being used to represent time.

One feature of interval graphs is that they are known to contain a "perfect elimination ordering". This is defined as an ordering of the vertices such that, for every vertex, all of its neighbours to the left of it in the ordering from a clique.

Theorem 3.3 *Every interval graph G has a perfect elimination ordering.*

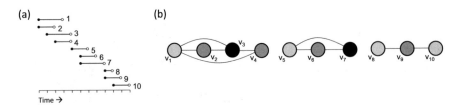

Fig. 3.7 An example set of intervals (**a**), and a corresponding interval graph (**b**). The left-to-right ordering of the vertices in this figure indicates a perfect elimination ordering $P = v_1, v_2, \ldots, v_{10}$

Proof To start, arrange the intervals of $\mathcal{I}$ in ascending order of start values, such that $a_1 \leq a_2 \leq \cdots \leq a_n$. Now label the vertices $v_1, v_2, \ldots, v_n$ to correspond to this ordering. This implies that for any vertex v_i, the corresponding intervals of all neighbours to its left in the ordering must contain the value a_i; hence all pairs of v_i's neighbours to the left must also share an edge, thereby forming a clique.   □

A perfect elimination ordering for our example interval graph is illustrated in Fig. 3.7. Here we see, for example, that v_4 forms a clique of size 3 with its neighbours v_1 and v_3, and that v_{10} forms a clique of size 2 with its neighbour v_9.

We now show that any graph featuring a perfect elimination ordering has a chromatic number $\chi(G) = \omega(G)$. In addition, all such graphs can be coloured optimally in polynomial time using the GREEDY algorithm.

Theorem 3.4 *Let G be a graph with a perfect elimination ordering $P = v_1, v_2, \ldots, v_n$. An optimal colouring for G is obtained by applying the GREEDY algorithm with the permutation P. Moreover, $\chi(G) = \omega(G)$.*

Proof During the execution of GREEDY, each vertex v_i is assigned to the lowest indexed colour not used by any of its neighbours preceding it in P. Clearly, each vertex has fewer than $\omega(G)$ neighbours that are already coloured. Hence at least one of the colours labelled 1 to $\omega(G)$ must be feasible for v_i. This implies that $\chi(G) \leq \omega(G)$. Since it is always necessary that $\omega(G) \leq \chi(G)$, this gives $\chi(G) = \omega(G)$. □

The optimal three-colouring provided in Fig. 3.7 shows the result of this process using the permutation $P = v_1, v_2, v_3, v_4, v_5, v_6, v_7, v_8, v_9, v_{10}$.

More generally, graphs featuring perfect elimination orderings are usually known as *chordal* graphs. Interval graphs are therefore a type of chordal graph. The problem of determining whether a graph is chordal or not can be achieved in polynomial time by algorithms such as lexicographic breadth-first search [3]. This implies that the problem of optimally colouring chordal graphs can be solved in polynomial time, making it a member of the problem class $\mathcal{P}$.

3.2.2   Upper Bounds

Upper bounds on the chromatic number can be derived by considering the number of edges in a graph and also the degrees of individual vertices. For instance, when a graph has a high density (that is, a high number of edges m in relation to the number of vertices n), a larger number of colours will often be needed because a greater proportion of the vertex pairs will need to be separated into different colour classes. This admittedly rather weak-sounding proposal gives rise to the following theorem.

Theorem 3.5 *Let G be a graph with chromatic number $\chi(G)$. Then*

$$\frac{\chi(G)(\chi(G) - 1)}{2} \leq m. \tag{3.4}$$

Proof Let $S = \{S_1, S_2, \ldots, S_{\chi(G)}\}$ be an optimal solution for G. In such a solution, there must be at least one edge between some vertex in S_i and some vertex in S_j, for all pairs $S_i, S_j \in S$. (If this was not the case, then S_i and S_j could be merged into a single independent set, meaning that S was not optimal). □

By solving this quadratic equation, this bound is more usefully stated as

$$\chi(G) \leq \frac{1}{2} + \sqrt{2m + \frac{1}{4}}. \tag{3.5}$$

Another upper bound can be derived by looking at the maximum degree of a graph.

Theorem 3.6 *Let G be a graph with maximal degree $\Delta(G)$ (that is, $\Delta(G) = \max\{\deg(v) : v \in V\}$). Then $\chi(G) \leq \Delta(G) + 1$.*

Proof Consider the behaviour of the GREEDY algorithm. Here, the ith vertex in the permutation P will be assigned to the lowest indexed colour class that contains none of its neighbouring vertices. Since each vertex has at most $\Delta(G)$ neighbours, no more than $\Delta(G) + 1$ colours will be needed to feasibly colour all the vertices of G. □

Another bound concerning vertex degrees can be calculated by examining all of a graph's subgraphs, identifying the minimal degree in each case, and then taking the maximum of these. For practical purposes this might be less useful than Theorem 3.6 for computing bounds quickly since the total number of subgraphs to consider might be prohibitively large. However, the following result is still useful, particularly when it comes to colouring planar graphs and graphs representing circuit boards (see Chap. 6).

> **Theorem 3.7** *Given a graph G, suppose that in every subgraph G' of G there is a vertex with degree less than or equal to a value δ. Then $\chi(G) \leq \delta + 1$.*

Proof We know there is a vertex with a degree of at most δ in G. Call this vertex v_n. We also know that there is a vertex of at most δ in the subgraph $G - \{v_n\}$, which we can label v_{n-1}. Next, we can label as v_{n-2} a vertex of degree of at most δ to form the graph $G - \{v_n, v_{n-1}\}$. Continue this process until all of the n vertices have been assigned labels. Now apply the GREEDY algorithm using the permutation $P = v_1, v_2, \ldots, v_n$. At each step of the algorithm, v_i will be adjacent to at most δ of the vertices $v_1, \ldots, v_{i-1}$ that have already been coloured; hence no more than $\delta + 1$ colours will be required.   □

Let us now examine some implications of the latter two theorems. It can be seen that Theorem 3.6 provides tight bounds for both complete graphs, where $\chi(K_n) = \Delta(K_n) + 1 = n$, and for odd cycles, where $\chi(C_n) = \Delta(C_n) + 1 = 3$. However, such accurate bounds will not always be so forthcoming. Consider, for example, the wheel graph comprising 100 vertices, W_{100}. This features a "central" vertex of degree $\Delta(W_{100}) = 99$, meaning that Theorem 3.6 merely informs us that the chromatic number of W_{100} is less than 100, even though it is actually just four! On the other hand, for any wheel graph, it is relatively easy to show that all of its subgraphs will contain a vertex with a degree of no more than 3 (i.e., $\delta = 3$). For wheel graphs where n is even, this allows Theorem 3.7 to return a tight bound since $\delta + 1 = \chi(W_n) = 4$.

3.2.2.1 Brooks' Bound

Beyond complete graphs and odd cycles, Theorem 3.6 can also be marginally strengthened due to the result of Brooks [4]. This proof is slightly more involved than that of Theorem 3.6 and requires some further definitions.

> **Definition 3.9** A *component* of a graph G is a set of vertices that are all connected by paths (that is, for every pair of vertices u, v in the component, there exists at least one uv-path). A *connected graph* is a graph comprising exactly one component.

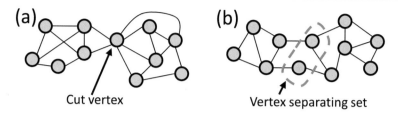

Fig. 3.8 Illustrations of a cut vertex and vertex separating set

Definition 3.10 A *cut vertex* (articulation point) v is a vertex whose removal from a graph G (together with all incident edges) increases the number of components. Thus a cut vertex of a connected graph is a vertex whose removal disconnects the graph. More generally, a *vertex separating set* of a graph G is a set of vertices whose removal increases the number of components.

Definition 3.11 A graph G is said to be *k-connected* if it remains connected whenever fewer than k vertices are removed. In other words, G will only become disconnected if a *vertex separating set* comprising k or more vertices is deleted.

Definition 3.12 A component of a graph is considered a *block* if it is 2-connected.

To illustrate these definitions, Fig. 3.8a shows a graph G comprising one component. Removal of the indicated cut vertex would split G into two components. Figure 3.8b can be considered a block in that it does not contain a cut vertex (i.e., it is 2-connected). However, it is not 3-connected, because removal of the two vertices in the indicated vertex separating set increases the number of components from one to two.

Having gone over the necessary terms, we are now in a position to state and prove Brooks' theorem. Observe that this theorem is concerned with connected graphs only. This is not restrictive, however, because if a graph G is composed of multiple components $G_1, \ldots, G_l$, then $\chi(G) = \max(\chi(G_1), \ldots, \chi(G_l))$; hence each component can be considered separately.

Theorem 3.8 (Brooks [4]) *Let G be a connected graph with maximal degree $\Delta(G)$. Suppose further that G is neither complete nor an odd cycle. Then $\chi(G) \leq \Delta(G)$.*

Proof The theorem is obviously correct for $\Delta(G) \leq 2$. For $\Delta(G) = 0$ and $\Delta(G) = 1$, the corresponding graphs will be the complete graphs K_1 and K_2, respectively, and are, therefore, not included in the theorem. For $\Delta(G) = 2$ on the other hand, G will be a path or even cycle (giving $\chi(G) = 2$) or will be an odd cycle, meaning it is not included in the theorem.

Assuming $\Delta(G) \geq 3$, let G be a counterexample with the smallest possible number of vertices for which the theorem does not hold, i.e., $\chi(G) > \Delta(G)$. We therefore assume that all graphs with fewer vertices than G can be feasibly coloured using $\Delta(G)$ colours.

Claim 1: G is connected. If G were not connected, then G's components would be a smaller counterexample, or all of G's components would be $\Delta(G)$-colourable.

Claim 2: G is 2-connected. If G were not 2-connected, then G would have at least one cut vertex v, and each block of G would be $\Delta(G)$-colourable. The colourings of each block could then be combined to form a feasible $\Delta(G)$-colouring.

Claim 3: G must contain three vertices v, u_1 and u_2 such that (a) u_1 and u_2 are nonadjacent; (b) both u_1 and u_2 are adjacent to v; and (c) $G - \{u_1, u_2\}$ is connected. Two cases must now be considered:

Case 1: G is 3-connected. Because G is not complete, there must be two vertices x and y that are nonadjacent. Let the shortest path between x and y in G be $x = v_0, \ldots, v_l = y$, where $l \geq 2$. Since this is the shortest xy-path, v_0 is not adjacent to v_2, so we can choose $u_1 = v_0$, $v = v_1$ and $u_2 = v_2$. This satisfies Claim 3.

Case 2: G is 2-connected but not 3-connected. In this case, there must exist two vertices u and v such that $G - \{u, v\}$ is disconnected. This means that the graph $G - \{v\}$ contains a cut vertex (i.e., u), but there is no cut vertex in G itself. In this case, v must be adjacent to at least one vertex in every block of the graph $G - \{v\}$. Let u_1 and u_2 be two vertices in two different end blocks of $G - \{v\}$ that are adjacent to v. The vertices u_1, u_2 and v now satisfy Claim 3.

Having proved Claims 1, 2, and 3, we now construct a permutation $P = p_1, p_2, \ldots, p_n$ of $G's$ vertices such that $p_1 = u_1$, $p_2 = u_2$, and $p_n = v$. The remaining parts of the permutation $p_3, \ldots, p_{n-1}$ are then formed such that, for $3 \leq i < j \leq n - 1$, the distance from p_n to p_i is greater than or equal to the distance from p_n to p_j (this can be determined using a breadth-first search tree rooted at v). If we now apply GREEDY using this permutation, the vertices $p_1 = u_1$ and $p_2 = u_2$ will first both be assigned to colour 1, because they are nonadjacent. Moreover, when we colour the vertices p_i (for $3 \leq i < n$), there will always be at least one colour $j \leq \Delta(G)$ that is feasible for p_i. Finally, when we come to colour vertex $p_n = v$, at most $\Delta(G) - 1$ colours will have been used to colour the neighbours of v (since its neighbours u_1 and u_2 have been assigned to the same colour) and so at least one of the $\Delta(G)$ colours will be feasible for v. This shows that $\chi(G) \leq \Delta(G)$ as required. $\qquad\square$

3.2.2.2 The Welsh–Powell Bound

One further upper bound on the chromatic number can be derived by considering the degree sequence of a graph. The proof of this bound is somewhat simpler than that of Brooks' bound.

Theorem 3.9 (Welsh and Powell [5]) *Given a graph G, assume that the vertices have been labelled such that* $\deg(v_1) \geq \deg(v_2) \geq \cdots \geq \deg(v_n)$. *Then,*

$$\chi(G) \leq \max_{i=1,\ldots,n} \min(\deg(v_i) + 1, i). \tag{3.6}$$

Proof Consider the behaviour of GREEDY using the permutation $P = v_1, v_2, \ldots, v_n$. When colouring vertex v_i, only vertices $v_1, \ldots, v_{i-1}$ have been coloured, so at most $i - 1$ colours are being used. Colour i is therefore available for v_i. Also, the number of different colours that are adjacent to v_i is at most $\deg(v_i)$. Hence GREEDY will assign v_i to a colour $j \leq \min(\deg(v_i) + 1, i)$. The maximum value for j across all vertices gives the upper bound. $\qquad\square$

To demonstrate this bound, consider the example graph G in Fig. 3.9 where, as required, vertices are labelled such that $\deg(v_1) \geq \deg(v_2) \geq \cdots \geq \deg(v_n)$. This gives rise to the following sequence of $(\deg(v_i) + 1, i)$ pairs:

$$(6, 1), (3, 2), (3, 3), (3, 4), (3, 5), (2, 6).$$

Taking the minimum of each pair leads to the sequence 1, 2, 3, 3, 3, 2, the maximum of which is 3. Hence $\chi(G) \leq 3$. In contrast, Brooks' bound only tells us that $\chi(G) \leq 5$ in this case.

Because the Welsh–Powell bound takes the minimum of each $(\deg(v_i) + 1, i)$ pair, it will always be at least as good as the upper bound $\chi(G) \leq \Delta(G) + 1$ seen in Theorem 3.6. Indeed, these two bounds will only be equal in situations where the graph G features at least $\Delta(G) + 1$ vertices with the maximum degree $\Delta(G)$. In these cases this implies the existence of a vertex v_i for which $i = \Delta(G) + 1$, meaning that $\min(\deg(v_i) + 1, i) = \Delta(G) + 1$. This occurs with cycle graphs, complete graphs, and (more generally) d-regular graphs.

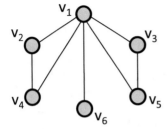

	v_1	v_2	v_3	v_4	v_5	v_6
deg	5	2	2	2	2	1

Fig. 3.9 Example graph G whose vertices are labelled such that $\deg(v_1) \geq \deg(v_2) \geq \cdots \geq \deg(v_n)$. The Welsh–Powell bound finds that $\chi(G) \leq \max_{i=1,...,n} \min(\deg(v_i) + 1, i) = 3$

WELSH-POWELL $(G = (V, E))$
(1) Relabel the vertices of G such that $\deg(v_1) \geq \deg(v_2) \geq ... \geq \deg(v_n)$
(2) Apply GREEDY using the permutation $P = v_1, v_2, ..., v_n$

Fig. 3.10 The WELSH–POWELL algorithm for graph colouring. The method GREEDY is defined in Fig. 3.2

The Welsh–Powell bound can also be used to define a simple constructive heuristic for graph colouring. This operates by arranging the vertices in non-ascending order of degree and then applying the GREEDY algorithm. The number of colours in solutions returned by this algorithm will never exceed the Welsh–Powell bound. The method is summarised in Fig. 3.10.

We have now analysed the behaviour of the GREEDY algorithm for graph colouring and reviewed several bounds on the chromatic number. In the next two sections, we consider three further heuristics for the graph colouring problem. As we will see, two of these algorithms, namely, DSATUR and RLF, are guaranteed to produce optimal solutions for some simple graph topologies. Often, they will also produce solutions that improve on the upper bounds mentioned above. Later, in Chaps. 4 and 5, we will see that these algorithms, along with GREEDY, can be used as building blocks in many of the more sophisticated algorithms available for graph colouring.

3.3 The DSATUR Algorithm

The DSATUR algorithm (abbreviated from "degree of saturation") was originally proposed by Brélaz [6]. Its behaviour is similar to the GREEDY algorithm in that it takes each vertex in turn and assigns it to the first suitable colour. The difference lies in the way that the vertex orderings are generated. As we have seen, with GREEDY the ordering is decided before any colouring takes place; on the other hand, for DSATUR the choice of which vertex to colour next is based on the characteristics of the current partial colouring and, particularly, the *saturation degrees* of the uncoloured vertices.

DSATUR $(G = (V, E))$
(1)   **for** $i \leftarrow 1$ **to** n **do**
(2)   Let v be the uncoloured vertex with the largest saturation degree. In case of ties, choose the vertex among these with the largest degree in the subgraph induced by the uncoloured vertices
(3) $c(v) \leftarrow j$, where j is the lowest colour label not assigned to any neighbours of v

Fig. 3.11   The DSATUR algorithm for graph colouring

Definition 3.13 Recall that $c(v)$ denotes the colour assigned to the vertex v, and let $c(v) = $ NULL for any vertex $v \in V$ not currently assigned to a colour. Given an uncoloured vertex v, the *saturation degree* of v, denoted by $\text{sat}(v)$, is the number of different colours assigned to adjacent vertices. That is, $\text{sat}(v) = |\{c(u) : u \in \Gamma(v) \wedge c(u) \neq \text{NULL}\}|$.

Pseudocode for the DSATUR algorithm is shown in Fig. 3.11. Much of the algorithm is the same as the GREEDY algorithm in that once a vertex has been selected, it is assigned to the lowest colour label not assigned to any of its neighbours.

Step (2) provides the main power behind the DSATUR algorithm in that it prioritises vertices that are seen to be the most "constrained"—that is, vertices that currently have the fewest colour options available to them. Consequently, these "more constrained" vertices are dealt with by the algorithm first, allowing the less constrained vertices to be coloured later.

Figure 3.12 shows an example application of DSATUR on a small graph. As shown, initially no vertices are coloured and so all vertices have a saturation degree equal to zero. The first vertex to be coloured is, therefore, v_4, which has the highest degree in the uncoloured subgraph. This is assigned to colour 1, as shown in Part (1). At this point, we now see that five vertices (v_1, v_3, v_5, v_6, v_7) feature the maximum saturation degree, so the next vertex to be chosen is the one among these with the highest degree in the uncoloured subgraph. This gives two options here, v_3 and v_7; consequently, one of these is chosen and then assigned to colour 2. The resultant colouring is shown in Part (2). This process continues in the same way until all vertices have been coloured.

Earlier we saw that the number of colours used in solutions produced by the GREEDY algorithm depends on the order that the vertices are fed into the procedure, with solution quality potentially varying a great deal. On the other hand, the DSATUR algorithm reduces this variance by generating the vertex ordering *during* a run according to its selection rules. As a result, DSATUR's performance is more predictable. One feature of the algorithm is that if a graph is composed of multiple components, then all vertices of a single component will be coloured before the other vertices are considered.

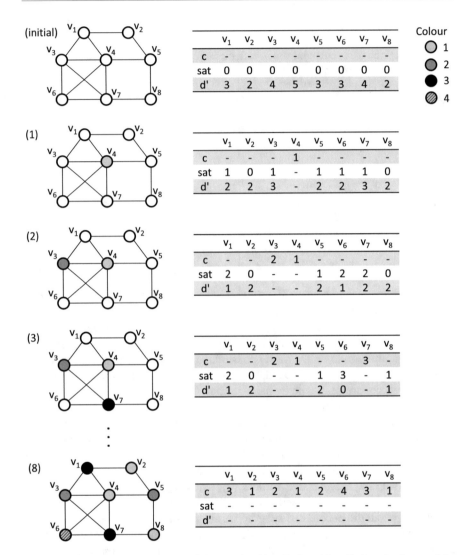

Fig. 3.12 Example application of the DSATUR algorithm. In the tables, d' gives the degree of the corresponding vertices in the uncoloured subgraph. Uncoloured vertices are shown in white

DSATUR is also exact for several elementary graph topologies. The first of these is the bipartite graph, and to prove this claim it is first necessary to show a well-known result on the structure of these graphs.

Theorem 3.10 *A graph is bipartite if and only if it contains no odd cycles.*

Proof Let G be a connected bipartite graph with vertex sets V_1 and V_2. (It is enough to consider G as being connected, as otherwise we could simply treat each component of G separately.) Let $(v_1, v_2, \ldots, v_l, v_1)$ be a cycle in G. We can also assume that $v_1 \in V_1$, $v_2 \in V_2$, $v_3 \in V_1$, and so on. Hence, a vertex $v_i \in V_1$ if and only if i is odd. Since $v_l \in V_2$, this implies l is even. Consequently G has no odd cycles.

Now suppose that G is known to feature no odd cycles. Choose any vertex v in the graph and let the set V_1 be the set of vertices such that the shortest path from each member of V_1 to v is of odd length, and let V_2 be the set of vertices where the shortest path from each member of V_2 to v is even. Observe now that there is no edge joining vertices of the same set V_i since otherwise G would contain an odd cycle. Hence G is bipartite.   □

This result allows us to prove the following theorem.

Theorem 3.11 (Brélaz [6]) *The* DSATUR *algorithm is exact for bipartite graphs.*

Proof Let G be a connected bipartite graph with $n \geq 3$. If G is not connected, it is enough to consider each component of G separately. For purposes of contradiction assume that one vertex v has a saturation degree of 2, meaning that v has two neighbours, u_1 and u_2, assigned to different colours. From these two neighbours, we can build two paths which, because G is connected, will have a common vertex u. Hence we have formed a cycle containing vertices v, u_1, u_2, u and perhaps others. Since G is bipartite, the length of this cycle must be even, meaning that the u_1 and u_2 must have the same colour, contradicting our initial assumption.   □

To illustrate the usefulness of this result, consider the bipartite graphs shown in Fig. 3.4 earlier. Here, many permutations of the vertices used in conjunction with the GREEDY algorithm will lead to colourings using more than two colours. Indeed, in the worst case they may even lead to $(n/2)$-colourings as demonstrated in the figure. In contrast, DSATUR is guaranteed to return the optimal solution for bipartite graphs, as it is for some further topologies.

Theorem 3.12 *The* DSATUR *algorithm is exact for cycle and wheel graphs.*

Proof Note that even cycles are two-colourable and are therefore bipartite. Hence they are dealt with by Theorem 3.11. However, it is useful to consider both even and odd cycles in the following.

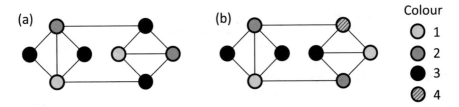

Fig. 3.13 Part **a** shows optimal three-colouring. Part **b** shows a suboptimal four-colouring produced by DSATUR

Let C_n be an uncoloured cycle graph. Since all degrees and saturation degrees are equal, the first vertex to be coloured, v, will be chosen arbitrarily by DSATUR. In the next $(n-2)$ steps, according to the behaviour of DSATUR a path of vertices of alternating colours will be constructed that extends from v in both clockwise and anticlockwise directions. At the end of this process, a path comprising $n-1$ vertices will have been formed, and a single vertex u will remain that is adjacent to both terminal vertices of this path. If C_n is an even cycle, $n-1$ will be odd, meaning that the terminal vertices have the same colour. Hence u can be coloured with the alternative colour. If C_n is an odd cycle, $n-1$ will be even, meaning that the terminal vertices will have different colours. Hence u will be assigned to a third colour.

For wheel graphs W_n a similar argument applies. Assuming $n \geq 5$, DSATUR will initially colour the central vertex v_n because it features the highest degree. Since v_n is adjacent to all other vertices in W_n, all remaining vertices $v_1, \ldots, v_{n-1}$ will now have a saturation degree of 1. The same colouring process as the cycle graphs C_{n-1} then follows.   □

Although, as these theorems show, DSATUR is exact for certain types of graph, the $\mathcal{NP}$-hardness of the graph colouring problem implies that it will be unable to produce optimal solutions for all graphs. Figure 3.13, for example, shows a small graph that, while actually three-colourable, will always be coloured using four colours by DSATUR. Janczewski et al. [7] have proved that this is the smallest such graph where this suboptimality occurs, but there are countless larger graphs where DSATUR will also not return the optimal. In other work, Spinrad and Vijayan [8] have also identified a graph topology of $\mathcal{O}(n)$ vertices that, despite being three-colourable, will be coloured using n different colours using DSATUR.

3.4 Colouring Using Maximal Independent Sets

We have seen that the GREEDY, WELSH–POWELL and DSATUR heuristics take vertices one at a time and assign them to the first colour seen to be feasible. An alternative strategy in graph colouring is to build solutions by constructing each *colour class* one at a time; that is, identify an independent set S in a graph, give these vertices the same colour, remove them, and then repeat these actions on the remaining subgraph until no graph remains.

To explore such a strategy, consider the following definitions.

Definition 3.14 Given a graph $G = (V, E)$, recall that an *independent set* is a set of vertices that are mutually nonadjacent. That is, $S \subseteq V$ is an independent set if and only if $\{u, v\} \notin E$ for all vertex pairs $u, v \in S$.

An independent set S is *maximal* if and only if, for all vertices $v \in V$, v is either in S, or is a neighbour of a vertex in S. A *maximum* independent set is the largest maximal independent set in G.

In Chap. 2 we saw that determining a maximum independent set in a graph is an $\mathcal{NP}$-hard problem. However, *maximal* independent sets can be determined in polynomial time. A simple algorithm for doing this is shown in the GET-MAXIMAL-I-SET procedure of Fig. 3.14. This operates by adding suitable vertices one at a time to the set S until no further vertices can be added. As shown, the algorithm operates by maintaining a set X. During a run, this contains all vertices that are not currently adjacent to any vertex in S. If desired, a second set Y can also be maintained to hold all vertices that *are* adjacent to vertices in S. When a vertex $u \in X$ is added to S, it is then removed from X together with all of its neighbours. These neighbours are also added to Y. The algorithm loops until X is empty, at which point S is a maximal independent set.

An example application of the GET-MAXIMAL-I-SET procedure is shown in Fig. 3.15.

A feasible colouring $\mathcal{S}$ of a graph G can be obtained through a series of calls to the GET-MAXIMAL-I-SET procedure. The method for doing this is shown in the procedure GREEDY-I-SET in Fig. 3.14. As shown, this operates by identifying a maximal independent set S and copying this to the solution $\mathcal{S}$. The vertices of S are then removed from G, and the process is repeated on the resultant subgraph, continuing until G is empty.

In its simplest form, GREEDY-I-SET can operate by randomly selecting vertices from X in each application of Step (3) of GET-MAXIMAL-I-SET. However, heuristic-based selection rules can also be applied to try and improve solution quality, as we now consider.

3.4.1 The RLF Algorithm

The Recursive Largest First (RLF) heuristic for graph colouring was originally proposed by Leighton [9]. It is a special case of the GREEDY-I-SET procedure in that it uses specific rules for selecting vertices in Step (3) of GET-MAXIMAL-I-SET. The intention behind these rules is to produce maximal independent sets containing many

---

GET-MAXIMAL-I-SET $(G = (V,E))$

---

(1) $X \leftarrow V, Y \leftarrow \emptyset, S \leftarrow \emptyset$
(2) **while** $X \neq \emptyset$ **do**
(3)   Remove a vertex u from X and insert it into S
(4)   **for all** $v \in \Gamma(u)$ **do**
(5)   Remove v from X (if present)
(6)   Insert v into Y (if not already present)
(7) **return** S

GREEDY-I-SET $(G = (V,E))$

---

(1) $\mathcal{S} = \emptyset$
(2) **while** $V \neq \emptyset$ **do**
(3) $S \leftarrow$ GET-MAXIMAL-I-SET(G)
(4) $\mathcal{S} \leftarrow \mathcal{S} \cup S$
(5)   Remove all vertices of S from G
(6) **return** $\mathcal{S}$

Fig. 3.14 The GREEDY-I-SET algorithm for graph colouring

vertices, thereby hopefully reducing the number of colours used in the final solution. These rules are as follows.

- In each application of GET-MAXIMAL-I-SET, when executing Step (3) for the first time, u is selected as the member of X that has the largest number of neighbours in X (that is, the vertex with the highest degree in the subgraph induced by X).
- In subsequent executions of Step (3), u is selected as the member of X that has the largest number of neighbours in Y. Ties can be broken by selecting the vertex among these that has the minimum number of neighbouring vertices in X.

As an example, executing the RLF algorithm on the graph from Fig. 3.15 results in vertices being selected in the following order: v_4, v_2 and then v_8 (colour 1); v_3 and then v_5 (colour 2); v_6 and then v_1 (colour 3); and then v_8 (colour 4). This gives the final (feasible) solution $\mathcal{S} = \{\{v_1, v_6\}, \{v_2, v_4, v_8\}, \{v_3, v_5\}, \{v_8\}\}$.

Like DSATUR, the RLF algorithm is also exact for several fundamental graph topologies.

Theorem 3.13 *The RLF algorithm is exact for bipartite graphs.*

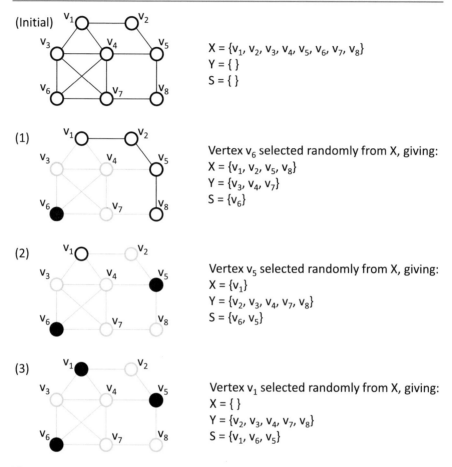

(Initial)

$X = \{v_1, v_2, v_3, v_4, v_5, v_6, v_7, v_8\}$
$Y = \{\,\}$
$S = \{\,\}$

(1)

Vertex v_6 selected randomly from X, giving:
$X = \{v_1, v_2, v_5, v_8\}$
$Y = \{v_3, v_4, v_7\}$
$S = \{v_6\}$

(2)

Vertex v_5 selected randomly from X, giving:
$X = \{v_1\}$
$Y = \{v_2, v_3, v_4, v_7, v_8\}$
$S = \{v_6, v_5\}$

(3)

Vertex v_1 selected randomly from X, giving:
$X = \{\,\}$
$Y = \{v_2, v_3, v_4, v_7, v_8\}$
$S = \{v_1, v_6, v_5\}$

Fig. 3.15 Example application of GET-MAXIMAL-I-SET using random vertex selection. Here, black-filled vertices are those currently assigned to the independent set S; black outlines show the subgraph induced by the vertices of X; and grey lines show the subgraph induced by the vertices of $Y \cup S$. At the end of this process, X will be empty and S will be a maximal independent set

Proof Let $G = (V_1, V_2, E)$ be a connected bipartite graph with $n \geq 3$. (As usual, if G is disconnected it is enough to consider each component separately.)

Assume without loss of generality that vertex $v \in V_1$ has the highest degree and is, therefore, assigned to the first colour. Consequently all neighbours of v (that is, $\Gamma(v) \subseteq V_2$) will be added to Y. It is now sufficient to show that the next $|V_1| - 1$ vertices assigned to the first colour will all be members of V_1. This is indeed the case because, in the next $|V_1| - 1$ steps, uncoloured vertices will be selected that have the largest number of adjacent vertices in the set Y. Since uncoloured vertices from the set $V_2 - Y$ will be nonadjacent to those in Y, only vertices from V_1 will be selected. This applies until all vertices from V_1 have been assigned to the first colour. At this point the subgraph induced by V_2 will have no edges, allowing RLF to colour all remaining vertices with the second colour.   □

Theorem 3.14 *The* RLF *algorithm is exact for cycle and wheel graphs.*

Proof Even cycles are two-colourable and are therefore dealt with by Theorem 3.13. However, for convenience we consider both even and odd cycles in the following. Let C_n be a cycle graph with vertices $V = \{v_1, \ldots, v_n\}$ and edges $E = \{\{v_1, v_2\}, \{v_2, v_3\}, \ldots, \{v_{n-1}, v_n\}, \{v_n, v_1\}\}$. For bookkeeping purposes, also assume that ties in the RLF selection rules are broken by taking the vertex with the lowest index. It is easy to see that this theorem holds without this restriction, however.

The degree of all vertices in C_n is 2, so the first vertex to be coloured will be v_1. Consequently, neighbouring vertices v_2 and v_{n-1} will be added to Y. According to the heuristics of RLF the next vertex to be coloured will be v_3, leading to v_4 being added to Y; then v_5, leading to v_6 being added to Y; and so on. At the end of this process, we will have colour class $S_1 = \{v_1, v_3, \ldots, v_{n-1}\}$ when n is even, and the colour class $S_1 = \{v_1, v_3, \ldots, v_{n-2}\}$ when n is odd. In the even case, this leaves an uncoloured subgraph with vertices $v_2, v_4, \ldots, v_n$ and no edges. Consequently RLF will assign all of these vertices to the second colour. In the odd case, we will be left with uncoloured vertices $v_2, v_4, \ldots, v_{n-1}, v_n$ together with a single edge $\{v_{n-1}, v_n\}$. Following the heuristic rules of RLF, all even-indexed vertices will then be assigned to the second colour, with v_n being assigned to the third.

For wheel graphs, W_n, similar reasoning applies. Assuming $n \geq 5$, the central vertex v_n will be coloured first because it has the highest degree. Since v_n is adjacent to all other vertices, no further vertices can be added to this colour, so the algorithm will move on to the second colour. The remaining uncoloured vertices now form the cycle graph C_{n-1}, and the same colouring process as above follows.   □

3.5 Empirical Comparison

In this section we now present a comparison of the five heuristics considered in this section, namely:

- The GREEDY algorithm, using randomly generated permutations of vertices (Sect. 3.1);
- The WELSH–POWELL algorithm (Sect. 3.2.2.2);
- The DSATUR algorithm (Sect. 3.3);
- The GREEDY-I-SET algorithm, using random vertex selection (Sect. 3.4); and
- The RLF algorithm (Sect. 3.4.1).

In the next subsection, we start by making some general points about good practice when empirically comparing algorithms. In Sect. 3.5.2 we then consider each of the five algorithms in turn and derive their complexities using big O notation. The results of the comparison are then discussed in Sect. 3.5.3. The implementations used in these experiments can be found in the online suite of graph colouring algorithms described in Sect. 1.4.1 and Appendix A.1.

3.5.1 Experimental Considerations

When new algorithms are proposed for the graph colouring problem, the quality of the solutions it produces will often be compared to those achieved on the same problem instances by other methods. A development in this regard occurred in 1992 with the organisation of the Second DIMACS Implementation Challenge (http://mat.gsia.cmu.edu/COLOR/instances.html), which resulted in a suite of differently structured graph colouring problems being placed into the public domain. Since this time, authors of graph colouring papers have often used this set (or a subset of it) and have concentrated on tuning their algorithms (perhaps by altering the underlying operators or run time parameters) to achieve the best possible results.

More generally, when testing and comparing the accuracy of two heuristic algorithms, an important question is:

> Are we attempting to show that Algorithm A produces better results than Algorithm B on (a) a particular problem instance? or (b) across a whole set of problem instances?

In some cases, we might be given a difficult practical problem that we only need to consider once, and whose efficient solution might save lots of money or other resources. Here, it would seem sensible to concentrate on answering question (a) and spend our time choosing the correct heuristics and parameters to achieve the best solution possible under the given time constraints. If our chosen algorithm involves stochastic behaviour (i.e., making random choices) multiple runs of the algorithm might then be performed on the problem instance to gain an understanding of its average performance with this case.

In most situations, however, it is more likely that when a new algorithm is proposed, the scientific community will be more interested in question (b) being answered—that is, we will want to understand and appreciate the performance of the algorithm across a whole set of problem instances, allowing more general conclusions to be drawn.

If we choose to try and answer (b) above, it is first necessary to decide what *types* of graphs (i.e., what population of problems instances) we wish to make statements about. For instance, this might be the set of all three-colourable graphs, or it could be the set of all graphs containing fewer than 1000 vertices. Typically, populations like these will be very large, or perhaps unlimited in size, and so it will be necessary to test our algorithms on randomly selected samples of these populations. Under appropriate experimental conditions, we might then be able to use the outcomes of

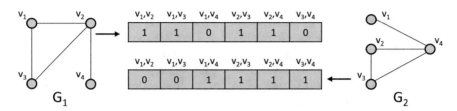

Fig. 3.16 Illustration of how different binary l-tuples can represent graphs that are isomorphic

these trials to make helpful statistical statements about the population itself, such as: "With $\geq 95\%$ confidence, Algorithm A produces solutions with fewer colours than Algorithm B on this particular graph type".

In this section, to compare the performance of our five heuristics, we make use of the following facts to define our population. Given a graph with n vertices there are a total of $l = \binom{n}{2}$ different pairs of vertices. Any graph with n vertices can therefore be represented by a binary l-tuple $\mathbf{b} = (b_1, b_2, \ldots, b_l)$ where an element $b_i = 1$ if the corresponding pair of vertices are adjacent, and $b_i = 0$ otherwise.

Now let $\mathcal{B}^{(n)}$ define the set of all possible binary l-tuples, where $l = \binom{n}{2}$. The size of this set is $|\mathcal{B}^{(n)}| = 2^l$ and can therefore be viewed as the set of all possible ways of connecting vertices in an n-vertex graph. However, we must be careful in this interpretation as it is not quite the same as saying that $\mathcal{B}^{(n)}$ represents the set of all *graphs* with n vertices (which it does not), because it fails to take into account the principle of graph isomorphisms.

Consider the example in Fig. 3.16, where we show two different binary 6-tuples and illustrate the graphs that they represent, called G_1 and G_2 here. Note that when we come to colour G_1 and G_2, their vertex labels are of little importance; indeed, without the labels these two graphs might be considered identical. In these circumstances G_1 and G_2 are considered *isomorphic* as there exists a way of converting one graph into the other by simply relabelling the vertices (in this example we can convert G_1 to G_2 by relabelling v_1 as v_2, v_2 as v_4, v_3 as v_3 and v_4 as v_1). Because the set $\mathcal{B}^{(n)}$ fails to take these isomorphisms into account, it must therefore be interpreted as the "set of all n-vertex graphs *and their isomorphisms*", as opposed to the set of all n-vertex graphs itself.

To generate a single member of the set $\mathcal{B}^{(n)}$ at random (i.e., to select an element of $\mathcal{B}^{(n)}$ such that each element is equally likely), it is simply necessary to generate an l-tuple $\mathbf{b}$ in which each element b_i is set to one with a probability of one half, and set to zero otherwise. This is the same process as producing a *random graph* with $p = 0.5$:

Definition 3.15 A *random graph*, denoted by $G_{n,p}$, is a graph that is generated by starting with n isolated vertices. Each pair of vertices u, v is then considered in turn, and the edge $\{u, v\}$ is added with probability p.

Observe that the degrees of the vertices in a random graph are binomially distributed; that is, $\deg(v) \sim B(n-1, p)$. As a result, they are also sometimes known as binomial graphs. Random graphs will be the focus of our algorithm comparison in this chapter, though we will also look at other types of graphs in later chapters.

3.5.2 Algorithm Complexities

Before comparing the quality of the solutions produced by our five heuristic algorithms, let us first determine their complexities using big O notation.

Recall from Sect. 3.1 that the GREEDY heuristic takes each vertex $v \in V$ in turn and assigns it to the colour $j \in \{1, 2, \ldots, \Delta(G)+1\}$ that represents the lowest colour label not being used by any of v's neighbours. An efficient way of determining a value for j is shown in Fig. 3.17. This operates using a $(\Delta(G)+1)$-tuple called *used* whose values are all initially set to false. Steps (1) to (3) of this procedure consider each neighbour u of v. If u is already coloured then $used(c(u))$ is set to true. On completion of this loop, false entries in *used* therefore correspond to colours not being used by the neighbours of v.

Steps (4) to (6) of this procedure then determine a value for j by simply identifying the index of the first false entry in *used*. Note that a false entry will always exist among the first $\deg(v) + 1$ elements. Finally, Steps (7) to (10) set the true elements of *used* back to false.

The GREEDY heuristic can be implemented by first setting $c(v)$ to NULL for each $v \in V$, and setting $used(i)$ to false for all $i \in \{1, 2, \ldots, \Delta(G) + 1\}$. This is clearly an $\mathcal{O}(n)$ process. The graph can then be coloured by making n separate calls to GET-LOWEST-FEASIBLE-COLOUR, one for each vertex $v \in V$. Since each of these calls has a complexity of $\mathcal{O}(\deg(v))$, the overall complexity of GREEDY is therefore $\mathcal{O}(n + \sum_{v \in V} \deg(v)) = \mathcal{O}(n + m)$.

As we saw in Sect. 3.2.2.2, the WELSH–POWELL heuristic operates in the same way as GREEDY except that vertices are first labelled such that $\deg(v_1) \geq \deg(v_2) \geq \cdots \geq \deg(v_n)$. This relabelling can be achieved in $\mathcal{O}(n \lg n)$ time by a standard sorting algorithm such as Merge Sort and Heap Sort. The overall complexity of WELSH–POWELL is therefore slightly more expensive than GREEDY at $\mathcal{O}(n \lg n + m)$.

Fig. 3.17 An $\mathcal{O}(\deg(v))$ algorithm for determining a feasible colour for a vertex v. Here, *used* is a $(\Delta(G) + 1)$-tuple whose values are initially false. As usual, $c(u)$ denotes the colour of a vertex u. If u has not yet been coloured, then $c(u) = $ NULL

GET-LOWEST-FEASIBLE-COLOUR (v)
(1) **for** $u \in \Gamma(v)$ **do**
(2)   **if** $c(u) \neq$ NULL **then**
(3) $used(c(u)) \leftarrow$ **true**
(4) **for** $i \leftarrow 1$ **to** $\Delta(G) + 1$ **do**
(5)   **if** $used(j) =$ **false then**
(6)   **break**
(7) **for** $u \in \Gamma(v)$ **do**
(8)   **if** $c(u) \neq$ NULL **then**
(9) $used(c(u)) \leftarrow$ **false**
(10) **return** j

DSATUR (G)
(1) $Q \leftarrow \emptyset$
(2) **for** $v \in V$ **do**
(3) $c(v) \leftarrow$ NULL
(4) $nc(v) \leftarrow \emptyset$
(5) $d(v) \leftarrow \deg(v)$
(6) $Q \leftarrow Q \cup \{(v, 0, d(v))\}$
(7) **while** $Q \neq \emptyset$ **do**
(8)   Let (u, x, y) be the element of Q with the maximum value for x, breaking ties with the maximum value for y
(9)   Remove the element (u, x, y) from Q
(10) $j \leftarrow$ GET-LOWEST-FEASIBLE-COLOUR(u)
(11) $c(u) \leftarrow j$
(12)   **for** $v \in \Gamma(u)$ such that $c(v) =$ NULL **do**
(13)   Remove the element $(v, |nc(v)|, d(v))$ from Q
(14) $nc(v) \leftarrow nc(v) \cup \{j\}$
(15) $d(v) \leftarrow d(v) - 1$
(16) $Q \leftarrow Q \cup \{(v, |nc(v)|, d(v))\}$

Fig. 3.18 An $\mathcal{O}((n + m) \lg n)$ specification of the DSATUR algorithm

We now turn our attention towards the DSATUR algorithm. In his original publication, Brélaz [6] states that the complexity of DSATUR is $\mathcal{O}(n^2)$. This can be achieved by performing n separate applications of an $\mathcal{O}(n)$ process that (a) identifies the next vertex to colour according to DSATUR's selection rules, and then (b) colours this vertex.

For sparse graphs, the complexity of DSATUR can be significantly improved by making use of a priority queue. During execution, this priority queue should store all vertices that are not yet coloured, together with their saturation degree and their degree in the subgraph induced by the uncoloured vertices. The priority queue should also allow the selection of the next vertex to colour (according to DSATUR's selection rules) in constant time.

A description of this approach is shown in Fig. 3.18. As shown, this procedure uses three n-tuples:

- $c(v)$ denotes the colour of vertex v, as before.
- $nc(v)$ stores the set of *neighbouring colours* of v (that is, colours currently being used by vertices adjacent to v). The saturation degree of a vertex v is therefore equal to $|nc(v)|$.
- $d(v)$ stores the degree of v in the subgraph induced by the uncoloured vertices.

In this pseudocode, the priority queue Q stores details about the uncoloured vertices. This information is in a 3-tuple containing the name of the vertex v, its saturation degree $|nc(v)|$, and $d(v)$ respectively.

GREEDY-I-SET (G)
(1) $i \leftarrow 1, X \leftarrow V, Y \leftarrow \emptyset$
(2) **for** $v \in V$ **do**
(3) $c(v) \leftarrow$ NULL
(4) **while** $X \neq \emptyset$ **do**
(5)   **while** $X \neq \emptyset$ **do**
(6)   Choose a random $u \in X$, remove it, and set $c(u) \leftarrow i$
(7)   **for** $v \in \Gamma(u)$ such that $c(v) =$ NULL **do**
(8)   Remove v from X and insert v into Y
(9)   Swap the contents of X and Y
(10) $i \leftarrow i+1$

Fig. 3.19 An $\mathcal{O}(n+m)$ specification of the GREEDY-I-SET algorithm

The contents of Q, c, nc, and d are initialised in Steps (1) to (6) of Fig. 3.18. In the remaining steps, an uncoloured vertex u is first removed from Q and coloured using the GET-LOWEST-FEASIBLE-COLOUR procedure. In Steps (12) to (16), the values of nc and d are then adjusted for each uncoloured neighbour v of u, and the contents of Q are updated accordingly.

The asymptotic running time of this version of DSATUR now depends on the data structures used for storing Q and each element of nc. An ideal option here is to use a binary heap or self-balancing binary tree since this allows vertex selection in constant time (Step (8)), with lookups, deletions, and insertions then being performed in logarithmic time.[2] Using these data structures, Steps (1) to (6) of Fig. 3.18 will take $\mathcal{O}(n \lg n)$ time. In the remaining steps, the neighbours of each vertex are considered once, and these are then updated using logarithmic-time operations, giving a run time of $\mathcal{O}(\sum_{v \in V} \deg(v) \lg n) = \mathcal{O}(m \lg n)$. This leads to an overall complexity for DSATUR of $\mathcal{O}((n+m) \lg n)$.

We now consider the two heuristics from Sect. 3.4 which, we recall, produce solutions by identifying and removing maximal independent sets from a graph.

An efficient version of GREEDY-I-SET using random vertex selection is shown in Fig. 3.19. Here it is not necessary for the contents of the sets X and Y to be ordered, so a suitable option is to use hash tables, which allow the addition and removal of elements in constant time.[3] Steps (1) to (3) of this procedure initialise the data structures and operate in $\mathcal{O}(n)$ time. The remaining steps then operate in $\mathcal{O}(\sum_{v \in V} \deg(v)) = \mathcal{O}(m)$ time. The overall complexity of GREEDY-I-SET is therefore equivalent to GREEDY at $\mathcal{O}(n+m)$.

Finally, we consider the complexity of the RLF algorithm. In his original publication [9] states this as $\mathcal{O}(n^3)$; however, this can be improved upon. As we saw in Sect. 3.4, RLF operates in much the same way as GREEDY-I-SET; indeed, the only

---

[2]In our C++ implementations we use the `set` container for these purposes. In most versions of C++, these are stored as self-balancing binary trees.

[3]In our C++ implementations we use the `unordered_set` container for storing X and Y.

Table 3.1 Summary of results produced by the GREEDY, GREEDY-I-SET, WELSH–POWELL, DSATUR and RLF algorithms on random graphs $G_{n,0.5}$

Algorithm[a]							
n	LB[b]	GREEDY	GREEDY-I-SET	WELSH–POWELL	DSATUR	RLF	UB[c]
100	9	21.38 ± 0.93	21.24 ± 1.03	19.99 ± 0.94	18.15 ± 0.76	17.13 ± 0.56	50.57 ± 0.82
1000	10	126.48 ± 1.31	126.35 ± 1.42	122.76 ± 1.28	115.26 ± 1.26	106.98 ± 1.03	500.55 ± 0.83
2000	10	224.08 ± 1.62	224.30 ± 1.79	219.21 ± 1.57	207.43 ± 1.16	194.25 ± 1.02	1000.45 ± 0.85

[a]Mean plus/minus standard deviation in number of colours, taken from runs across 100 graphs
[b]Lower bound. This is the largest value x for which Eq. (3.3) is greater than or equal to 0.99
[c]The Welsh–Powell upper bound (Theorem 3.9). Mean taken across 100 graphs

difference occurs in Step (6) of the pseudocode in Fig. 3.19, where the selection of a vertex $v \in X$ is made according to heuristic rules as opposed to at random. The use of these rules does indeed increase the complexity of the procedure because each time a vertex u is selected, the number of neighbours in X and the number of neighbours in Y need to be recalculated for each uncoloured vertex. These calculations can be performed in $\mathcal{O}(m)$ time, meaning that the overall complexity of RLF is $\mathcal{O}(mn)$. Note that this is the highest complexity of the five heuristics considered in this section.

3.5.3   Results and Analysis

Table 3.1 summarises the number of colours used[4] by the solutions returned by the five heuristics. These trials were carried out using random graphs of edge probability $p = 0.5$. For each value of n, one hundred graphs were generated and each algorithm was executed on it once. In applications of the GREEDY algorithm, the vertex permutation P was generated randomly. For GREEDY-I-SET, vertices were selected from X at random.

The results in Table 3.1 show that the two simplest algorithms, GREEDY and GREEDY-I-SET, produce the poorest solutions overall. There are also no significant differences between these results.[5] As we move from left to right in the table we see that the results returned by the corresponding algorithms improve. Each of these improvements was also seen to be significant for the three values of n. We can therefore conclude that, out of these five constructive heuristics, RLF produces the best

---

[4]Mean plus/minus standard deviation in number of colours, taken from runs across 100 graphs.
[5]The samples collected for each algorithm and value of n were not generally found to be derived from an underlying normal distribution according to a Shapiro–Wilk test; consequently, statistical significance is claimed here according to the results of a nonparametric related samples Wilcoxon Signed Rank test at the 0.1% significance level.

solutions across the set of all graphs and their isomorphisms for $n = 100$, 1000, and 2000.

The data in Table 3.1 also reveals that the generated lower and upper bounds seem to be some distance from the number of colours ultimately used by the algorithms. This indicates that the Welsh–Powell bound tends to provide a rather inaccurate upper bound for random graphs of density 0.5. It also suggests two factors concerning the lower bound: (a) that the probabilistic bound determined by Eq. (3.3) is quite inaccurate and/or (b) that all five heuristic algorithms are producing solutions whose numbers of colours are some distance from the chromatic number.

The graphs shown in Fig. 3.20 expand upon the results of Table 3.1 by considering random graphs across a range of different values for p. Bounds are also indicated by the shaded areas. We see that the unshaded areas of these graphs are generally quite wide, with the algorithms' results falling in a fairly narrow band within these. This again indicates the inadequacy of the upper bounds considered in this chapter, particularly for larger values of n. The Welsh–Powell bound appears to be the most accurate for these graphs overall.

The differences in solution quality between these five algorithms are presented more clearly in Fig. 3.21. Here, the bars in the charts show the number of colours used in solutions produced by GREEDY. The lines then indicate the percentage of this figure used by the remaining four algorithms. We see that WELSH–POWELL, DSATUR and RLF achieve percentages of less than 100% across all of the tested values for p, indicating their superior performance across the set of random graphs, from sparse to dense. We also see that RLF consistently produces the lowest percentages, once again indicating its general superiority over the other methods.

We now consider the implications of the computational complexity of the five heuristics. Figure 3.22 shows the strong correlation that exists between the number of checks performed by each algorithm and its subsequent run time. They also show that greater amounts of computation are required for dense graphs, which is what we should expect when we consider the complexities of these algorithms, as discussed in Sect. 3.5.2.

The scales of the charts in Fig. 3.22 also give us further information. We see that the number of checks performed by GREEDY, GREEDY-I-SET, and WELSH–POWELL are very similar, indicating that the additional $\mathcal{O}(n \lg n)$ sorting operation required by WELSH–POWELL has a negligible effect. On the other hand, we see that the CPU time of GREEDY-I-SET is higher than the other two methods. This is due to the additional overheads of using the C++ container `unordered_set` for storing X and Y in the implementation. Finally, we also see that the computational requirements of RLF are the highest of the five algorithms. This is to be expected due to its higher complexity of $\mathcal{O}(mn)$.

Fig. 3.20 Number of colours in solutions produced by the GREEDY, GREEDY-I-SET, WELSH–POWELL, DSATUR, and RLF algorithms on random graphs $G_{n,p}$ with various values of p, using $n = 100, 1000,$ and 2000, respectively. The shaded curves at the bottom right indicates the lower bound derived from Eq. (3.3). The shaded curves at the top left indicate, from top to bottom, the upper bounds derived from Theorems 3.5, 3.6, and 3.9, respectively. All points are the mean across 100 graphs

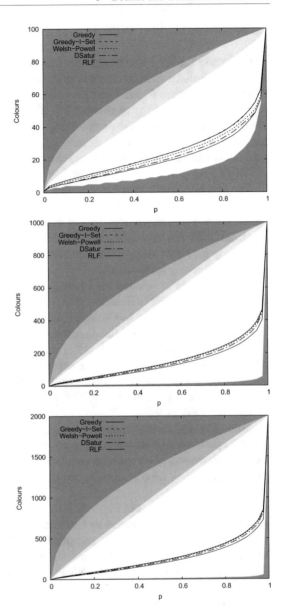

3.6 Chapter Summary and Further Reading

In this chapter, we have reviewed a number of bounds for the graph colouring problem. We have also compared and contrasted five constructive heuristics. A summary of these bounds and heuristics is shown in Table 3.2. For random graphs of different sizes and densities (including sets of graphs and their isomorphisms), we have seen that the RLF algorithm generally produces the solutions with the fewest colours, though this comes at the expense of added computation time.

In the next two chapters, we will analyse techniques that seek to improve upon the solutions produced by these constructive heuristics. We now end this chapter by providing points of reference for further work on bounds for the chromatic number.

Reed [10] has shown that Brooks' bound (Theorem 3.8) can be improved by one colour when a graph G has a sufficiently large value for $\Delta(G)$ and also has no cliques of size $\Delta(G)$. Specifically:

Theorem 3.15 (Reed [10]) *There exists some value δ such that if $\Delta(G) \geq \delta$ and $\omega(G) \leq \Delta(G) - 1$ then $\chi(G) \leq \Delta(G) - 1$.*

In this work, a sufficient value for δ is shown to be 10^{14}. This is obviously a very large number, though it is suggested that "a more careful analysis could bring the bound down to 1000".

A further conjecture, now known as Reed's Conjecture, is also stated in this paper. This proposes that for any graph G,

$$\chi(G) \leq \left\lceil \frac{1 + \Delta(G) + \omega(G)}{2} \right\rceil. \tag{3.7}$$

A good survey on these issues can be found in the work of Cranston and Rabern [11].

In Sect. 3.2.1.1 of this chapter, we saw that interval graphs (and more generally chordal graphs) always feature chromatic numbers $\chi(G)$ equal to their clique numbers $\omega(G)$. Chordal graphs form part of a larger family of graphs known as *perfect* graphs which, in addition to satisfying this criterion, are also known to maintain this property when any of its vertices are removed.

Definition 3.16 A graph $G = (V, E)$ is *perfect* if, for every subgraph $G' \subseteq G$, $\chi(G') = \omega(G')$.

Defining the structures needed for a graph to be perfect has been the subject of much research in the field of graph theory and was eventually settled by Chudnovsky et al. [12], who proved the earlier conjecture of Berge [13], which stated that a graph is perfect if and only if it contains no odd hole and no odd antihole. (A hole is an induced subgraph which is a cycle of length at least 4; an antihole is the complement.) See the work of Mackenzie [14] for further details.

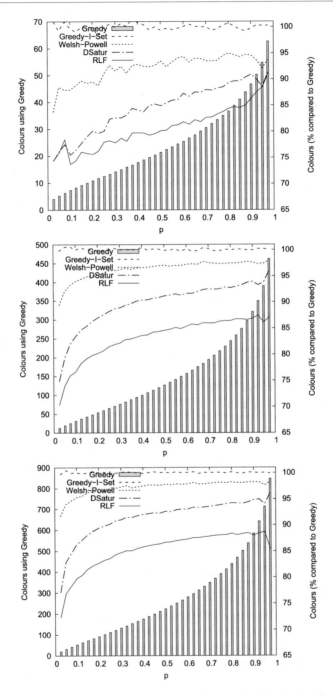

Fig. 3.21 Mean quality of solutions achieved on random graphs $G_{n,p}$ by the GREEDY-I-SET, WELSH–POWELL, DSATUR and RLF algorithms in comparison to GREEDY. All points are the mean across 100 graphs using $n = 100$, 1000, and 2000 respectively

Table 3.2 Summary of the bounds and heuristic algorithms considered in this chapter

Bound	Notes
$\chi(G) \geq \omega(G)$	Involves calculating the size of the largest clique $\omega(G)$, which is $\mathcal{NP}$-hard
$\chi(G) \geq \lceil n/\alpha(G) \rceil$	Involves calculating the size of the largest independent set $\alpha(G)$, which is $\mathcal{NP}$-hard
$\chi(G) \leq 1/2 + \sqrt{2m+1/4}$	Derived from Theorem 3.5
$\chi(G) \leq \Delta(G) + 1$	Can be observed due to the behaviour of the GREEDY algorithm. Brooks' bound strengthens this to $\chi(G) \leq \Delta(G)$, providing G is not a complete graph or an odd cycle
$\chi(G) \leq \max_{i=1,\dots,n} \min(\deg(v_i) + 1, i)$	Assumes vertices are labelled such that $\deg(v_1) \geq \deg(v_2) \geq \dots \geq \deg(v_n)$ At least as good as the previous bound

Algorithm	Complexity	Notes
GREEDY	$\mathcal{O}(n + m)$	Different vertex permutations can result in solutions of varying quality. Produces an optimal solution with the correct permutation
GREEDY-I-SET	$\mathcal{O}(n + m)$	Similar to GREEDY. Extracts maximal independent sets one at a time
WELSH-POWELL	$\mathcal{O}(n \lg n + m)$	Applies GREEDY using a permutation of vertices in descending order of degree
DSATUR	$\mathcal{O}(n^2)$	Exact for bipartite, cycle, and wheel graphs. Can be implemented with complexity $\mathcal{O}((n + m) \lg n)$, which is more favourable for sparse graphs
RLF	$\mathcal{O}(mn)$	Exact for bipartite, cycle, and wheel graphs. More expensive than DSATUR, but produces better results with random graphs

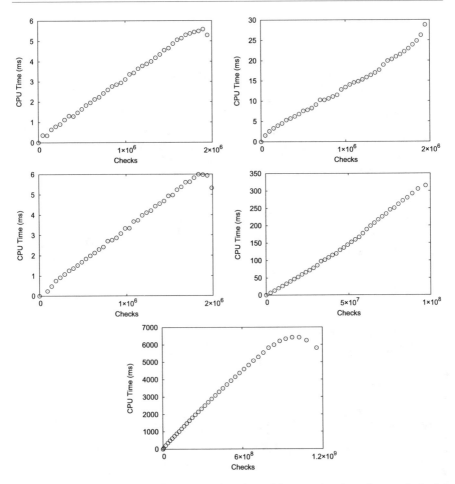

Fig. 3.22 Relationship between the number of checks and the execution time of, respectively, the
GREEDY, GREEDY-I-SET, WELSH–POWELL, DSATUR and RLF algorithms. Each point in the figure
is the mean taken from 100 random graphs $G_{1000,p}$. Moving from left to right in each figure, these
probabilities are $p = 0.025, 0.05, 0.075, \ldots, 0.975$. All experiments were performed on a 3.0 GHz
Windows 7 PC with 3.87 GB RAM

Looking at other topologies, bounds on the chromatic number of random graphs
have also been determined by Bollobás [15], who states that with very high probability, a random graph $G_{n,p}$ will have a chromatic number $\chi(G_{n,p})$:

$$\frac{n}{s} \leq \chi(G_{n,p}) \leq \frac{n}{s} + \left(1 + \frac{3\log\log n}{\log n}\right), \tag{3.8}$$

where

$$s = \lceil 2\log_d n - \log_d \log_d n + 2\log_d(e/2) + 1\rceil \tag{3.9}$$

with $q = 1 - p$ and $d = 1/q$. The finding is based on calculating the expected
number of disjoint cliques within a random graph.

Further bounds on general graphs have also been given by Berge [16], who finds

$$\frac{n^2}{n^2 - 2m} \leq \chi(G), \tag{3.10}$$

and Hoffman [17], who has shown

$$1 - \frac{\lambda_1(G)}{\lambda_n(G)} \leq \chi(G), \tag{3.11}$$

where $\lambda_1(G)$ and $\lambda_n(G)$ are the biggest and smallest eigenvalues of the adjacency matrix of G. Both of these usually give very loose lower bounds in practice, however.

Note that the five constructive methods in this section should be classed as *heuristic* algorithms as opposed to approximation algorithms. Unlike heuristics, approximation algorithms are usually associated with provable bounds on the quality of solutions they produce compared to the optimal. So for the graph colouring problem, using $A(G)$ to denote the number of colours used in a feasible solution produced by algorithm A with graph G, a good approximation algorithm should feature an approximation ratio $A(G)/\chi(G)$ as close to 1 as possible. To date, the best-known approximation ratio for a graph colouring algorithm is $\mathcal{O}(n(\log \log n)^2/(\log n)^3)$ due to Halldórsson [18]. This method operates by randomly selecting independent sets which are then allocated a colour and removed from the graph. The process repeats until the graph is empty. Those seeking an algorithm with a low approximation ratio for the graph colouring problem, however, should take note of the following theorem:

Theorem 3.16 (Garey and Johnson [19]) *If, for some constant $r < 2$ and constant d, there exists a polynomial-time graph colouring algorithm A which is guaranteed to produce $A(G) \leq r\chi(G) + d$, then there also exists an algorithm A' which guarantees $A'(G) = \chi(G)$.*

In other words, this states that we cannot hope to find an approximation algorithm A for the graph colouring problem that, for all graphs, produces $A(G) < 2\chi(G)$ unless $\mathcal{P} = \mathcal{NP}$.

References

1. Bollobás B (1998) Modern graph theory. Springer
2. Mycielski J (1955) Sur le coloriage des graphes. Colloq Math 3:161–162
3. Rose D, Lueker G, Tarjan R (1976) Algorithmic aspects of vertex elimination on graphs. SIAM J Comput 5(2):266–283
4. Brooks R (1941) On colouring the nodes of a network. Math Proc Cambridge Philos Soc 37:194–197
5. Welsh D, Powell M (1967) An upper bound for the chromatic number of a graph and its application to timetabling problems. Comput J 12:317–322

6. Brélaz D (1979) New methods to color the vertices of a graph. Commun ACM 22(4):251–256
7. Janczewski R, Kubale M, Manuszewski K, Piwakowski K (2001) The smallest hard-to-color graph for algorithm DSatur. Discret Math 236:151–165
8. Spinrad J, Vijayan G (1984) Worst case analysis of a graph coloring algorithm. Discret Appl Math 12:89–92
9. Leighton F (1979) A graph coloring algorithm for large scheduling problems. J Res Natl Bur Stand 84(6):489–506
10. Reed B (1999) A strengthening of Brooks' theorem. J Comb Theory Ser B 76(2):136–149
11. Cranston D, Rabern L (2014) Brooks' theorem and beyond. J Graph Theory. https://doi.org/10.1002/jgt.21847
12. Chudnovsky M, Robertson N, Seymour P, Thomas R (2006) The strong perfect graph theorem. Ann Math 164(1):51–229
13. Berge C (1960) Les problémes de coloration en théorie des graphes. Publ Inst Stat Univ Paris 9:123–160
14. Mackenzie D (2002) Graph theory uncovers the roots of perfection. Science 38:297
15. Bollobás B (1988) The chromatic number of random graphs. Combinatorica 8(1):49–55
16. Berge C (1970) Graphs and hypergraphs. North-Holland
17. Hoffman A (1970) On eigenvalues and colorings of graphs. In: Graph theory and its applications, Proc. Adv. Sem., Math., Research Center, University of Wisconsin, Madison, WI, 1969. Academic Press, New York, pp 79–91
18. Halldórsson M (1993) A still better performance guarantee for approximate graph coloring. Inf Process Lett 45(1):19–23. https://www.sciencedirect.com/science/article/pii/0020019093902466
19. Garey M, Johnson D (1976) The complexity of near-optimal coloring. J Assoc Comput Mach 23(1):43–49

Advanced Techniques for Graph Colouring

4

In this chapter, we review many of the algorithmic techniques that can be used for the graph colouring problem. The intention is to give the reader an overview of the different strategies available, including both exact and heuristic methods. As we will see, a variety of different approaches are available, including backtracking algorithms, integer programming, column generation, evolutionary algorithms, neighbourhood search, and other metaheuristics. Full descriptions of these techniques are provided as they arise in the text. We also describe ways in which graph colouring problems can be reduced in size and/or broken up, helping to improve algorithm performance in many cases.

4.1 Exact Algorithms

Exact algorithms are those that, given sufficient time, will always determine the optimal solution to a computational problem. As discussed in Chap. 2, one way of exactly solving an $\mathcal{NP}$-hard combinatorial problem such as graph colouring is to exhaustively search the space of all possible candidate solutions; however, as problem sizes grow, the running times for such brute-force methods soon become too large, making them impractical.

Despite this, it is still possible to design exact algorithms that are significantly faster than exhaustive search, though still not operating in polynomial time. Often we can also choose to impose computation limits on these algorithms, allowing good quality (though not-necessarily-optimal) solutions to be returned within reasonable time frames. Three alternatives are considered in the following subsections, namely, backtracking, integer programming, and column generation.

© The Author(s), under exclusive license to Springer Nature Switzerland AG 2021   77
R. M. R. Lewis, *Guide to Graph Colouring*, Texts in Computer Science,
https://doi.org/10.1007/978-3-030-81054-2_4

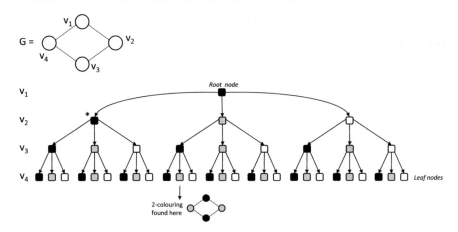

Fig. 4.1 The complete search tree that results when attempting to colour the graph G with a maximum of three colours

4.1.1   Backtracking Approaches

Backtracking is a type of algorithm that, given enough time, is guaranteed to achieve an optimal solution to a combinatorial problem. Where required, the algorithm can also be modified to find *all* optimal solutions. Backtracking works by systematically building up partial solutions into complete solutions. During this construction process, as soon as evidence is gained telling us that there is no way of completing the current partial solution to produce an optimal solution, the algorithm *backtracks* to try and find suitable ways of adjusting this partial solution.

A useful way of describing backtracking is to consider a *search tree* in which internal nodes represent partial solutions and leaf nodes represent complete solutions. Each path from the root node to a leaf node in this tree then describes the individual assignments that make up a single, complete solution.

The search tree for a small graph colouring problem is given in Fig. 4.1. In this particular case, we are seeking to optimally colour the four-vertex graph G shown at the top-left. We are also allowing ourselves a maximum of three colours—black, grey, and white here—because $\chi(G) \leq \Delta(G) + 1 = 3$ in this case. Finally, we are also assuming that vertex v_1 is fixed with the colour black. This is permitted due to the solution symmetries exhibited in graph colouring, as noted in Sect. 2.2.

As shown in Fig. 4.1, each level of this search tree signifies a different colour assignment to a particular vertex. The leftmost path in the tree therefore represents the solution in which all four vertices are coloured black. Because there are three choices of colour at every node in this tree, this leads to $3^{n-1} = 3^3 = 27$ different root-to-leaf paths. Together, these paths signify all possible candidate solutions to this problem. Evaluating all of these paths will therefore allow us to identify an optimal solution.

At this point, a method that evaluates every root-to-leaf path in this search tree will be no different to the exhaustive techniques reviewed in Sect. 2.2. However, backtracking techniques will often allow us to disregard large sections of the search tree, thereby improving execution times. This is usually referred to as *pruning*. To see this, consider a parse of this tree from the root down in depth-first order (that is, we are navigating the search tree such that the leaf nodes are considered from left to right).

1. First, at each node x in the tree, a test can be conducted to see whether the current partial solution can be completed to make a feasible solution. If it cannot, then the whole of the subtree rooted at x can be pruned, and therefore ignored by the algorithm.

   An example of this occurs at the node marked by the asterisk (*) in Fig. 4.1. Here we see that the assignment of the colour black to v_2 will result in a clash with its neighbour v_1, which is also black. As a result, there is no need to consider the subtree rooted at x because all of its leaf nodes define infeasible solutions. Instead, the algorithm can *backtrack* to the parent node and consider its other branches.

2. Second, consider the situation where solutions are evaluated according to a cost function that we want to minimise. Suppose further that the best feasible solution observed so far, S, has a cost of $f(S)$, but that the algorithm has not yet completed its parse of the search tree. Now, let x be a node in the search tree such that all complete solutions S' stemming from x are known to have a cost $f(S') \geq f(S)$. In this case, the whole of the subtree rooted at x can be pruned because none of its solutions will improve on the best solution seen so far. Instead, the algorithm can again backtrack to the parent node.

   As we know, for graph colouring the cost of a feasible solution is given by the number of different colours it is using. This means that once a feasible k-colouring has been achieved, there is no need to consider paths in the search tree that involve using k or more colours. More specifically, given a node x in the search tree, if the path from the root to x uses k or more colours, then the subtree rooted at x can be ignored. In Fig. 4.1, for example, we see that a feasible two-colouring is established at the eleventh leaf node from the left. From this point onwards, there is no need to consider subtrees rooted at any node x where the root-to-x path uses two or more colours.

The application of these two rules allows the backtracking algorithm to remove many parts of the search tree, thereby improving performance. Of course, due to the $\mathcal{NP}$-hardness of graph colouring, the time requirements of this approach may still be excessively large, and executions will often need to be terminated prematurely, perhaps leaving the user with a suboptimal solution. On the other hand, the systematic construction of different solutions, together with the way that the algorithm can ignore large swathes of the solution space, means that backtracking is usually far more efficient than brute-force enumeration.

Continuing with our example, Fig. 4.2 shows the parts of the search tree from Fig. 4.1 that are ultimately considered by the backtracking algorithm for graph colouring. As shown, each path through the tree terminates in one of three ways: by an

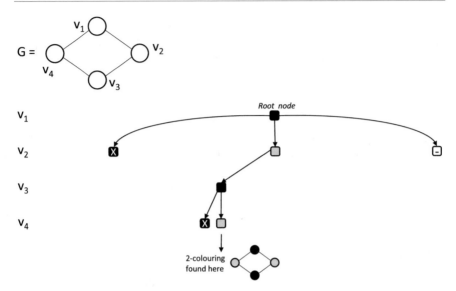

Fig. 4.2 The search tree considered by the backtracking algorithm when trying to colour G using a maximum of three colours. Nodes in the tree marked by "X" indicate applications of Rule 1. Nodes marked by "–" indicate applications of Rule 2

application of Rule 1 above, an application of Rule 2, or at a leaf node. If the entire search tree is parsed in this way, the returned solution is guaranteed to be optimal.

Kubale and Jackowski [1] have reviewed several ways in which backtracking algorithms for graph colouring can be enhanced. Suggested ideas include the following:

- *Using appropriate orderings of the vertices.* In the example of Fig. 4.1, the vertices are considered in label order; however, this is an arbitrary choice. Better performance has been noted if vertices are sorted by decreasing degree, or by ordering vertices such that those with the fewest available colours are coloured first (in a similar fashion to the DSATUR algorithm).
- *Using better branching rules.* In our current example, each branch is considered in order from left to right, that is, the vertex is first considered for assignment to black, then grey, then white. However, better performance might be gained by making more informed decisions. For example, we may choose to explore branches that involve assigning vertices to the largest colour classes first (the rationale being that forming large colour classes leads to an overall reduction in the number of colour classes). Similarly, we might choose to prioritise colour classes that contain large numbers of high-degree vertices.

Another improvement for backtracking with graph colouring can be achieved by first employing a procedure to find a large clique C in G, and then relabelling the vertices such that $C = \{v_1, v_2, \ldots, v_l\}$. Since all of the vertices in this clique need to be assigned to different colours, we can then (a) fix the colours of these vertices as $c(v_i) = i$ (for all $v_i \in C$) and (b) prevent any branching at nodes corresponding to these vertices in the search tree. In addition to reducing the size of the search tree, this method also provides a lower bound on $\chi(G)$; consequently, if a feasible l-colouring is achieved during a run, the algorithm can halt immediately and give the user a provably optimal solution.

A backtracking algorithm using an effective combination of these schemes is considered further in Chap. 5.

4.1.2 Integer Programming

Another way of achieving an exact algorithm for graph colouring is to use a special type of linear programming model called *integer programming* (IP). Linear programming (LP) is a general methodology for achieving optimal solutions to linear mathematical models. Such models consist of variables, linear constraints, and a linear objective function. The variables take on numerical values, and the constraints are used to define feasible ranges for these variables. The objective function is then used to measure the quality of a solution and to define the particular assignment of values to variables that is considered optimal.

In general, the variables of an LP are continuous in the sense that they are permitted to be fractional. On the other hand, IP models are those in which the variables are restricted to integer values. Though this might seem like a subtle difference, this insistence on integer-valued variables greatly increases the number of problems that can be modelled. Indeed, IP models can be used to solve a wide variety of combinatorial problems, including supply chain design, resource management, timetabling, employee rostering, and, as we will see presently, graph colouring.

Whereas algorithms such as the well-known simplex method are known to be effective for solving LPs, there is no single preferred technique for solving integer programs. Instead, various exact methods are available, including branch-and-bound, cutting-plane, branch-and-price, as well as various hybrid techniques. Because of their wide applicability, several off-the-shelf software applications have also been developed in recent decades for solving linear and integer programming models. These include commercial packages such as Xpress, CPLEX, and Gurobi, and free open-source applications such as the SCIP optimisation suite and Coin-OR. Such packages allow users to input their particular model (in terms of variables, constraints, and an objective function) and then simply click a button, at which point the software goes on to produce solutions using the methods just mentioned.

In this subsection, we will focus on how branch-and-bound can be used to solve the graph colouring problem. Readers interested in finding out about other methods for integer programming are invited to consult the textbook of Wolsey [2], which provides a thorough overview of the subject.

For now, we will consider a very basic IP formulation of the graph colouring problem. As usual, let $G = (V, E)$ be a graph with n vertices and m edges. We now define two binary matrices $\mathbf{X}_{n \times n}$ and $\mathbf{Y}_n$ that will hold the variables of this problem.

These are interpreted as follows:

$$X_{ij} = \begin{cases} 1 \text{ if vertex } v_i \text{ is assigned to colour } j, \\ 0 \text{ otherwise,} \end{cases} \tag{4.1}$$

$$Y_j = \begin{cases} 1 \text{ if at least one vertex is assigned to colour } j, \\ 0 \text{ otherwise.} \end{cases} \tag{4.2}$$

Note that the elements of $\mathbf{X}$ and $\mathbf{Y}$ are not only required to be integers here but also binary. This is a common restriction in integer programming models and it is necessary in this formulation because only two options are available: vertex v_i is either assigned to colour j, or it is not. The objective in this model is to now minimise the number of colours being used according to the objective function

$$\min \sum_{j=1}^{n} Y_j. \tag{4.3}$$

The following constraints now need to be satisfied:

$$X_{ij} + X_{lj} \leq Y_j \qquad \forall \{v_i, v_l\} \in E, \ \forall j \in \{1, \ldots, n\} \tag{4.4}$$

$$\sum_{j=1}^{n} X_{ij} = 1 \qquad \forall v_i \in V \tag{4.5}$$

$$X_{ij} \in \{0, 1\} \qquad \forall v_i \in V, \ \forall j \in \{1, \ldots, n\} \tag{4.6}$$

$$Y_j \in \{0, 1\} \qquad \forall j \in \{1, \ldots, n\}. \tag{4.7}$$

Here, Constraints (4.4) ensure that pairs of adjacent vertices are not assigned to the same colour and that $Y_j = 1$ if and only if some vertex is assigned to colour j. Consequently, $Y_j = 1$ indicates that colour j is being used in a solution. Constraints (4.5) specify that each vertex should be assigned to exactly one colour. Constraints (4.6) and (4.7) then ensure that only binary values are permitted in $\mathbf{X}$ and $\mathbf{Y}$. These are known as the *integrality* constraints of the model.

To illustrate this IP model, consider the small four-vertex graph shown in Fig. 4.3. An optimal solution to this problem is as follows:

$$\mathbf{X} = \begin{pmatrix} 0 \ 0 \ 0 \ 1 \\ 1 \ 0 \ 0 \ 0 \\ 0 \ 0 \ 1 \ 0 \\ 0 \ 0 \ 0 \ 1 \end{pmatrix} \tag{4.8}$$

$$\mathbf{Y} = \begin{pmatrix} 1 \ 0 \ 1 \ 1 \end{pmatrix}.$$

As can be seen, the cost of this optimal solution is $\sum_{j=1}^{n} Y_j = 3$, which is the chromatic number for this graph. Specifically in this solution, the colours labelled 1, 2, and 4 are being used, and these are assigned to the four vertices as follows: $c(v_1) = 4$, $c(v_2) = 1$, $c(v_3) = 3$, and $c(v_4) = 4$. This solution is also considered an *integer* solution, in that all values in $\mathbf{X}$ and $\mathbf{Y}$ satisfy the integrality constraints of the IP.

4.1.2.1   The Branch-and-Bound Algorithm

To achieve an optimal integer solution like the one shown in (4.8), we can use the branch-and-bound algorithm. Branch-and-bound operates by considering *relaxations* of the IP model in which some or all of the integrality constraints are dropped. This allows the decision variables to assume fractional values.

For example, consider the case where we remove the requirement that the elements of $\mathbf{X}$ need to be binary (that is, we ignore (4.6)). This relaxation now has the following optimal solution:

$$\mathbf{X} = \begin{pmatrix} 0 & 1/2 & 0 & 1/2 \\ 0 & 1/2 & 0 & 1/2 \\ 0 & 1/2 & 0 & 1/2 \\ 0 & 1/2 & 0 & 1/2 \end{pmatrix} \tag{4.9}$$

$$\mathbf{Y} = \begin{pmatrix} 0 & 1 & 0 & 1 \end{pmatrix}.$$

We can see that the cost of this solution is $\sum_{j=1}^{n} Y_j = 2$; however, it is not valid from the point of view of graph colouring because each vertex is specified as being (nonsensically) "half assigned" to colour 2 and "half assigned" to colour 4. That said, this solution is still helpful because it provides us with a lower bound on the optimal cost: here it tells us that any integer solution for this problem must have a cost of at least two (colours).

Branch-and-bound uses these ideas by first removing the integrality constraints of the problem. The resultant problem, known as the LP *relaxation* of the original IP, can then be solved using the simplex method. If the optimal solution to this relaxation also happens to satisfy the integrality constraints, then we have found the optimal solution to the original IP and can finish. More likely, however, is that the solution to this LP will feature some fractional values where integer values are required (as with (4.9)). In cases where binary values are required for the decision variables, at this point we can now select one of these fractional-valued variables X and impose two new constraints, $X = 0$ and $X = 1$.

If the original LP relaxation is denoted by LP_1, let us now denote these two new problems by LP_2, where $X = 0$ is imposed, and LP_3, where $X = 1$ is imposed. In this case, the variable X is called the "branching variable", and we are said to have *branched on* X to produce two new subproblems LP_2 and LP_3. This means that we have now replaced LP_1 with two, more restricted problems. We can now apply the same ideas to these new LPs by solving them and, if necessary, performing further branching. By following this process we therefore construct a *search tree* with the root node LP_1. The other nodes in this tree are then the various LPs generated by branching, and the leaves of this tree are the nodes that have not yet been branched on.

Now suppose that we have just solved a particular node in the search tree. If all of the integrality constraints of the original IP are satisfied in this solution, then we have found a new integer (feasible) solution. This node can now be labelled as *fathomed*, meaning that there is no need to branch on it. It therefore forms a permanent leaf in the search tree. In addition, we can also check to see if this integer solution is the best integer solution seen so far during the run. If this is the case, then this solution,

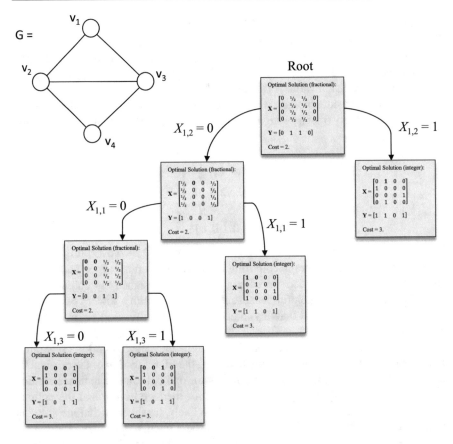

Fig. 4.3 Example branch-and-bound search tree for the IP specified in (4.1)–(4.7). In this case, the root problem is defined by dropping the integrality constraints for **X**. The branching variable is always selected as the first fractional value in **X**

together with its cost, should be stored. There are also two further ways in which a node can become fathomed. First, if the LP specified at this node admits no feasible solution; second, if the optimal solution to the LP has a cost that is worse than the best integer solution observed so far.

An example of how the branch-and-bound process operates with our IP formulation for graph colouring is shown in Fig. 4.3. To begin, we have defined the root problem by removing the integrality constraints for **X**. An optimal solution for this relaxed problem appears at the top of the figure and features a cost of two. Since the solution contains fractional values, we now branch on this node. This results in two new problems, one with the additional constraint $X_{1,2} = 0$ and one where $X_{1,2} = 1$. Continuing this process leads to the tree shown. Note that in this particular case, all of the leaf nodes result in integer solutions; hence, all leaf nodes are fathomed and there is no need for further branching. The best observed integer solution in this tree

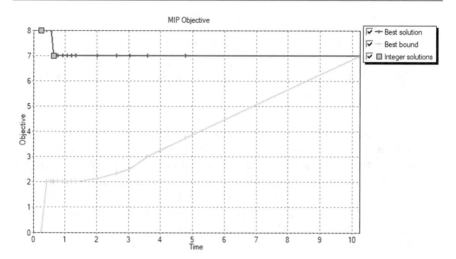

Fig. 4.4 Screenshot of the application XPress IVE. In this case, we are using branch-and-bound to optimally colour a small random graph $G_{30,0.5}$ using the IP given in (4.1)–(4.7). Here, "best solution" indicates the upper bound and "best bound" indicates the lower bound. The algorithm halts (with a certificate of optimality) when these values have become equal

corresponds to an optimal solution to the original IP. The cost of this solution also corresponds to the chromatic number of the graph (three in this case).

During the execution of the branch-and-bound algorithm, note that two important values are stored. The first of these is the cost of the best integer solution observed so far. In cases where we are attempting to minimise the cost function (such as here), this value gives an upper bound, telling us that we will never need to accept a solution with a cost that is worse (higher) than this value. The second value is a lower bound and is obtained by taking the best (minimum) cost across all of the unfathomed leaf nodes in the current search tree. This tells us that there is no integer solution to the original IP that has a cost that is better (lower) than this value. If an integer solution is obtained whose cost equals this lower bound, this tells us that an optimal solution has been found. In this case, the branch-and-bound algorithm can halt immediately, providing the user with a certificate of optimality.

Figure 4.4 illustrates how branch-and-bound can refine these upper and lower bounds during a run using a small random graph $G_{30,0.5}$. As shown, in this particular case, the algorithm has produced integer solutions with costs of eight, and then seven, in just under a second. In the remainder of the run, this latter solution is not improved upon, so the upper bound does not change further; however, the expansion of the search tree allows the lower bound to be improved. At around 10 seconds, the upper and lower bounds are seen to be equal, proving that an optimal solution to the problem (using seven colours) has been obtained.

Fig. 4.5 Example graph
with $n = 8$ vertices and
$m = 12$ edges

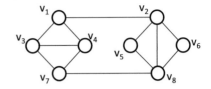

Note that our descriptions of the branch-and-bound framework in this section
have left various components unspecified. These include the search strategy (i.e.,
the order in which nodes in the search tree are generated and solved), the branching
strategy (i.e., how branching variables are selected), and the pruning rules (i.e., the
rules that prevent the exploration of suboptimal subtrees). A survey of the different
options available is due to Morrison et al. [3]. Nowadays, commercial solvers, such as
Xpress, CPLEX, and Gurobi, also contain various other mechanisms for improving
the performance of branch-and-bound. These include the use of parallelism, pre-
solve routines, cutting planes, and heuristics.

The fact that $\mathcal{NP}$-hard problems can be formulated using IP implies that branch-
and-bound will not always be able to determine an optimal solution in reasonable
time; indeed, in some cases, it may not even be able to determine an *integer* solution
in acceptable run time. However, if an integer solution has been found, then the gap
between the generated upper and lower bounds will provide the user with useful
information about its quality.

There will also often be more than one way in which a combinatorial problem can
be expressed using IP. The task of finding a "good" formulation is very important but
may sometimes be quite empirical in nature. Avoiding features such as "symmetry"
in the formulation will often help, as we now discuss.

4.1.2.2 Avoiding Solution Symmetry

One of the major drawbacks with the IP model defined in the previous section is
its underlying symmetry. Indeed, it is noticeable in this model that any feasible k-
colouring of a graph can be expressed in $^{n}P_k = \frac{n!}{(n-k)!}$ different ways by simply
permuting the columns of the **X** matrix. This has the potential to make branch-and-
bound very inefficient as it can drastically increase the size of the search tree.

One way to alleviate this problem might be to introduce further constraints as
follows:

$$Y_j \geq Y_{j+1} \qquad\qquad \forall j \in \{1, \ldots, n-1\}. \qquad (4.10)$$

These constraints ensure that a k-coloured solution will only ever use the columns
labelled 1 to k in **X**. For example, consider the graph shown in Fig. 4.5. In our original

model (4.1)–(4.7), the returned optimal solution is

$$\mathbf{X} = \begin{pmatrix} 0\,0\,1\,0\,0\,0\,0\,0 \\ 0\,0\,0\,0\,0\,1\,0\,0 \\ 0\,0\,0\,0\,0\,1\,0\,0 \\ 1\,0\,0\,0\,0\,0\,0\,0 \\ 0\,0\,1\,0\,0\,0\,0\,0 \\ 0\,0\,1\,0\,0\,0\,0\,0 \\ 0\,0\,1\,0\,0\,0\,0\,0 \\ 1\,0\,0\,0\,0\,0\,0\,0 \end{pmatrix}$$

$$\mathbf{Y} = \begin{pmatrix} 1\,0\,1\,0\,0\,1\,0\,0 \end{pmatrix}.$$

In this case, we see that, out of the eight available colours, this solution is using colours 1, 3, and 6. However, by adding (4.10) to the IP, we now get the following optimal solution:

$$\mathbf{X} = \begin{pmatrix} 1\,0\,0\,0\,0\,0\,0\,0 \\ 0\,0\,1\,0\,0\,0\,0\,0 \\ 0\,0\,1\,0\,0\,0\,0\,0 \\ 0\,1\,0\,0\,0\,0\,0\,0 \\ 1\,0\,0\,0\,0\,0\,0\,0 \\ 1\,0\,0\,0\,0\,0\,0\,0 \\ 1\,0\,0\,0\,0\,0\,0\,0 \\ 0\,1\,0\,0\,0\,0\,0\,0 \end{pmatrix}$$

$$\mathbf{Y} = \begin{pmatrix} 1\,1\,1\,0\,0\,0\,0\,0 \end{pmatrix}.$$

This solution actually gives us the same partition of the vertices as the previous one (namely, $\{\{v_1, v_5, v_6, v_7\}, \{v_2, v_3\}, \{v_4, v_8\}\}$); however, it is now only using the first $k = 3$ columns of $\mathbf{X}$.

Although this extra constraint provides an improvement on the first model, note that there are still equivalent solutions that arise due to the $k!$ ways in which the first k colours can be permuted. To eliminate some of these equivalent solutions, we might instead choose to replace (4.10) with the following:

$$\sum_{i=1}^{n} X_{ij} \geq \sum_{i=1}^{n} X_{ij+1} \qquad\qquad \forall j \in \{1, \ldots, n-1\}. \qquad (4.11)$$

These new constraints ensure that the number of vertices assigned to colour j is always greater than or equal to the number of vertices assigned to colour $j + 1$. This again presents improvements over the preceding model, but there still exists some symmetry due to colour classes of the same size being interchangeable. One final improvement suggested by Méndez-Díaz and Zabala [4] can therefore be obtained by replacing (4.11) with the following two constraints:

$$X_{ij} = 0 \qquad\qquad \forall v_i \in V, \ j \in \{i + 1, \ldots, n\} \qquad (4.12)$$

$$X_{ij} \leq \sum_{l=j-1}^{i-1} X_{lj-1} \qquad \forall v_i \in V - \{v_1\}, \ \forall j \in \{2, \ldots, i - 1\}. \qquad (4.13)$$

Together, these constraints specify a unique permutation of the first k columns for each possible k-colouring. Specifically, vertex v_1 must be assigned to colour 1, v_2 must be assigned to either colour 1 or colour 2, and so on. (Or, in other words, the columns are sorted according to the minimally labelled vertex in each colour class.) Under these constraints the optimal solution to our example problem is now:

$$
\mathbf{X} = \begin{pmatrix}
1 & 0 & 0 & 0 & 0 & 0 & 0 & 0 \\
0 & 1 & 0 & 0 & 0 & 0 & 0 & 0 \\
0 & 1 & 0 & 0 & 0 & 0 & 0 & 0 \\
0 & 0 & 1 & 0 & 0 & 0 & 0 & 0 \\
1 & 0 & 0 & 0 & 0 & 0 & 0 & 0 \\
1 & 0 & 0 & 0 & 0 & 0 & 0 & 0 \\
1 & 0 & 0 & 0 & 0 & 0 & 0 & 0 \\
0 & 0 & 1 & 0 & 0 & 0 & 0 & 0
\end{pmatrix}
$$

$$
\mathbf{Y} = \begin{pmatrix} 1 & 1 & 1 & 0 & 0 & 0 & 0 & 0 \end{pmatrix}
$$

as required.

4.1.2.3 Other Improvements

In addition to eliminating symmetries, there are also other ways in which the performance of the branch-and-bound algorithm for graph colouring might be improved.

First, note that our current IP model uses the matrices $\mathbf{X}_{n \times n}$ and $\mathbf{Y}_n$. This allows up to n colours to be used by the solution, though fewer colours will usually be sufficient. The size of the branch-and-bound search tree might therefore be reduced by first determining an upper bound k on the chromatic number $\chi(G)$, and then using $\mathbf{X}_{n \times k}$ and $\mathbf{Y}_k$ instead. A suitable value for k could be found using one of the upper bounds seen in Chap. 3, or by using a heuristic to construct a feasible solution $\mathcal{S}$, and then setting k to $|\mathcal{S}|$.

As with the backtracking method, improvements might also be seen by first identifying a large clique C in G and then relabelling the vertices such that $C = \{v_1, v_2, \ldots, v_l\}$. Since all of the vertices in this clique need to be assigned to different colours, extra constraints of the form

$$ X_{ii} = 1 \qquad\qquad \forall v_i \in C \qquad\qquad (4.14) $$

can then be added to the model. Once again, the intention behind this modification is to reduce the size of the search tree and therefore improve the performance of branch-and-bound. Whether this makes a difference in practice will usually need to be checked empirically, however.

4.1.2.4 Branch-and-Bound Performance

In this section, we now provide some empirical data on how well the branch-and-bound method is able to cope with different graph colouring problem instances. To do this, we use our previous IP model with the additional antisymmetry constraints defined by (4.12) and (4.13). Given the (binary) matrices $\mathbf{X}_{n \times n}$ and $\mathbf{Y}_n$, the full IP model is therefore

$$ \min \sum_{j=1}^{n} Y_j \qquad\qquad (4.15) $$

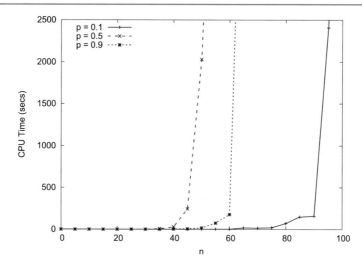

Fig. 4.6 Time required to solve (to optimality) various random graphs $G_{n,p}$ using branch-and-bound

subject to:

$$X_{ij} + X_{lj} \leq Y_j \qquad\qquad \forall \{v_i, v_l\} \in E, \ \forall j \in \{1, \ldots, n\} \qquad (4.16)$$

$$\sum_{j=1}^{n} X_{ij} = 1 \qquad\qquad \forall v_i \in V \qquad (4.17)$$

$$X_{ij} = 0 \qquad\qquad \forall v_i \in V, \ j \in \{i+1, \ldots, n\} \qquad (4.18)$$

$$X_{ij} \leq \sum_{l=j-1}^{i-1} X_{lj-1} \qquad \forall v_i \in V - \{v_1\}, \ \forall j \in \{2, \ldots, i-1\} \qquad (4.19)$$

$$X_{ij} \in \{0, 1\} \qquad\qquad \forall v_i \in V, \ \forall j \in \{1, \ldots, n\} \qquad (4.20)$$

$$Y_j \in \{0, 1\} \qquad\qquad \forall j \in \{1, \ldots, n\}. \qquad (4.21)$$

In these trials, we used the software XPress IVE (v. 8.11) for specifying and solving this IP model. A full listing of this code is given in Appendix A.5. All trials were conducted on a 3.2 GHz Windows machine with 8 GB of RAM using a time limit of 1 hour.

Figure 4.6 shows the times required for solving various random graphs $G_{n,p}$. For $p = 0.5$, we see that branch-and-bound can solve graphs of up to 40 vertices in very short amounts of time. Beyond this, however, the time requirements increase drastically—indeed, no graphs with $n > 50$ were solved within the 1-hour time limit. For the other values of p, similar results occur, though slightly larger values of n can be tolerated. That said, no graphs with $n \geq 100$ have been solved within the time limit here.

Fig. 4.7 The bars in these charts show the time requirements of branch-and-bound to solve random graphs $G_{n,p}$ to optimality (for $n = 25$, 50, and 100, respectively). The lines in the charts show the upper and lower bounds achieved within the 1-hour time limit. In cases where these bounds are equal, the problem has been solved to optimality; in the remaining cases, branch-and-bound has terminated without finding a provably optimal solution

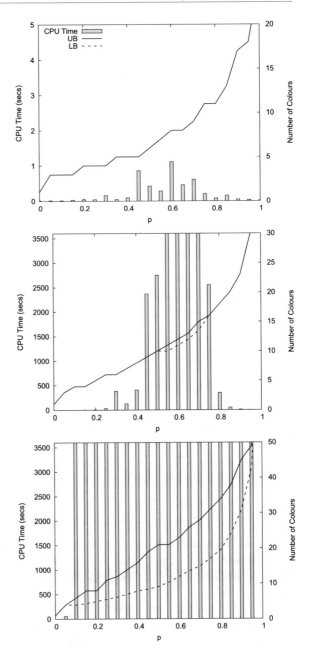

Similar results to these are shown in Fig. 4.7. In the top chart (where $n = 25$), all problems are solved to optimality in a matter of seconds. For the larger random graphs $G_{50,p}$, we see that most problems are solved within the time limit, but that difficulties arise for values of p between 0.55 and 0.7. In these cases, there exists a gap between the upper and lower bounds, indicating that a provably optimal solution has not been established within the 1-hour time limit. These patterns are even more striking for $G_{100,p}$ where only $p = 0.05$ is solved within the time limit.

4.1.3   Minimum Coverings and Column Generation

An exact algorithm for graph colouring can also be created using a technique known as *column generation*. The idea in column generation is to make use of integer programming, but to also avoid considering all variables of the problem explicitly; instead, variables are only introduced when they have the potential to improve the objective function. This will often make problem sizes more manageable, allowing larger problems to be tackled.

To apply column generation with graph colouring, we can make use of the minimum set covering problem. This is defined as follows.

> **Definition 4.1** Let $U = \{1, 2, \dots, n\}$ be a set of integers known as the "universe" and let $\mathbb{S} = \{S_1, S_2, \dots, S_l\}$ be a set whose elements are subsets of this universe. The *minimum set covering problem* involves identifying the smallest subset of $\mathbb{S}$ whose union equals the universe. That is, we are seeking a subset $\mathcal{S} \subseteq \mathbb{S}$ for which $\bigcup_{S \in \mathcal{S}} S = U$ and where $|\mathcal{S}|$ is minimal.

For example, given $U = \{1, 2, 3, 4\}$ and $\mathbb{S} = \{\{1\}, \{1, 2\}, \{1, 3\}, \{3, 4\}, \{4\}\}$, one example covering is $\{\{1, 2\}, \{1, 3\}, \{4\}\}$, which contains three elements of $\mathbb{S}$ and all members of U. However, the *minimal* covering in this case is $\{\{1, 2\}, \{3, 4\}\}$, which contains just two elements of $\mathbb{S}$.

The minimum set covering problem can be formulated by the following integer program:

$$\min \sum_{S \in \mathbb{S}} X_S \tag{4.22}$$

subject to:

$$\sum_{S:u \in S} X_S \geq 1 \qquad\qquad \forall u \in U \tag{4.23}$$

$$X_S \in \{0, 1\} \qquad\qquad \forall S \in \mathbb{S}. \tag{4.24}$$

In this formulation, the variable $X_S = 1$ if the subset $S \in \mathbb{S}$ is being used in the covering; else $X_S = 0$. We are therefore seeking to minimise the number of sets being used for the covering (4.22), while covering every element u of the universe U (4.23). The integrality constraints are given by (4.24).

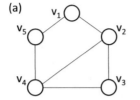 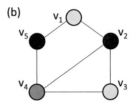

Fig. 4.8 Part **a** shows a small graph with $n = 5$ vertices and $m = 6$ edges. As a set covering problem, this graph colouring problem is defined by the universe $U = V = \{v_1, v_2, v_3, v_4, v_5\}$ and the set of all maximal independent sets $\mathbb{S} = \{\{v_1, v_3\}, \{v_1, v_4\}, \{v_2, v_5\}, \{v_3, v_5\}\}$. In this case, a minimum set covering of the universe is given by $S = \{\{v_1, v_3\}, \{v_1, v_4\}, \{v_2, v_5\}\}$, which has a size of three. If S contains any duplicates (as is the case with v_1 here), these can be removed arbitrarily to form a feasible colouring. An optimal three-colouring in this example is therefore $S = \{\{v_1, v_3\}, \{v_4\}, \{v_2, v_5\}\}$, as shown in Part **b**

To solve the graph colouring problem using set covering, we can take the set $\mathbb{S}$ of all maximal independent sets of a graph G (see Definition 3.14). An optimal colouring of G can then be found by simply identifying a minimum set covering S of the universe $U = V = \{v_1, v_2, \ldots, v_n\}$. An example of this process is shown in Fig. 4.8.

Attempting to solve the graph colouring problem using a set covering formulation brings two difficulties, however.

- First, the set covering problem is itself an $\mathcal{NP}$-hard problem [5], implying that it cannot be solved in polynomial time in general.
- Second, the number of maximal independent sets in a graph has the potential to grow exponentially in relation to graph size, meaning that the task of constructing a set $\mathbb{S}$ that contains *all* maximal independent sets will often be beyond our means.

To illustrate the second point, Table 4.1 shows the number of maximal independent sets that exist in different random graphs $G_{n,p}$. As n is increased, we see that the number of maximal independent sets grows quickly. This is particularly the case for sparser graphs where, in our case, there is often insufficient memory to complete the computation. On the other hand, for dense graphs, the presence of so many edges in the graphs makes the maximal independent sets quite small, meaning that they can be enumerated quickly, even for fairly large values of n.

If we want to use set covering principles to solve the graph colouring problem, an obvious approach is to populate the set $\mathbb{S}$ with all maximal independent sets of G and then seek to solve the integer program specified in (4.22)–(4.24). As we have seen, however, this has the potential to involve huge numbers of variables. In addition, it is likely that most of these variables will assume values of zero, meaning that the corresponding members of $\mathbb{S}$ are not needed to optimally colour the graph. Column generation seeks to resolve these issues by using a smaller number of variables, and then only adding further variables to the IP when it is deemed necessary.

Table 4.1 Number of maximal independent sets in random graphs $G_{n,p}$. These figures were generated using the NetworkX command `nx.find_cliques(H)`, where H is the complement of the random graph $G_{n,p}$. Further details on the NetworkX library are given in Appendix A.3. The number of seconds required to execute these operations is shown in italics. Missing values indicate that the operations did not complete due to an Out of Memory error. All trials were conducted on a 3.2 GHz Windows machine with 8 GB RAM

n	Edge probability p				
	0.1	0.25	0.5	0.75	0.9
50	43,815	9015	1021	205	87
	0.26	*0.06*	*<0.01*	*<0.01*	*<0.01*
100	–	1,314,202	14,841	1227	337
		9.75	*0.11*	*<0.01*	*<0.01*
250	–	–	1,578,449	22,925	2599
			13.82	*0.16*	*<0.01*
500	–	–	–	263,957	15,903
				1.95	*0.09*
1000	–	–	–	4,144,959	99,646
				37.97	*0.60*
2000	–	–	–	–	747,296
					6.80

In more detail, column generation operates by starting with a relatively small set $\mathbb{S}$. The contents of $\mathbb{S}$ might be generated at random or via specialised heuristics—at this stage it is only necessary that $\mathbb{S}$ covers the universe of the set covering problem. In the next stage, the linear relaxation of the set covering IP is solved. The dual values from this optimal solution are then used to define a so-called *pricing problem* that needs to be solved. The solution to this pricing problem can then be used to determine if further variables need to be added to the IP. If this is the case then $\mathbb{S}$ is updated, the linear relaxation is solved again, and the process is repeated. Otherwise, an integer solution to the IP gives an optimal solution to the original problem.

The original application of column generation to graph colouring is due to Mehrotra and Trick [6], who identified the pricing problem as the maximum weighted independent set problem. This is defined as follows.

Definition 4.2 Let $G = (V, E)$ be a graph with weights $w(v)$ for each vertex $v \in V$. The *maximum weighted independent set problem* involves identifying an independent set of vertices $V' \subseteq V$ whose weight $\sum_{v \in V'} w(v)$ is maximal among all independent sets of G.

In this work, the authors also consider a relaxed version of the set covering IP in which the integrality constraints are replaced by the requirement that each variable

X_S should be nonnegative. This LP is as follows:

$$\min \sum_{S \in \mathbb{S}} X_S \qquad (4.25)$$

subject to:

$$\sum_{S:e \in S} X_S \geq 1 \qquad \forall e \in U \qquad (4.26)$$

$$X_S \geq 0 \qquad \forall S \in \mathbb{S}. \qquad (4.27)$$

The overall column generation algorithm of Mehrotra and Trick [6] now operates using the following four steps. An example run of this algorithm is shown in Fig. 4.9.

1. Given a graph $G = (V, E)$, let the universe $U = V$. Also, let $\mathbb{S} = \{S_1, S_2, \ldots, S_l\}$ be a set whose elements are maximal independent sets of G, and for which $\bigcup_{S \in \mathbb{S}} S = U$.
2. Solve the LP given by (4.25)–(4.27). The optimal solution to this LP gives a dual value π_i for each vertex $v_i \in V$.
3. For each $v_i \in V$, let $w(v_i) = \pi_i$. Now solve the maximum weighted independent set problem. If the weight of this solution $\sum_{v \in V'} w(v) > 1$, then add V' to $\mathbb{S}$ and return to Step 2. Otherwise proceed to Step 4.
4. If the current solution to the linear relaxation is an integer solution then stop: we have determined an optimal colouring of G. Otherwise, solve the original set covering IP (4.22)–(4.24) using $\mathbb{S}$. An optimal solution to this IP corresponds to an optimal colouring of G.

Note that in each application of Step 3, a solution to the maximum weighted independent set problem is required. This problem is $\mathcal{NP}$-hard, so attempting to solve it exactly has the potential to slow the overall algorithm considerably.[1] To help mitigate this, Mehrotra and Trick [6] note that it is not necessary to find the *optimal* solution in each case; instead, it suffices to find *any* independent set whose weight is greater than one. Options considered by the authors include a backtracking procedure and heuristics. They also propose a modification in which these methods seek to find independent sets with weights that exceed a target of 1.1, thereby encouraging better quality independent sets to be added to $\mathbb{S}$ at earlier stages of the run. In empirical tests, their algorithm is shown to solve random graphs $G_{n,0.5}$ of up to 70 vertices in less than 1 min. For other topologies, graphs with many hundreds of vertices are solved.

In more recent work, Gualandi and Malucelli [7] have sought to enhance this column generation algorithm by using other methods for solving the maximum weight

---

[1] The problem is $\mathcal{NP}$-hard because it is a generalisation of the (unweighted) maximum independent set problem seen in Sect. 2.3, which is also $\mathcal{NP}$-hard. Any instance of the latter can be converted into a corresponding instance of the former by simply allocating a weight of one to all vertices of the graph G. A maximum weighted independent set of this graph is then equivalent to a maximum independent set of G.

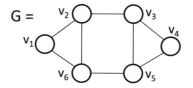

Iteration 1:

$\mathbb{S}$ $= \{\{v_1, v_3\}, \{v_1, v_4\}, \{v_1, v_5\}, \{v_2, v_4\}, \{v_4, v_6\}\}$

Optimal $= (1, 0, 1, 1, 1)$

Duals $= (0, 1, 1, 0, 1, 1)$

The maximum weight independent set $\{v_2, v_5\}$ has a weight of two. Add this to $\mathbb{S}$.

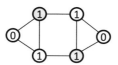

Iteration 2:

$\mathbb{S}$ $= \{\{v_1, v_3\}, \{v_1, v_4\}, \{v_1, v_5\}, \{v_2, v_4\}, \{v_4, v_6\}, \{v_2, v_5\}\}$

Optimal $= (1, 0, 0, 0, 1, 1)$

Duals $= (0, 0, 1, 0, 1, 1)$

The maximum weight independent set $\{v_3, v_6\}$ has a weight of two. Add this to $\mathbb{S}$.

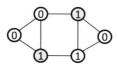

Iteration 3:

$\mathbb{S}$ $= \{\{v_1, v_3\}, \{v_1, v_4\}, \{v_1, v_5\}, \{v_2, v_4\}, \{v_4, v_6\}, \{v_2, v_5\}, \{v_3, v_6\}\}$

Optimal $= (1, 0, 0, 0, 1, 1, 0)$

Duals $= (0, 0, 1, 1, 1, 0)$

The maximum weight independent set has a weight of one. An optimal (integer) solution to this set covering problem therefore corresponds to an optimal colouring of G.

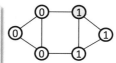

Fig. 4.9 Example run of the column generation algorithm for graph colouring. In Iteration 1, the contents of $\mathbb{S}$ are maximal independent sets that, together, cover the vertex set of G. At the end of Iteration 3, an optimal colouring $\{\{v_1, v_3\}, \{v_2, v_5\}, \{v_4, v_6\}\}$ for G has been determined. Note that in this particular example the variables in the optimal solutions have assumed binary values; however, this is not enforced. Instead, each variable only needs to assume a nonnegative value, as defined in the LP given in (4.25)–(4.27)

independent set problem. Options considered include high-performance exact methods and constraint programming. Some variants of their algorithm are shown to be able to solve random graphs $G_{n,0.5}$ of up to 250 vertices within 1 h.

4.2 Inexact Heuristics and Metaheuristics

Another promising approach to graph colouring is to make use of heuristic- and metaheuristic-based approaches. In contrast to the exact methods that we have considered, these approaches are *inexact* and, as such, are not guaranteed to return an optimal solution, even when granted excess time. Also, even if they do happen to achieve an optimal solution, these methods will often fail to recognise this and will continue to execute until some user-defined stopping criterion is met.

On the other hand, heuristics and metaheuristics have the advantage of being highly adaptable to different problems. As a result, various algorithms of this type have been shown to produce excellent results on a large range of different graph colouring problems. Often, heuristics and metaheuristics are also considered to produce better solutions than exact algorithms for larger instances, though this depends on the makeup of the graph, together with the amount of time the user is willing to wait.

In this section, we discuss several ideas stemming from these algorithms. In our descriptions, we will often refer to different types of metaheuristics and will describe these in more detail as they arise in the text. For now, it suffices to say that metaheuristics can be considered types of *higher level* algorithmic frameworks that are intended to be applicable to a variety of different optimisation problems with relatively few modifications needing to be made in each case. Examples of metaheuristics include evolutionary algorithms, simulated annealing, tabu search, and ant colony optimisation. Typically they operate by navigating their way through a large set of candidate solutions (the *solution space*), attempting to optimise an objective function that reflects each member's quality. In some cases, metaheuristics will attempt to make educated decisions about which parts of the solution space to explore based on the characteristics of candidate solutions already observed. This allows them to "learn" about the solution space during a run and therefore hopefully make good decisions about where high-quality best solutions can be found.

In terms of the graph colouring problem, it is helpful to classify heuristic and metaheuristic methods according to how their solution spaces are defined. Specifically, we can consider feasible-only solution spaces, spaces of complete improper k-colourings, and spaces of incomplete proper k-colourings. Each of these categories is now considered in turn.

4.2.1   Feasible-Only Solution Spaces

The first set of algorithms we consider are those that operate in the space of feasible colourings. Approaches of this type seek to identify solutions within this space that feature the minimum number of colours. Often these methods make use of the GREEDY algorithm to construct solutions; hence, they are concerned with identifying good permutations of the vertices. (Recall from Theorem 3.2 that, for any graph, a permutation of the vertices always exists that decodes into an optimal solution through the application of GREEDY.)

One early example of this type of approach was the *iterated Greedy* algorithm of Culberson and Luo [8]. This rather elegant algorithm exploits the findings of Theorem 3.1, namely, that, given a feasible colouring $\mathcal{S}$, a permutation of the vertices can be formed that, when fed back into the GREEDY algorithm, results in a new feasible solution $\mathcal{S}'$ such that $|\mathcal{S}'| \leq |\mathcal{S}|$. To start, DSATUR is used to produce an initial feasible solution. Then, at each iteration, the current solution $\mathcal{S} = \{S_1, \ldots, S_{|\mathcal{S}|}\}$ is taken and its colour classes are reordered to form a new permutation of the vertices. This permutation is then used with GREEDY to produce a new feasible solution before the process is repeated indefinitely.

Culberson and Luo [8] suggest several ways in which reorderings of the colour classes can be achieved at each iteration. These include the following:

- *Largest first*: Arranging the colour classes in order of decreasing size;
- *Reverse*: Reversing the order of the colour classes in the current solution; and
- *Random*: Rearranging the colour classes randomly.

The largest first heuristic is used in an attempt to construct large independent sets in the graph, while the reverse heuristic encourages vertices to be mixed among different colour classes. The random heuristic is then used to try and prevent the algorithm from cycling among the same set of solutions, allowing new regions of the solution space to be explored. Culberson and Luo [8] ultimately recommend selecting these heuristics randomly at each iteration according to the ratio 5:5:3, respectively.

Two further algorithms that operate within the feasible-only space are the evolutionary algorithms (EAs) of Mumford [9] and Erben [10]. EAs are a type of metaheuristic inspired by biological evolution. They operate by maintaining a *population* of candidate solutions that represent a sample of the solution space. During a run, EAs then attempt to improve the quality of members within this population using the following operators:

- *Recombination*. This aims to create new solutions by combining different parts of existing members of the population. Often these "parts" are referred to as "building blocks", and the existing and new solutions are known as "parent" and "offspring" solutions, respectively.
- *Mutation*. This makes changes to a candidate solution to allow new regions of the solution space to be explored. These changes might be made randomly, or perhaps via some sort of local search operator.

	Parent S_1	Parent S_2	Offspring S'	Comments
1)	$\{\{v_1,v_2,v_3\},$ $\{v_4,v_5,v_6,v_7\},$ $\{v_8,v_9,v_{10}\}\}$	$\{\{v_3,v_4,v_5,v_7\},$ $\{v_1,v_6,v_9\},$ $\{v_2,v_8\},$ $\{v_{10}\}\}$	$\{\{v_1,v_2,v_3\},$ $\{v_4,v_5,v_6,v_7\},$ $\{v_8,v_9,v_{10}\}\}$	S' is initially a copy of Parent S_1
2)	$\{\{v_1,v_2,v_3\},$ $\{v_4,v_5,v_6,v_7\},$ $\{v_8,v_9,v_{10}\}\}$	$\{\{v_3,v_4,v_5,v_7\},$ $\{v_1,v_6,v_9\},$ $\{v_2,v_8\},$ $\{v_{10}\}\}$	$\{\{v_1,v_2,v_3\},$ $\{v_4,v_5,v_6,v_7\},$ $\{v_8,v_9,v_{10}\},$ $\{v_2,v_8\},$ $\{v_{10}\}\}$	Randomly select come colour classes from S_2 and copy them to S'. The sets $\{v_2,v_8\}$ and $\{v_{10}\}$ are chosen here
3)	$\{\{v_1,v_2,v_3\},$ $\{v_4,v_5,v_6,v_7\},$ $\{v_8,v_9,v_{10}\}\}$	$\{\{v_3,v_4,v_5,v_7\},$ $\{v_1,v_6,v_9\},$ $\{v_2,v_8\},$ $\{v_{10}\}\}$	$\{\{v_4,v_5,v_6,v_7\},$ $\{v_2,v_8\},$ $\{v_{10}\}\}$	Remove classes from S' that came from S_1 and that contain duplicates. In this case vertices v_1, v_3 and v_9 and are now missing from S'
4)	$\{\{v_1,v_2,v_3\},$ $\{v_4,v_5,v_6,v_7\},$ $\{v_8,v_9,v_{10}\}\}$	$\{\{v_3,v_4,v_5,v_7\},$ $\{v_1,v_6,v_9\},$ $\{v_2,v_8\},$ $\{v_{10}\}\}$	$\{\{v_4,v_5,v_6,v_7\},$ $\{v_2,v_8,v_9,v_1\},$ $\{v_{10},v_3\}\}$	Reinsert missing vertices into S' using GREEDY to form a feasible colouring.

Fig. 4.10 Example application of the recombination operator of Erben [10]

- *Evolutionary pressure.* As with biological evolution, EAs usually also exhibit some bias towards keeping good candidate solutions in the population and rejecting bad ones. Hence, high-quality solutions are more likely to be used for generating new offspring, and weaker solutions are more likely to be deleted from the population.

The evolution, and hopefully improvement, of an EA's population takes place through the repeated application of the above operators; however, it is also usually necessary to design specialised recombination and mutation operators that can suitably exploit the underlying structures of the problem at hand.

The recombination operator of Erben [10] seeks to do this by considering the *colour classes* of a solution as the underlying building blocks. Specifically, it operates by first taking two feasible parent solutions S_1 and S_2. A subset of colour classes is then selected from S_2 and copied into a copy of S_1 to form a new offspring solution called S'. At this point, S' will contain multiple occurrences of some vertices, and so the algorithm goes through the colour classes of S' and deletes all colour classes that came from parent S_1 and that contain a duplicate. This operation results in an offspring solution that is proper, but most likely partial; thus any missing vertices are randomly permuted and reinserted back into S' using the GREEDY algorithm to form a feasible solution. An example of this process is shown in Fig. 4.10.

One notable feature of this recombination operator is that, before the GREEDY algorithm is used to reinsert missing vertices, each colour class in the offspring will be a copy of a colour class existing in at least one of the parents. That is:

$$S_i \in S' \implies S_i \in (S_1 \cup S_2). \tag{4.28}$$

In particular, if a colour class is seen to exist in both parents then this colour class will be guaranteed to be present in the offspring solution. Features such as these are generally considered desirable in a recombination operator in that they provide a mechanism by which useful building blocks (colour classes in this case) can be passed from parents to offspring. For this particular operator, however, there is also the possibility that an offspring might inherit all of its colour classes from the second parent due to the policy of deleting colour classes originating from the first parent.

The mutation operator of this EA works in a similar fashion to recombination by deleting some randomly selected colour classes from a solution, randomly permuting these vertices, and then reinserting them into the solution via GREEDY. The following heuristic-based objective function is also proposed for gauging the quality of a solution $\mathcal{S}$:

$$f_1(\mathcal{S}) = \frac{\sum_{S_i \in \mathcal{S}} \left(\sum_{v \in S_i} \deg(v) \right)^2}{|\mathcal{S}|}. \tag{4.29}$$

Here, $\sum_{v \in S_i} \deg(v)$ gives the total degree of all vertices assigned to the colour class S_i. The aim is to maximise f_1 by making increases to the numerator (by forming large colour classes containing high-degree vertices) and/or decreases to the denominator (by reducing the number of colour classes). It is also suggested that this objective function allows evolutionary pressure to be sustained in a population for longer during a run (compared to the more obvious choice of using the number of colours $|\mathcal{S}|$), because it allows greater distinction between individuals.

The EA of Mumford [9] also seeks to construct offspring solutions by combining the colour classes of parent solutions. In this research, two recombination operators are suggested: the Merge Independent Sets (MIS) operator and the Permutation One Point (POP) operator. The MIS operator starts by taking two feasible parent solutions, $\mathcal{S}_1$ and $\mathcal{S}_2$, and constructs two permutations. As with the iterated Greedy algorithm, vertices within the same colour classes are put into adjacent positions in these permutations. For example, the two solutions

$$\mathcal{S}_1 = \{\{v_1, v_2, v_3\}, \{v_4, v_5, v_6, v_7\}, \{v_8, v_9, v_{10}\}\}$$
$$\mathcal{S}_2 = \{\{v_1, v_6, v_9\}, \{v_2, v_8\}, \{v_3, v_4, v_5, v_7\}, \{v_{10}\}\}$$

might result in the following two vertex permutations:

$$P_1 = (v_1, v_2, v_3 : v_4, v_5, v_6, v_7 : v_8, v_9, v_{10})$$
$$P_2 = (\mathbf{v_1, v_6, v_9} : \mathbf{v_2, v_8} : \mathbf{v_3, v_4, v_5, v_7} : \mathbf{v_{10}}).$$

(For convenience, colons are used in these permutations to mark boundaries between different colour classes.) In the next step of the operator, the two permutations are merged randomly such that the boundaries between the colour classes are maintained. For example, we might merge the above examples to get

$$(\mathbf{v_1, v_6, v_9} : v_1, v_2, v_3 : v_4, v_5, v_6, v_7 : \mathbf{v_2, v_8} : \mathbf{v_3, v_4, v_5, v_7} : v_8, v_9, v_{10} : \mathbf{v_{10}}).$$

Finally, two offspring permutations are formed by using the first occurrence of each vertex for the first offspring, and the second occurrence for the second offspring:

$$P_1' = (\mathbf{v_1, v_6, v_9}, v_2, v_3, v_4, v_5, v_7, \mathbf{v_8}, v_{10})$$
$$P_2' = (v_1, v_6, \mathbf{v_2, v_3, v_4, v_5, v_7}, v_8, v_9, \mathbf{v_{10}}).$$

These permutations are then converted into feasible offspring solutions using the GREEDY algorithm.

The POP operator of Mumford [9] follows a similar scheme by first forming two permutations, P_1 and P_2, as above. A random cut point is then chosen, and the first portion of P_1 up to the cut point becomes the first portion of the first offspring permutation P_1'. The remainder of P_1' is then obtained by copying the vertices absent from P_1' in the order that they occur in P_2. The second offspring permutation P_2' is found in the same way, but with the roles of the parents reversed. For example, using "|" to signify the cut point, the permutations

$$P_1 = (v_1, v_2, v_3, v_4 \mid v_5, v_6, v_7, v_8, v_9, v_{10})$$

$$P_2 = (\mathbf{v_1, v_6, v_9, v_2} \mid \mathbf{v_8, v_3, v_4, v_5, v_7, v_{10}})$$

result in the following new permutations:

$$P_1' = (v_1, v_2, v_3, v_4, \mathbf{v_6, v_9, v_8, v_5, v_7, v_{10}}) \text{ and}$$

$$P_2' = (\mathbf{v_1, v_6, v_9, v_2}, v_3, v_4, v_5, v_7, v_8, v_{10}).$$

As before, two offspring solutions are then formed by feeding these new permutations into the GREEDY algorithm.

As we have seen, the recombination operators used in the EAs of both Mumford [9] and Erben [10] attempt to provide mechanisms by which good colour classes within a population can be propagated, thereby hopefully allowing good offspring solutions to be formed. However, the overall performance of their algorithms as presented in their papers does not seem as strong as that of other algorithms reported in the literature. That said, we will see later that excellent results can occur when evolutionary-based algorithms are hybridised with local search-based procedures.

Moving away from EAs, the technique of Lewis [11] also operates in the space of feasible solutions. The suggested algorithm makes use of operators based on the iterated Greedy algorithm for making large changes to a solution. These are then combined with other specialised operators that make small alterations to a solution while ensuring that it remains feasible at all times. These latter operators are the so-called *Kempe chain interchange* and *pair swap* operators, defined as follows:

Definition 4.3 Let $S = \{S_1, \ldots, S_k\}$ be a feasible solution. Given an arbitrary vertex $v \in S_i$ and a second colour class $S_{j \neq i}$, a *Kempe chain* is defined as a connected subgraph that contains v, and that only comprises vertices coloured with colours i and j. The set of vertices involved in such a chain is denoted by KEMPE(v, i, j). A *Kempe chain interchange* involves taking a Kempe chain and swapping the colours of all vertices contained within it.

Definition 4.4 Let the Kempe chains KEMPE(u, i, j) and KEMPE(v, j, i) both contain just one vertex each (therefore implying that u and v are nonadjacent.) A *pair swap* involves swapping the colours of u and v.

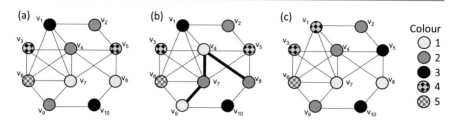

Fig. 4.11 a An example five-colouring; **b** the result of a Kempe chain interchange using KEMPE(v_7, 1, 2); **c** the result of a pair swap using v_1 and v_5

Figure 4.11 shows examples of these operators. In Fig. 4.11a, we see that KEMPE(v_7, 1, 2) = $\{v_4, v_7, v_8, v_9\}$. Interchanging the colours of these vertices gives the colouring shown in Fig. 4.11b. For an example pair swap, observe that in Fig. 4.11a the Kempe chains identified by KEMPE(v_1, 3, 4) and KEMPE(v_5, 4, 3) both contain just one vertex each. Hence, a pair swap can be performed using v_1 and v_5, as shown in Fig. 4.11c.

The fact that applications of these operators will always preserve the feasibility of a solution is due to the following theorem.

Theorem 4.1 *Given a proper solution $S = \{S_1, \ldots, S_k\}$, an application of a Kempe chain interchange or a pair swap will result in a new solution S' that is also proper.*

Proof For the Kempe chain interchange operator, consider the situation where S is proper but S' is not. Because a Kempe chain interchange involves two colours i and j, S' must feature a pair of adjacent vertices u and v that are assigned to the same colour i or the same colour j. Without loss of generality, assume this to be colour i. This tells us that u and v must have both been in the Kempe chain used for the interchange since they are adjacent, implying that u and v were both assigned to colour j in S. However, this is impossible since S is known to be proper.

According to the conditions given in Definition 4.4, v cannot be adjacent to any vertex coloured with j, and u cannot be adjacent to any vertex coloured with i. Hence, swapping the colours of u and v will also ensure that S' is proper. □

The method of Lewis [11] has been shown to outperform those of both Culberson and Luo [8] and Erben [10] on a variety of different graphs. Consequently, this forms one of the case-study algorithms discussed in Chap. 5.

4.2.2 Spaces of Complete, Improper k-Colourings

Many algorithms proposed for graph colouring have been designed to explore the space of complete improper colourings. Such methods typically start by proposing a fixed number of colours k, with each vertex then being assigned to one of these colours using heuristics, or possibly at random. During this assignment there may be vertices that cannot be assigned to any of the k colours without inducing a clash; however, these will be assigned to one of the colours anyway. (Recall that a clash occurs when a pair of adjacent vertices are assigned to the same colour—see Definition 1.4.)

The above assignment process leaves us with a k-partition of the vertices $\mathcal{S} = \{S_1, \ldots, S_k\}$ that represents a complete, but most likely improper k-colouring. A natural way to measure the quality of this solution is to now count the number of clashes. This can be achieved via the following objective function:

$$f_2(\mathcal{S}) = \sum_{\forall \{u,v\} \in E} g(u, v) \tag{4.30}$$

where

$$g(u, v) = \begin{cases} 1 \text{ if } c(u) = c(v) \\ 0 \text{ otherwise.} \end{cases}$$

The aim of an algorithm that uses this solution space is to make alterations to the k-partition so that the number of clashes is reduced to zero. If this is achieved, k might then be reduced and the process restarted. Alternatively, if all clashes cannot be eliminated, k can be increased.

Perhaps the first algorithm to make use of the above strategy was due to Chams et al. [12], who made use of the simulated annealing metaheuristic. Soon after this, Hertz and de Werra [13] proposed a similar algorithm called TABUCOL based on the tabu search metaheuristic of Glover [14]. Simulated annealing and tabu search are types of metaheuristics based on the concept of local search. In essence, local search algorithms make use of *neighbourhood operators* which are simple schemes for changing (or disrupting) a particular candidate solution. In the examples just cited, this operator is simply:

- Take a vertex v that is currently assigned to colour i, and assign it to a new colour j (where $1 \leq i \neq j \leq k$).

Given a particular candidate solution $\mathcal{S}$, the *neighbourhood* of $\mathcal{S}$, denoted by $N(\mathcal{S})$, is then defined as the set of all candidate solutions that can be produced from $\mathcal{S}$ via a single application of this operator.

A very simple way of performing local search with such a neighbourhood operator is to make use of the elementary *random descent* method. This starts by generating an initial solution $\mathcal{S}$ by some means and then evaluating it according to the chosen objective function $f(\mathcal{S})$, which we are seeking to minimise. At each iteration of the algorithm, a *move* in the solution space is attempted by randomly applying the neighbourhood operator to the incumbent solution $\mathcal{S}$ to form a new solution $\mathcal{S}'$ (that is, the new solution $\mathcal{S}'$ is chosen randomly from the set $N(\mathcal{S})$). If this new solution

Fig. 4.12 The simulated
annealing algorithm. Here, a
random value for r in the
range [0, 1] is generated in
each iteration. All other
notation is described in the
accompanying text

$$
\begin{array}{ll}
\multicolumn{2}{c}{\text{SIMULATED-ANNEALING } (i \leftarrow 0)} \\
\hline
(1) & \text{Produce an initial solution } S \\
(2) & \text{Choose an initial value for } t \\
(3) & \textbf{while (not} \text{ stopping condition) } \textbf{do} \\
(4) & \quad \text{Randomly choose } S' \in N(S) \\
(5) & \quad \textbf{if } f(S') \leq f(S) \textbf{ then} \\
(6) & \quad\quad S \leftarrow S' \\
(7) & \quad \textbf{else if } r \leq \exp(-\delta/t) \textbf{ then} \\
(8) & \quad\quad S \leftarrow S' \\
(9) & \quad i \leftarrow i+1 \\
(10) & \quad \textbf{if } i \ (\text{mod } z) = 0 \textbf{ then} \\
(11) & \quad\quad t \leftarrow \alpha t \\
\hline
\end{array}
$$

is seen to be better than the incumbent (i.e., $f(S') < f(S)$) then it is set as the incumbent for the next iteration (i.e., $S \leftarrow S'$); otherwise, no changes occur. The algorithm can then be left to run indefinitely or until some user-defined stopping criterion is met.

Although the random descent method is very intuitive, it is highly susceptible to getting caught at local optima within the solution space. This occurs when all neighbours of the incumbent solution feature an equal or inferior cost—that is, $\forall S' \in N(S)$, $f(S') \geq f(S)$. It is obvious that if a random descent algorithm reaches such a point in the solution space, then no further improvements (or changes to the solution) will be possible.

The *simulated annealing* algorithm is a generalisation of random descent which offers a mechanism by which this issue can be avoided. In essence, the main difference between the two methodologies lies in the criterion used for deciding whether to perform a move or not. As noted, for random descent this criterion is simply $f(S') < f(S)$. Simulated annealing uses this criterion, but it also accepts a move to a worse solution with probability $\exp(-\delta/t)$, where $\delta = |f(S) - f(S')|$ gives the proposed change in cost and t is a parameter known as the *temperature*. Typically, t is set to a relatively high value at the beginning of execution. This results in nearly all moves being accepted, meaning that the exploration method closely resembles a random walk through the solution space. During a run, the temperature t is then slowly reduced, meaning that the chances of accepting a worsening move become increasingly less likely. This causes the algorithm's behaviour to approach that of the random descent method. This additional acceptance criterion allows the algorithm to escape from local optima, helping SA to explore a greater span of the solution space.

A pseudocode description of the simulated annealing algorithm is given in Fig. 4.12. In this case, the temperature t is reduced every z iterations by multiplying it by a *cooling rate* α in the range (0, 1). Many other cooling schemes are possible, however. Since its introduction by Kirkpatrick et al. [15], simulated annealing has become a well-known and often very successful method for combinatorial optimisation problems, including applications in areas such as scheduling [16], university timetabling [17], packing problems [18], and bridge construction [19]. Methods

Fig. 4.13 The steepest
descent algorithm. Once
Step (7) is reached, a local
optimum has been reached
and the algorithm terminates

STEEPEST-DESCENT ()
(1) Produce an initial solution S
(2) **while (true) do**
(3)   Choose $S' \in N(S) : \forall S'' \in N(S), f(S') \leq f(S'')$
(4)   **if** $f(S') < f(S)$ **then**
(5) $S \leftarrow S'$
(6)   **else**
(7)   **break**

based on simulated annealing were also the winners of the first two International
Timetabling Competitions held in 2003 and 2007 (see Chap. 9).

One potentially problematic feature of the simulated annealing metaheuristic is
that it does not maintain any memory of the solutions previously observed. Indeed,
during a run, it may visit the same solution multiple times, or could even spend significant amounts of time cycling within the same subset of solutions. In contrast to this,
the *tabu search* metaheuristic contains mechanisms that are intended to help avoid
cycling, therefore encouraging the algorithm to enter new regions of the solution
space.

In the same way that simulated annealing can be considered an extension of
random descent, tabu search can be seen as an extension of the *steepest descent*
methodology. Steepest descent acts similarly to random descent in that it starts with
an initial solution S and then repeatedly applies a neighbourhood operator to try to
make improvements. In contrast, however, at each iteration of the steepest descent
algorithm *all* solutions in the neighbourhood are evaluated, with the best of these
then being chosen as the next incumbent. A pseudocode description of this process
is given in Fig. 4.13.

One advantage of using steepest descent over random descent is that it is abun-
dantly clear when a local optimum has been reached (the algorithm will not be able
to identify any neighbouring solution S' that is better than S). Tabu search extends
steepest descent by offering a mechanism for escaping these local optima. It does
this by also allowing worsening moves to be made when they are seen to be the
best available in the current neighbourhood. To avoid cycling, tabu search then also
makes use of a memory structure called a tabu list that keeps track of previously
visited solutions and bans the algorithm from returning to these for a certain period
of time. This encourages the algorithm to enter new parts of the solution space.

As we have discussed, the papers of Chams et al. [12] and Hertz and de Werra
[13] suggested some time ago that both the simulated annealing and tabu search
metaheuristics are suitable for graph colouring problems. A tabu search method
called TABUCOL, in particular, has proved to be very popular, both when used in
isolation and when used as an improvement procedure as part of broader algorithmic
schemes. This algorithm is discussed further in Chap. 5.

In more recent years, many other methods for exploring the space of complete
improper k-colourings have been proposed, including techniques based on

- Evolutionary algorithms [20–23];
- Iterated local search [24,25];
- GRASP algorithms [26];
- Variable neighbourhood search [27];
- Ant colony optimisation [28].

Two of the most notable examples from the above list, particularly due to the quality of results they produce, are the hybrid evolutionary algorithm of Galinier and Hao [23] and the ant colony optimisation algorithm of Thompson and Dowsland [28]. Both of these algorithms make use of population-based methods combined with the TABUCOL algorithm. The idea behind this hybridisation is to use the population-based elements of the algorithms to guide the search over the long term, gently directing it towards favourable regions of the solution space, with the TABUCOL element then being used to identify high-quality solutions within these regions. Both of these algorithms are considered further in Chap. 5.

4.2.3 Spaces of Partial, Proper k-Colourings

A further strategy for graph colouring involves exploring the space of proper *partial* solutions. This scheme again involves stipulating a fixed number of colours k at the outset. In this case, however, when vertices are encountered that cannot be feasibly assigned to a colour, they are transferred to a set of uncoloured vertices U. A solution S is therefore defined by a set of k feasible colour classes (independent sets) $\{S_1, \ldots, S_k\}$ together with a set of uncoloured vertices U such that $\left(\bigcup_{j=1,\ldots,k} S_j\right) \cup U = V$. The aim is to then make changes to the colour classes so that all vertices in U can be feasibly coloured, resulting in $U = \emptyset$. If this goal is achieved, k can then be reduced and the algorithm repeated, as with the previous scheme.

An effective example of this strategy is the PARTIALCOL algorithm of Blöchliger and Zufferey [29]. This approach uses tabu search and operates in a very similar fashion to the TABUCOL algorithm, albeit with a different neighbourhood operator. Specifically, a move in the solution space is achieved as follows:

- Select an uncoloured vertex $v \in U$ and assign it to a colour class S_j. Next, take all vertices $u \in S_j$ that are adjacent to v and move them from S_j to U.

In their work, Blöchliger and Zufferey [29] make use of the simple objective function $f_3 = |U|$ to evaluate solutions. A second objective function, $f_4 = \sum_{v \in U} \deg(v)$, is also suggested but is found to only give better solutions in a small number of cases. This algorithm is discussed in more detail in Chap. 5.

An earlier algorithm using this same scheme is due to Morgenstern and Shapiro [30]. This uses the same neighbourhood operator and objective function in conjunction with simulated annealing. However, it also employs an additional operator that is periodically applied to the partial solution to help reinvigorate the search process. Specifically, this mechanism shuffles vertices between colour classes in the partial solution while not introducing any clashes. This has the effect of moving the algorithm into different parts of the solution space while not changing its objective function value.

High-quality results based on exploring the space of partial proper k-colourings have also been reported by Malaguti et al. [31]. This algorithm is similar to the hybrid evolutionary algorithm of Galinier and Hao [23] and uses an analogous recombination operator together with a local search procedure based on PARTIALCOL. Their approach also makes use of the objective function f_4 in an attempt to sustain evolutionary pressure during execution. One feature of this work is that, during a run of the EA, a set is maintained containing various independent sets encountered by the algorithm. Upon termination of the EA, this set is then used in conjunction with a set covering IP to try and make further improvements to the quality of solution returned by the algorithm.

4.2.4 Combining Solution Spaces

Interesting work has also been carried out by Hertz et al. [32], who propose a method for operating in *different* solution spaces during different stages of execution. Specifically, TABUCOL is used to explore the space of complete improper k-colourings, and PARTIALCOL is used for the space of partial, proper solutions. The main idea is that a local optimum in one solution space is not necessarily a local optimum in another. Hence, when the search is deemed to have stagnated in one space, a procedure is used to alter the incumbent solution so that it becomes a member of another space. (For example, a complete improper solution formed by TABUCOL is converted into a partial proper solution by considering clashing vertices in a random order, and moving them into the set U until no clashes remain.) The search can then be continued in this new space where further improvements might be made, with the process being repeated as long as necessary. The authors also propose a third solution space based on the idea of assigning orientations to edges in the graph and then trying to minimise the length of the longest paths within the resultant directed graph (see also the work of Gendron et al. [33]). The authors note, however, that improvements are rarely achieved during exploration of the latter space, but that its inclusion is still useful because it tends to make large alterations to a solution, helping to diversify the search.

4.2.5 Problems Related to Graph Colouring

Concluding this review of different algorithms for the graph colouring problem, it is relevant to note that many of the schemes mentioned above are also commonly used in algorithms that tackle related problems. For example, we can observe the existence of timetabling algorithms that use constructive heuristics with backtracking [34]; algorithms that allow additional timeslots (colours) in a timetable and then only deal with feasible solutions [10,35–37]; methods that fix the number of timeslots in advance and then allow constraints to be violated (i.e., clashes to occur) [38–40]; and also algorithms that deal with partial timetables, never allowing constraint violations to occur [17,41,42]. Similar examples can also be noted in other related problems such as the frequency assignment problem [43,44].

4.3 Reducing Problem Size

This section now looks at ways in which the size of a graph colouring problem can be reduced. In some cases, this can lead to shorter run times and/or more accurate results.

4.3.1 Removing Vertices and Splitting Graphs

Given a graph $G = (V, E)$, it is sometimes possible to remove certain vertices and edges to create a smaller subgraph G'. If we then establish an optimal colouring of G', we can then reinstate the missing vertices and assign them to appropriate colours, giving an optimal colouring for G. Two options for doing this are now outlined.

1. Let $v \in V$ such that $\Gamma(v) = V - \{v\}$. In this case, v is adjacent to all other vertices in G implying that, in any feasible colouring, v will always assume its own unique colour. Now let $G' = G - \{v\}$ and assume we have established a feasible colouring of G'. The vertex v can now be allocated to a new colour, giving a feasible colouring to G that is using one more colour than G'.
2. Let $u, v \in V$ such that $\Gamma(u) \subseteq \Gamma(v)$. This implies that u and v are nonadjacent. Now let $G' = G - \{u\}$. A feasible colouring of G can be established by taking a feasible colouring of G' and simply assigning u to the same colour as v.

The above steps can be applied repeatedly, creating a series of smaller graphs, until neither condition holds.

In some cases, it is also possible to split a graph into a set of different subgraphs that can each be coloured separately. This can be done as follows.

3. Given, $G = (V, E)$, let $C \subseteq V$ be a subset of vertices such that (a) C is a clique in G and (b) the vertices of C are a vertex separating set (see Definition 3.10). Now label as $G_1, \ldots, G_l$ the components that are formed by deleting C from G, and let G'_i be the subgraph induced by $G_i \cup C$, for all $i \in \{1, \ldots, l\}$. Feasible colourings of the smaller subgraphs $G'_1, \ldots, G'_l$ can now be produced separately and then merged into a feasible colouring of G.

As an example of Item 2, consider Fig. 4.14a where, as required, $\Gamma(u) \subseteq \Gamma(v)$. It is clear from this figure that u can always be assigned to the same colour as v.

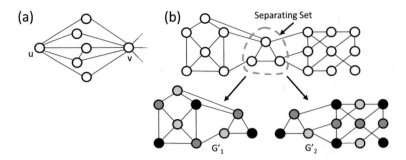

Fig. 4.14 Examples of graphs that can be reduced in size before colouring

Hence, u can be removed from the graph together with all its incident edges. These can be reinstated once the remaining vertices have been coloured.

To illustrate Item 3, consider Fig. 4.14b. As indicated, this graph contains a vertex separating set of size 3 that is also a clique. In this case, the two smaller subgraphs G'_1 and G'_2 can be coloured separately. If the vertices in the separating set are not allocated to the same colours in each subgraph (as is the case here), then a colour relabelling can be applied to make this so. The subgraphs can then be merged to form a complete feasible colouring for G. Note that, by definition, this feature includes cases where a graph G is disconnected (giving $|C| = 0$) or where G contains a cut vertex or bridge ($|C| = 1$). Also note that $\chi(G) = \max\{\chi(G'_1), \ldots, \chi(G'_l)\}$.

In practice, it is easy to check whether vertices exist in a graph that satisfy the conditions in Items 1 and 2 and, depending on the topology of the graph, it might be possible to remove many vertices before applying a graph colouring algorithm. The problem of identifying vertex separating sets is also solvable by various polynomial-time algorithms (such as the approach of Kanevsky [45]), and it only takes the addition of a simple checking step to determine whether these separating sets also constitute cliques or not.

In addition to these steps, in situations where we are trying to solve the decision variant of the graph colouring problem (given an integer k, identify whether a feasible k-colouring exists), it is also permissible to delete all vertices with degrees of less than k. That is, we can reduce the size of G by removing all vertices belonging to the set $\{v \in V : \deg(v) < k\}$. This is allowed since, obviously, vertices with fewer than k adjacent vertices will always have a feasible colour from the set $\{1, \ldots, k\}$ to which they can be assigned. Hence, colours can be allocated to these vertices once the remaining subgraph has been coloured.

4.3.2 Extracting Independent Sets

A further method for reducing the size of a graph involves the identification and removal of independent sets. A suitable process can be summarised as follows.

Given a graph $G = (V, E)$:

1. First let $G' = G$. Now, identify an independent set I_1 in G' and remove it. Repeat this step on G' a further $l - 1$ times to form a set of l disjoint independent sets $\{I_1, I_2, \ldots, I_l\}$. Call G' the residual graph.
2. Next, use any graph colouring algorithm to find a feasible colouring for the residual graph G'. Call this solution $\mathcal{S}' = \{S_1, \ldots, S_k\}$.
3. A feasible $(k + l)$-colouring for the original graph G is now obtained by setting $\mathcal{S} = \mathcal{S}' \cup \{I_1, I_2, \ldots, I_l\}$.

For Step 1, it is usually helpful to identify large maximal independent sets because this will leave us with a smaller residual graph. Recall, however, that the problem of identifying the *maximum* independent set in a graph is itself an $\mathcal{NP}$-hard problem. Methods for identifying large independent sets in a graph range from simple heuristics such as the RLF algorithm (Sect. 3.4.1) to advanced metaheuristic algorithms such as the tabu search approach of Wu and Hao [46].

A simple local search-based scheme for identifying an independent set of size q in a graph $G = (V, E)$ might operate as follows. First select a subset of q vertices $I \subseteq V$ and evaluate it according to the following cost function:

$$f_5(I) = \sum_{\forall u, v \in I} g(u, v), \tag{4.31}$$

where

$$g(u, v) = \begin{cases} 1 \text{ if } \{u, v\} \in E \\ 0 \text{ otherwise.} \end{cases}$$

In other words, f_5 is simply a count on the number of edges in the subgraph induced by I. If $f_5(I) = 0$ then I is an independent set of size q; otherwise, alter the contents of I using a suitable neighbourhood operator. One such operator is to swap a vertex $u \in I$ with a vertex $v \in (V - I)$. Once an independent set has been established it can then be removed from the graph. We might also seek to fulfill additional criteria, such as identifying an independent set that, when removed, leaves a residual graph with the fewest number of edges, thereby hopefully giving it a lower chromatic number. This might involve making suitable changes to the neighbourhood operator and cost function.

Note that if we choose to reduce a problem's size by extracting independent sets, a suitable balance will need to be struck between the time dedicated to this task and the time spent colouring the residual graph itself. We should also be mindful of the fact that extracting the *wrong* independent sets may also prevent us from being able to identify the optimal solution to the original graph colouring problem.

Finally, recall from Chap. 2 that the problem of identifying a maximum independent set in a graph G is equivalent to identifying a maximum clique in G's complement graph $\bar{G} = (V, \bar{E})$ (where $\bar{E} = \{\{u, v\} : \{u, v\} \notin E\}$). A helpful survey on heuristic algorithms for both of these problems is provided by Pelillo [47].

4.4 Chapter Summary

In this chapter, we have examined several different algorithmic techniques for the graph colouring problem. Initial sections considered three different ways to achieve *exact* algorithms for this problem, namely, using backtracking, integer programming (via branch-and-bound) and column generation. Given the $\mathcal{NP}$-hard nature of the graph colouring problem, all of these algorithms feature exponential growth rates in the worst case; however, they still offer significant improvements over methods based on brute-force enumeration (such as that seen in Sect. 2.2).

In the later sections of this chapter, we have also described some different heuristic algorithms for graph colouring. Some of these also make use of metaheuristic techniques like evolutionary algorithms, local search, and ant colony optimisation. Unlike exact methods, these heuristics will rarely prove solution optimality, even if granted excess time. However, in many cases, they are still able to provide very high-quality solutions in comparison to exact algorithms. Evidence for this will be presented in the next chapter, where we will present a comparison of six contrasting graph colouring algorithms.

References

1. Kubale M, Jackowski B (1985) A generalized implicit enumeration algorithm for graph coloring. Commun ACM 28(28):412–418
2. Wolsey L (2020) Integer programming, 2nd edn. Wiley. ISBN 978-1119606536
3. Morrison D, Jacobson S, Sauppe J, Sewell E (2016) Branch-and-bound algorithms: a survey of recent advances in searching, branching, and pruning. Discret Optim 19:79–102
4. Méndez-Díaz I, Zabala P (2008) A cutting plane algorithm for graph coloring. Discret Appl Math 156:159–179
5. Karp M (1972) Complexity of computer computations. In: Reducibility among combinatorial problems. Plenum, New York, pp 85–103
6. Mehrotra A, Trick M (1996) A column generation approach for graph coloring. INFORMS J Comput 8(4):344–354
7. Gualandi S, Malucelli F (2012) Exact solution of graph coloring problems via constraint programming and column generation. INFORMS J Comput 24(1)
8. Culberson J, Luo F (1996) Exploring the k-colorable landscape with iterated greedy. In: Cliques, coloring, and satisfiability: second DIMACS implementation challenge, vol 26. American Mathematical Society, pp 245–284
9. Mumford C (2006) New order-based crossovers for the graph coloring problem. In: Parallel problem solving from nature (PPSN) IX. LNCS, vol 4193. Springer, pp 880–889
10. Erben E (2001) A grouping genetic algorithm for graph colouring and exam timetabling. In: Practice and theory of automated timetabling (PATAT) III. LNCS, vol 2079. Springer, pp 132–158
11. Lewis R (2009) A general-purpose hill-climbing method for order independent minimum grouping problems: a case study in graph colouring and bin packing. Comput Oper Res 36(7):2295–2310
12. Chams M, Hertz A, Dubuis O (1987) Some experiments with simulated annealing for coloring graphs. Eur J Oper Res 32:260–266

13. Hertz A, de Werra D (1987) Using Tabu search techniques for graph coloring. Computing 39(4):345–351
14. Glover F (1986) Future paths for integer programming and links to artificial intelligence. Comput Oper Res 13(5):533–549
15. Kirkpatrick S, Gelatt C, Vecchi M (1983) Optimization by simulated annealing. Science 220(4598):671–680
16. Sekiner S, Kurt M (2007) A simulated annealing approach to the solution of job rotation scheduling problems. Appl Math Comput 188(1):31–45
17. Lewis R, Thompson J (2015) Analysing the effects of solution space connectivity with an effective metaheuristic for the course timetabling problem. Eur J Oper Res 240:637–648
18. Egeblad J, Pisinger D (2009) Heuristic approaches for the two- and three-dimensional knapsack packing problem. Comput Oper Res 36(4):1026–1049
19. Perea C, Alcaca J, Yepes V, Gonzalez-Vidosa F, Hospitaler A (2008) Design of reinforced concrete bridge frames by heuristic optimization. Adv Eng Softw 39(8):676–688
20. Dorne R, Hao J-K (1998) A new genetic local search algorithm for graph coloring. In: Eiben A, Back T, Schoenauer M, Schwefel H (eds) Parallel problem solving from nature (PPSN) V. LNCS, vol 1498. Springer, pp 745–754
21. Eiben A, van der Hauw J, van Hemert J (1998) Graph coloring with adaptive evolutionary algorithms. J Heurist 4(1):25–46
22. Fleurent C, Ferland J (1996) Genetic and hybrid algorithms for graph colouring. Ann Oper Res 63:437–461
23. Galinier P, Hao J-K (1999) Hybrid evolutionary algorithms for graph coloring. J Comb Optim 3:379–397
24. Chiarandini M, Stützle T (2002) An application of iterated local search to graph coloring. In: Proceedings of the computational symposium on graph coloring and it's generalizations, pp 112–125
25. Paquete L, Stützle T (2002) An experimental investigation of iterated local search for coloring graphs. In: Cagnoni S, Gottlieb J, Hart E, Middendorf M, Raidl G (eds) Applications of evolutionary computing, proceedings of EvoWorkshops2002: EvoCOP, EvoIASP, EvoSTim. LNCS, vol 2279. Springer, pp 121–130
26. Laguna M, Marti R (2001) A grasp for coloring sparse graphs. Comput Optim Appl 19:165–78
27. Avanthay C, Hertz A, Zufferey N (2003) A variable neighborhood search for graph coloring. Eur J Oper Res 151:379–388
28. Thompson J, Dowsland K (2008) An improved ant colony optimisation heuristic for graph colouring. Discret Appl Math 156:313–324
29. Blöchliger I, Zufferey N (2008) A graph coloring heuristic using partial solutions and a reactive tabu scheme. Comput Oper Res 35:960–975
30. Morgenstern C, Shapiro H (1990) Coloration neighborhood structures for general graph coloring. In: Proceedings of the first annual ACM-SIAM symposium on discrete algorithms, San Francisco, California, USA. Society for Industrial and Applied Mathematics, pp 226–235
31. Malaguti E, Monaci M, Toth P (2008) A metaheuristic approach for the vertex coloring problem. INFORMS J Comput 20(2):302–316
32. Hertz A, Plumettaz M, Zufferey N (2008) Variable space search for graph coloring. Discret Appl Math 156(13):2551–2560
33. Gendron B, Hertz A, St-Louis P (2007) On edge orienting methods for graph coloring. J Comb Optim 13(2):163–178
34. Carter M, Laporte G, Lee SY (1996) Examination timetabling: algorithmic strategies and applications. J Oper Res Soc 47:373–383
35. Burke E, Elliman D, Weare R (1995) Specialised recombinative operators for timetabling problems. In: The artificial intelligence and simulated behaviour workshop on evolutionary computing, vol 993. Springer, pp 75–85

36. Cote P, Wong T, Sabourin R (2005) Application of a hybrid multi-objective evolutionary algorithm to the uncapacitated exam proximity problem. In: Burke E, Trick M (eds) Practice and theory of automated timetabling (PATAT) V. LNCS, vol 3616. Springer, pp 294–312
37. Lewis R, Paechter B (2007) Finding feasible timetables using group based operators. IEEE Trans Evol Comput 11(3):397–413
38. Carrasco M, Pato M (2001) A multiobjective genetic algorithm for the class/teacher timetabling problem. In: Burke E, Erben W (eds) Practice and theory of automated timetabling (PATAT) III. LNCS, vol 2079. Springer, pp 3–17
39. Colorni A, Dorigo M, Maniezzo V (1997) Metaheuristics for high-school timetabling. Comput Optim Appl 9(3):277–298
40. Di Gaspero L, Schaerf A (2002) Multi-neighbourhood local search with application to course timetabling. In: Burke E, De Causmaecker P (eds) Practice and theory of automated timetabling (PATAT) IV. LNCS, vol 2740. Springer, pp 263–287
41. Burke E, Newall J (1999) A multi-stage evolutionary algorithm for the timetable problem. IEEE Trans Evol Comput 3(1):63–74
42. Paechter B, Rankin R, Cumming A, Fogarty T (1998) Timetabling the classes of an entire university with an evolutionary algorithm. In: Baeck T, Eiben A, Schoenauer M, Schwefel H (eds) Parallel problem solving from nature (PPSN) V. LNCS, vol 1498. Springer, pp 865–874
43. Aardel K, van Hoesel S, Koster A, Mannino C, Sassano A (2002) Models and solution techniques for the frequency assignment problems. 4OR: Q J Belgian, French and Italian Oper Res Soc 1(4):1–40
44. Valenzuela C (2001) A study of permutation operators for minimum span frequency assignment using an order based representation. J Heurist 7:5–21
45. Kanevsky A (1993) Finding all minimum-size separating vertex sets in a graph. Networks 23:533–541
46. Wu Q, Hao J-K (2012) Coloring large graphs based on independent set extraction. Comput Oper Res 39:283–290
47. Pelillo M (2009) Encyclopedia of optimization. In: Heuristics for maximum clique and independent set, 2nd edn. Springer, pp 1508–1520

Algorithm Case Studies

5

In this chapter, we present detailed descriptions of six different high-performance algorithms for the graph colouring problem. We also compare the performance of these algorithms over a wide range of graphs to gauge their relative strengths and weaknesses. The considered problem instances include random, flat, planar, and scale-free graphs, together with some real-world graphs from the fields of timetabling and social networking.

Our six case-study algorithms are described in the following six sections. Implementations of these can be found in the online suite of graph colouring algorithms described in Appendix A.1.

5.1 The TABUCOL Algorithm

As we mentioned in the previous chapter, TABUCOL has been used as a local search subroutine in a number of high-performing hybrid algorithms, including those of Avanthay et al. [1], Dorne and Hao [2], Galinier and Hao [3] and Thompson and Dowsland [4]. The specific version of TABUCOL that we consider here is the so-called "improved" variant, which was originally used by Galinier and Hao [3].

As we have seen, TABUCOL operates in the space of complete improper k-colourings using an objective function that simply counts the number of clashes (i.e., objective function f_2 from Eq. (4.30)). Given a candidate solution $\mathcal{S} = \{S_1, \ldots, S_k\}$, moves in the solution space are performed by selecting a vertex $v \in S_i$ whose assignment to colour class S_i is currently causing a clash, and then switching it to a new colour class S_j (where $i \neq j$). Note that previous incarnations of this algorithm also allowed nonclashing vertices to be moved between colours, though this is generally thought to worsen performance [5].

The tabu list of this algorithm is stored using a matrix $\mathbf{T}_{n \times k}$. If at iteration l of the algorithm, the neighbourhood operator transfers a vertex v from S_i to S_j, then the

© The Author(s), under exclusive license to Springer Nature Switzerland AG 2021   113
R. M. R. Lewis, *Guide to Graph Colouring*, Texts in Computer Science,
https://doi.org/10.1007/978-3-030-81054-2_5

element T_{vi} is set to $l + t$, where t is a positive integer that will be defined presently. This signifies that the transfer of v back to colour class S_i is *tabu* (i.e., disallowed) for the next t iterations of the algorithm (or, in other words, that v cannot be moved back to S_i until at least iteration $l + t$). Note that this has the effect of making *all* solutions containing the assignment of vertex v to S_i tabu for t iterations.

As is typical in applications of tabu search, in each iteration of TABUCOL, the entire set of neighbouring solutions is considered. That is, the cost of moving each clashing vertex into all other $k - 1$ colour classes is evaluated. This process consumes the majority of the algorithm's execution time; however, it can be sped up considerably through the use of appropriate data structures. To explain, let x denote the number of vertices involved in a clash in the current solution $\mathcal{S}$. This leads to $x(k-1)$ members in the set of neighbouring solutions $N(\mathcal{S})$. (Obviously, there is a strong positive correlation between x and the objective function, so better solutions will tend to have smaller neighbourhoods.) A naïve implementation of the TABUCOL algorithm would now set about separately performing the $x(k - 1)$ different neighbourhood moves and evaluating all of the resulting solutions. However, this is not necessary, particularly because only two colour classes are affected by each neighbourhood move.

A more efficient approach involves making use of an additional matrix $\mathbf{C}_{n \times k}$ where, given the current solution $\mathcal{S} = \{S_1, \ldots, S_k\}$, element C_{vj} denotes the number of vertices in colour class S_j that are adjacent to vertex v. When an initial solution is generated, all elements in $\mathbf{C}$ will first need to be calculated. This can be done using the $\mathcal{O}(nk + m)$ procedure POPULATE-C shown in Fig. 5.1. In each subsequent iteration of TABUCOL, the act of moving a vertex v from S_i to S_j will result in a new solution $\mathcal{S}'$ whose cost is simply:

$$f_2(\mathcal{S}') = f_2(\mathcal{S}) + C_{vj} - C_{vi}. \tag{5.1}$$

Since $f_2(\mathcal{S})$ will already be known, this means that the cost of all neighbouring solutions can be determined by simply reading through each row of $\mathbf{C}$ corresponding to the clashing vertices in $\mathcal{S}$. In a solution with x clashing vertices, this action has a complexity of $\mathcal{O}(xk)$. Once a move has been selected and performed (i.e., once v has been moved from S_i to S_j), the matrix $\mathbf{C}$ is then updated using the $\mathcal{O}(|\Gamma(v)|)$ UPDATE-C procedure shown in Fig. 5.1. As shown in this pseudocode, neighbours of v are now marked as being adjacent to one fewer vertex in colour class S_i and one additional vertex in colour class S_j. The complexity of each individual iteration of TABUCOL is therefore $\mathcal{O}(xk + |\Gamma(v)|)$, which is $\mathcal{O}(nk + m)$ in the worst case.

Having evaluated all neighbouring solutions, in each iteration, TABUCOL selects and performs the non-tabu move that brings about the largest decrease (or failing that the smallest increase) in cost. Any ties in this criterion are broken randomly. In addition, TABUCOL employs an *aspiration criterion* which allows tabu moves to be performed on occasion. Specifically, tabu moves are permitted if they are seen to improve on the best solution found so far during the run. This is particularly helpful if a tabu move is seen to lead to a solution with zero cost, at which point the algorithm can halt. Finally, if *all* moves are seen to be tabu, then a vertex $v \in V$ is selected at random and moved to a new randomly selected colour class. The tabu list is then updated as usual.

POPULATE-C ()

(1) $C_{vj} \leftarrow 0 \; \forall v \in V, j \in \{1, 2, \ldots, k\}$
(2)   **forall** $\{u, v\} \in E$ **do**
(3) $C_{u,c(v)} \leftarrow C_{u,c(v)} + 1$
(4) $C_{v,c(u)} \leftarrow C_{v,c(u)} + 1$

UPDATE-C (v, i, j)

(1)   **forall** $u \in \Gamma(v)$ **do**
(2) $C_{ui} \leftarrow C_{ui} - 1$
(3) $C_{uj} \leftarrow C_{uj} + 1$

Fig. 5.1 Procedures for populating and updating the matrix **C** used with TABUCOL. In POPULATE-C, $c(v)$ gives the colour of vertex v in the current solution. UPDATE-C is used when TABUCOL has moved vertex v from colour i to colour j. $\Gamma(v)$ denotes the set of all vertices adjacent to vertex v

In the version of TABUCOL used here, an initial candidate solution is constructed by taking a random ordering of the vertices and applying a modified version of the GREEDY algorithm in which only k colours are permitted. Thus, if vertices are encountered that cannot be assigned to any of the k colours without inducing a clash, these are assigned to one of the existing colours randomly. Of course, we could use more sophisticated constructive methods here, but it is stated by both Galinier and Hertz [5] and Blöchliger and Zufferey [6] that the method of initial solution generation is not critical in TABUCOL's performance.

Finally, concerning the *tabu tenure*, Galinier and Hao [3] have suggested making t a random variable that is proportional to the incumbent solution's cost. The idea here is that when the incumbent solution is poor, its high cost will lead to large values for t, which will hopefully force the algorithm into different regions of the solution space where better solutions can be found. On the other hand, when the incumbent solution has a low cost, the algorithm should focus on the current region by using low values for t. For a current solution S, Galinier and Hao [3] suggest using $t = 0.6 \times f_2(S) + r$, where r is selected randomly from the set $\{0, 1, 2, \ldots, 9\}$. These particular settings have been used in various other applications of TABUCOL [3, 4, 6] and are generally thought to give good results; however, it should be noted that other schemes for determining t may well be more appropriate for certain graphs.

5.2   The PARTIALCOL Algorithm

The PARTIALCOL algorithm of Blöchliger and Zufferey [6] operates in a similar fashion to TABUCOL in that it uses the tabu search metaheuristic to seek a proper k-colouring. However, in contrast to TABUCOL, PARTIALCOL does not consider improper solutions; instead, vertices that cannot be assigned to any of the k colours without causing a clash are left uncoloured. A solution $S = \{S_1, \ldots, S_k, U\}$ is

Fig. 5.2 Procedure for
updating **C** once
PARTIALCOL has moved
vertex v from the set U to the
colour class S_j

UPDATE-C (v, j)
(1) **forall** $u \in \Gamma(v)$ **do**
(2) $\quad C_{uj} \leftarrow C_{uj} + 1$
(3) $\quad$ **if** $c(u) = j$ **then**
(4) $\quad\quad$ **forall** $w \in \Gamma(u)$ **do**
(5) $\quad\quad\quad C_{wj} \leftarrow C_{wj} - 1$

therefore made up of k independent sets $S_1, \ldots, S_k$ together with a set of uncoloured vertices U. The aim of PARTIALCOL is to then make alterations to the independent sets so that U can be emptied, giving $U = \emptyset$ (and therefore a feasible k-colouring).

Because of its use of partial, proper solutions, the neighbourhood operator of PARTIALCOL is somewhat different to that of TABUCOL. Specifically, a move in the solution space is achieved by selecting an uncoloured vertex $v \in U$ and assigning it to a colour class S_j. The move is then completed by taking all vertices $u \in S_j$ that are adjacent to v and transferring them from S_j into U. This ensures that S_j continues to be an independent set. Having performed such a move, all corresponding elements T_{uj} in the tabu list are then marked as tabu for the next t iterations of the algorithm.

In each iteration of PARTIALCOL, the complete set of $|U| \times k$ neighbouring solutions is examined. The move to be performed is then chosen using the same criteria as TABUCOL, but using the cost function $f_3(\mathcal{S}) = |U|$. As with TABUCOL the matrix **C** can again be used to speed up the process of evaluating the set of neighbouring solutions. The contents of C are first initialised using the $\mathcal{O}(nk + m)$ procedure POPULATE-C from Fig. 5.1. The act of moving vertex v from U to colour class S_j then leads to a new solution $\mathcal{S}'$ whose cost is simply:

$$f_3(\mathcal{S}') = f_3(\mathcal{S}) + C_{vj} - 1. \tag{5.2}$$

The costs of all neighbouring solutions can be determined by scanning the rows of **C** corresponding to vertices in U. This has a complexity of $\mathcal{O}(|U|k)$. Once a move has been performed (i.e., the vertex $v \in U$ has been transferred to S_j and all vertices in the set $\{u \in S_j : u \in \Gamma(v)\}$ have been moved to U), the **C** matrix is updated using the $\mathcal{O}(m)$ procedure given in Fig. 5.2. Like TABUCOL, each iteration of PARTIALCOL therefore has a worst-case complexity of $\mathcal{O}(nk + m)$.

An initial solution to PARTIALCOL is generated using a Greedy process analogous to that of TABUCOL. The only difference is that when vertices are encountered for which there exist no clash-free colours, these are put into the set U. The only other operational difference between the two algorithms relates to the calculation of the tabu tenure t. In their original paper, Blöchliger and Zufferey [6] use an algorithm variant known as FOO-PARTIALCOL. Here, FOO abbreviates "Fluctuation Of the Objective-function", and indicates their use of a mechanism that alters t based on the algorithm's search progress. In essence, if during a run the objective function has not altered for a lengthy period of time, it is assumed that the search has stagnated in a particular region of the solution space and so t is increased to try to encourage the algorithm to leave this region. Similarly, when the objective function is seen to be fluctuating, t is slowly reduced, counteracting these effects. Note that this scheme requires values to be assigned to several parameters, the meanings of which are described by Blöchliger and Zufferey [6]. In our case, we choose to use settings recommended by the authors and these are included in our source code of this algorithm. We are perfectly at liberty to use other simpler schemes for calculating t if required, however.

5.3   The Hybrid Evolutionary Algorithm (HEA)

The third algorithm that we consider is the hybrid evolutionary algorithm (HEA) of Galinier and Hao [3]. The HEA operates by maintaining a population of candidate solutions that are evolved via a problem-specific recombination operator and a local search method. Like TabuCol, the HEA operates in the space of complete improper k-colourings using cost function f_2 (Eq. (4.30)).

The algorithm begins by creating an initial population of candidate solutions. Each member of this population is formed using a modified version of the DSATUR algorithm for which the number of colours k is fixed at the outset. To provide diversity between members, the first vertex is selected at random and assigned to the first colour. The remaining vertices are then taken in sequence according to the maximum saturation degree (with ties being broken randomly) and assigned to the lowest indexed colour class S_i seen to be feasible (where $1 \leq i \leq k$). When vertices are encountered for which no feasible colour class exists, these are kept to one side and are assigned to random colour classes at the end of this process. Upon construction of this initial population, an attempt is then made to improve each member by applying the local search routine, defined below.

As is typical for an evolutionary algorithm, for the remainder of the run the algorithm evolves the population using recombination, mutation, and evolutionary pressure. In each iteration, two parent solutions S_1 and S_2 are selected from the population at random, and copies of these are used in conjunction with the recombination operator to produce one offspring solution S'. This offspring is then improved using local search and inserted into the population by replacing the weaker of its two parents. Note that there is no bias towards selecting fitter parents for recombination; rather evolutionary pressure only exists due to the offspring replacing their weaker parent (regardless of whether the parent has a better cost than its offspring).

The recombination operator proposed Galinier and Hao [3] is the so-called Greedy partition crossover (GPX). The idea behind GPX is to construct offspring using large colour classes inherited from the parent solutions. A demonstration of how this is done is given in Fig. 5.3. As shown, the largest (not necessarily proper) colour class from the parents is first selected and copied into the offspring (ties are broken randomly). To avoid duplicate vertices occurring in the offspring at a later stage, these copied vertices are then removed from both parents. To form the next colour, the other (modified) parent is then considered and, again, the largest colour class is selected and copied into the offspring, before these vertices are removed from both parents. This process is continued by alternating between the parents until the offspring's k colour classes have been formed.

	Parent S_1	Parent S_2	Offspring S'	Comments
1)	$\{\{v_1,v_2,v_3\},$ $\{v_4,v_5,v_6,v_7\},$ $\{v_8,v_9,v_{10}\}\}$	$\{\{v_3,v_4,v_5,v_7\},$ $\{v_1,v_6,v_9\},$ $\{v_2,v_8,v_{10}\}\}$	$\{\}$	To start, the offspring solution $S' = \emptyset$.
2)	$\{\{v_1,v_2,v_3\},$ $\{v_8,v_9,v_{10}\}\}$	$\{\{v_3\},$ $\{v_1,v_9\},$ $\{v_2,v_8,v_{10}\}\}$	$\{\{v_4,v_5,v_6,v_7\}\}$	Select the colour class containing the most vertices and copy it into S'. (Class $\{v_4,v_5,v_6,v_7\}$ from S_1 in this case.) Delete the copied vertices from both S_1 and S_2.
3)	$\{\{v_1,v_3\},$ $\{v_9\}\}$	$\{\{v_3\},$ $\{v_1,v_9\}\}$	$\{\{v_4,v_5,v_6,v_7\},$ $\{v_2,v_8,v_{10}\}\}$	Select the largest colour class in S_2 and copy it into S'. Delete the copied vertices from both S_1 and S_2.
4)	$\{\{v_9\}\}$	$\{\{v_9\}\}$	$\{\{v_4,v_5,v_6,v_7\},$ $\{v_2,v_8,v_{10}\},$ $\{v_1,v_3\}\}$	Select the largest colour class in S_1 and copy it into S'. Delete the copied vertices from both S_1 and S_2.
5)	$\{\{v_9\}\}$	$\{\{v_9\}\}$	$\{\{v_4,v_5,v_6,v_7\},$ $\{v_2,v_8,v_{10},v_9\},$ $\{v_1,v_3\}\}$	Having formed k colour classes, assign any missing vertices to random colours to form a complete but not necessarily proper offspring solution S'.

Fig. 5.3 Example application of the Greedy partition crossover (GPX) operator of Galinier and Hao [3], using $k = 3$

At this point, each colour class in the offspring will be a subset of a colour class existing in at least one of the parents. That is:

$$\forall S_i \in S' \; \exists S_j \in (S_1 \cup S_2) \; : \; S_i \subseteq S_j, \qquad (5.3)$$

where S', S_1, and S_2 represent the offspring, and the first and second parents, respectively. However, some vertices may also be missing in the offspring (as is the case with vertex v_9 in Fig. 5.3). This issue is resolved by assigning the missing vertices to random colour classes. Finally, local search is executed on the offspring before inserting it into the population.

In this algorithm, TABUCOL is used for the local search routine, executing it for a fixed number of iterations I and using the same tabu tenure scheme as described in Sect. 5.1. In their original paper, Galinier and Hao [3] manually tune I for different problem instances. In our case, we choose not to follow this strategy and require a setting for I to be determined automatically by the algorithm. We also need to be wary that if I is set too low, then insufficient local search will be carried out on each newly created solution, while an I that is too high will result in too much effort being placed on local search as opposed to the global search carried out by the evolutionary operators. Ultimately, we choose to settle on $I = 16n$, which roughly corresponds to the settings used in the most successful runs reported by Galinier and Hao [3]. In all cases reported here, we also use a population size of 10, as recommended by the authors.

5.4 The ANTCOL Algorithm

Like the HEA, the ANTCOL algorithm of Thompson and Dowsland [4] is another metaheuristic-based method that combines global and local search operators, in this case using the ant colony optimisation (ACO) metaheuristic.

ACO is an algorithmic framework that was originally inspired by how real ants determine efficient paths between food sources and their colonies. In their natural habitat, when no food source has been identified, ants tend to wander about rather randomly. However, when a food source is found, the discovering ants will take some of this back to the colony leaving a pheromone trail in their wake. When other ants discover this pheromone, they are less likely to continue wandering at random, but may instead follow the trail. If they go on to discover the same food source, they will then follow the pheromone trail back to the nest, adding their own pheromone in the process. This encourages further ants to follow the trail. In addition to this, pheromones on a trail also tend to evaporate over time, reducing the chances of an ant following it. The longer it takes for an ant to traverse a path, the more time the pheromones have to evaporate; hence, shorter paths tend to see a more rapid build-up of pheromone, making other ants more likely to follow it and deposit their own pheromone. This positive feedback eventually leads to all ants following a single, efficient path between the colony and food source.

As might be expected, initial applications of ACO were aimed towards problems such as the travelling salesman problem and vehicle routing problems, where we seek to identify efficient paths and cycles in graphs (see, for example, the work of Dorigo et al. [7] and Rizzoli [8]). However, applications to many other problems have also been made.

The idea behind the ANTCOL algorithm is to use virtual "ants" to produce individual candidate solutions. During a run each ant produces its solution in a nondeterministic manner, using probabilities based on heuristics and also on the quality of solutions produced by previous ants. In particular, if previous ants have identified features that are seen to lead to better-than-average solutions, the current ant is more likely to include these features in its solution, generally leading to a reduction in the number of colours during the course of a run.

A full description of the ANTCOL algorithm is provided in Fig. 5.4. As shown in the pseudocode, in each cycle of the algorithm (Steps (4) to (16)), several ants each produce a complete, though not necessarily feasible, solution. In Step (11), the details of each of these solutions are then added to a trail update matrix δ. At the end of a cycle, the contents of δ are used together with an evaporation rate ρ to update the global trail matrix t (Step (15)).

As shown, at the start of each cycle, an individual ant constructs a candidate solution S using the procedure BUILDSOLUTION. This procedure is based on the GREEDY-I-SET algorithm seen in Chap. 3 (Fig. 3.14) and operates by building up each colour class in a solution one at a time. Recall that during the construction of each colour class $S_i \in S$, GREEDY-I-SET makes use of two sets: X, which contains the uncoloured vertices that can currently be added to S_i without causing a clash; and Y, which holds the uncoloured vertices that *cannot* be feasibly added to S_i. The modifications that BUILDSOLUTION employs are as follows:

- In the procedure, a maximum of k colour classes are permitted. Once these have been constructed, any remaining vertices are left uncoloured.

ANTCOL $(G = (V, E))$

---

(1) $t_{uv} \leftarrow 1 \ \forall u, v \in V : u \neq v$
(2) $k \leftarrow n$
(3) $S_{best} \leftarrow \{\{v_1\}, \{v_2\}, \ldots, \{v_n\}\}$
(4) **while (not** stopping condition) **do**
(5) $\delta_{uv} \leftarrow 0 \ \forall u, v \in V : u \neq v$
(6)   **for** $(ant \leftarrow 1$ **to** $nants)$ **do**
(7) $S \leftarrow$ BUILDSOLUTION(k)
(8)   **if** $(S$ is a partial solution) **then**
(9)   Randomly assign the uncoloured vertices to colour classes in S
(10)   Run TABUCOL on S
(11) $\delta_{uv} \leftarrow \delta_{uv} + F(S) \ \forall u, v : c(u) = c(v) \wedge u \neq v$
(12)   **if** $(S$ is feasible) **then**
(13)   **if** $(|S| < |S_{best}|)$ **then** $S_{best} \leftarrow S$
(14)   **break**
(15) $t_{uv} \leftarrow \rho \times t_{uv} + \delta_{uv} \ \forall u, v \in V : u \neq v$
(16) $k \leftarrow |S_{best}| - 1$

Fig. 5.4 The ANTCOL algorithm. At termination, the best feasible solution found is S_{best}, using $|S_{best}| = k + 1$ colours

- The first vertex to be assigned to a colour class S_i $(1 \leq i \leq k)$ is chosen randomly from the set X.
- In remaining cases, each vertex v is then assigned to colour S_i with probability

$$P(v, i) = \begin{cases} \dfrac{\tau_{vi}^{\alpha} \times \eta_{vi}^{\beta}}{\sum_{u \in X}(\tau_{ui}^{\alpha} \times \eta_{ui}^{\beta})} & \text{if } v \in X \\ 0 & \text{otherwise} \end{cases}, \tag{5.4}$$

where τ_{vi} is calculated

$$\tau_{vi} = \frac{\sum_{u \in S_i} t_{uv}}{|S_i|}. \tag{5.5}$$

Note that the calculation of τ_{vi} makes use of the global trail matrix t, meaning that higher values are associated with combinations of vertices that have been assigned the same colour in high-quality, previously observed solutions. The value η_{vi}, meanwhile, is associated with a heuristic rule which, in this case, is the degree of vertex v in the graph induced by the set of currently uncoloured vertices $X \cup Y$. Larger values for τ_{vi} and η_{vi} thus contribute to larger values for $P(v, i)$, encouraging vertex v to be assigned to colour class S_i. The parameters α and β are used to control the relative strengths of τ and η in the equation.

The ANTCOL algorithm also makes use of a "multi-sets" operator within the BUILDSOLUTION procedure. Since the process of constructing a colour class is probabilistic in this case, the operator makes v separate attempts to construct each colour class. It then selects the one that results in the minimum number of edges in the graph induced by the set of remaining uncoloured vertices Y (since such graphs will tend to feature lower chromatic numbers).

On completion of BUILDSOLUTION, the generated solution S will be proper, but could be partial. If the latter is true, all uncoloured vertices are assigned to random colour classes to form a complete, improper solution, and TABUCOL is run for I iterations. Details on the solution are then written to the trail update matrix δ using the evaluation function

$$F(S) = \begin{cases} 1/f_2 & \text{if } f_2 > 0 \\ 3 & \text{otherwise,} \end{cases} \tag{5.6}$$

as shown in Step (11) of Fig. 5.4. This means that higher quality solutions contribute larger values to δ, encouraging their features to be included in solutions produced by future ants.

The parameters used in our application, and recommended by Thompson and Dowsland [4], are as follows: $\alpha = 2$, $\beta = 3$, $\rho = 0.75$, $nants = 10$, $I = 2n$, and $v = 5$. The tabu tenure scheme of TABUCOL is the same as in previous descriptions.

5.5 The Hill-Climbing (HC) Algorithm

In contrast to the preceding four algorithms, the hill-climbing (HC) algorithm of Lewis [9] operates in the space of feasible solutions. The initial solution for this approach is formed using the DSATUR heuristic. During a run, the algorithm then operates on a single feasible solution $S = \{S_1, \ldots, S_k\}$ (where each $S_i \in S$ is nonempty), with the aim of minimising k.

A single cycle of this algorithm operates by first moving a small number of colour classes (independent sets) from S into a second set T. This gives two partial proper solutions such that $|S| + |T| = k$. A specialised local search procedure is then run for I iterations. This procedure attempts to feasibly transfer vertices from colour classes in T into colour classes in S while ensuring that all elements in S and T continue to be independent sets. If successful, this has the effect of increasing the cardinality of the colour classes in S and may also empty some of the colour classes in T, reducing the total number of colours being used. At the end of the local search procedure, all nonempty colour classes in T are then moved back into S to form a complete, feasible solution.

The first iteration of the local search procedure operates by considering each vertex v in T and checking whether it can be feasibly transferred into any of the colour classes in S (i.e., without causing a clash). If this is the case, such transfers are performed. The remaining $I - 1$ iterations of the procedure then operate as follows. First, an alteration is made to a randomly selected pair of colour classes $S_i, S_j \in S$ using either a Kempe chain interchange or a pair swap (see Definitions 4.3 and 4.4). Because this will alter the makeup of S_i and S_j, this then raises the possibility that other vertices in T can now also be moved into these colour classes. Again, these transfers are made if they are seen to retain feasibility. The local search procedure continues in this fashion until I iterations have been performed.[1]

On completion of the local search procedure, the colour classes in $\mathcal{T}$ are moved back into $\mathcal{S}$ to form a complete feasible solution. The independent sets in $\mathcal{S}$ are then ordered according to some (possibly random) heuristic, and an updated solution is formed by constructing a permutation of the vertices in the same manner as that of the iterated Greedy algorithm (see Sect. 4.2.1) and then applying the GREEDY algorithm. This completes a single cycle of the HC algorithm.

The HC algorithm performs a series of cycles until a user-defined computation limit is reached. The application of GREEDY in each cycle is intended to generate large alterations to the incumbent solution, which is then passed back to the local search procedure for further optimisation. Note that none of the stages of this algorithm allow the number of colour classes being used to increase, thus providing its hill-climbing characteristics.

As with the previous algorithms, several parameters need to be set with the HC algorithm, each that can influence its performance. The values used in our experiments here were determined in preliminary tests and according to those reported by Lewis [9]. For the local search procedure, independent sets are moved into $\mathcal{T}$ by considering each $S_i \in \mathcal{S}$ in turn and transferring it with probability $1/|\mathcal{S}|$. The local search procedure is then run for $I = 1000$ iterations and, in each iteration, the Kempe chain and swap neighbourhoods are called with probabilities 0.99 and 0.01, respectively. Finally, when constructing the permutation of the vertices for passing to the GREEDY algorithm, the independent sets are ordered using the same 5:5:3 ratio as detailed in Sect. 4.2.1.

5.6 The Backtracking Algorithm

The sixth and final algorithm considered in this chapter is the backtracking approach of Korman [10]. Essentially, this operates in the same manner as the basic backtracking approach discussed in Sect. 4.1.1, though with the following modifications:

- Given the graph $G = (V, E)$, initially vertices are relabelled such that $\deg(v_1) \geq \deg(v_2) \geq \cdots \geq \deg(v_n)$, breaking ties randomly.
- When performing a forward step, the next vertex to be coloured is chosen as the uncoloured vertex with the smallest number of feasible colours to which it can currently be assigned. Ties are broken using the vertex among these with the lowest index.

---

[1]Note that in some cases a Kempe chain will contain all vertices in both colour classes, that is, the graph induced by $S_i \cup S_j$ will form a connected bipartite graph. Kempe chains of this type are known as *total*, and interchanging their colours serves no purpose since this only results in the two colour classes being relabelled. Consequently, total Kempe chains are ignored by the algorithm.

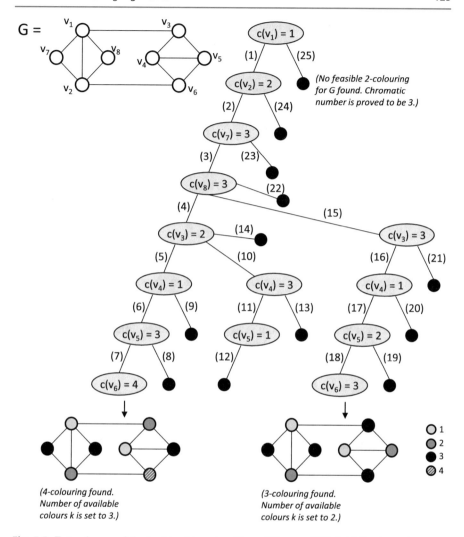

Fig. 5.5 Example run of the backtracking algorithm of Korman [10]. Initially, k can be set to $\Delta(G) + 1$. As usual, the notation $c(v) = j$ means that vertex v is assigned to colour i

An example run of this algorithm is illustrated in Fig. 5.5. Each grey node in this search tree represents a *decision* (an assignment of a colour to a vertex) and grey leaf nodes correspond to a feasible solution. For clarity, the order in which these nodes are visited is shown by the numbers next to the corresponding links in the tree. This corresponds to a depth-first parse of the search tree.

As seen in Steps (1) to (7) of Fig. 5.5, the algorithm starts by performing a series of assignments. This results in the four-colouring of G shown at the bottom left. At each node along this path, the above rules have been used to select the next vertex to colour. The lowest feasible colour from the set $\{1, \ldots, k\}$ has then been assigned to the vertex. At this point, we are now interested in producing a solution using fewer colours, so we set k to be equal to the number of colours in this current solution minus one and then backtrack. At Step (8), the algorithm now attempts to assign v_6 to the colours 1, 2, and 3. However, all of these colours are infeasible, so no further branches need exploring. This is signified by the black node in the search tree. At Step (9), the algorithm then backtracks to try and identify a new colour for v_5. In this case, colours 1 and 2 are not feasible, and colour 3 has already been tried so, again, no further branching is necessary.

At Step (15), vertex v_3 is assigned to colour 3, producing a new grey node. As shown, the subtree rooted at this node contains a three-colouring. Once this is discovered, k can be set to two. At this point, no further branching is possible, meaning that once the algorithm has backtracked to the root of the search tree, the discovered three-colouring is the guaranteed optimum.

Note that several parameters can also be set when applying this algorithm, some of which might alter the performance quite drastically. These include limiting the number of branches that can be considered at each node of the search tree and also prohibiting branching at certain levels. In practice, it is not obvious how these settings might be chosen *a priori* for individual graphs, so, in our case, we opt for the most natural configuration, which is to simply attempt a complete exploration of the search tree.[2] This means that the algorithm is exact under excess time, though, of course, such run-lengths will not be possible in many cases.

5.7 Algorithm Comparison

In this section, we now compare the performance of the above six algorithms using a variety of different graph types. As with our comparison of constructive algorithms in Chap. 3, we begin by considering random graphs. We then go on to look at flat graphs, scale-free graphs, and planar graphs. We will also examine sets of graphs arising from two real-world practical problems, namely, university timetabling and social networking.

As with our previous experiments, computational effort for these algorithms is measured by counting the number of constraint checks (see Sect. 1.4.1). Solution quality is measured by recording the smallest number of colours being used in any feasible solution observed during a run. Note that because TABUCOL, PARTIALCOL, and the HEA operate using infeasible solutions, settings for k are required which will need to be modified during a run. In our case, initial values are determined by executing DSATUR on each instance and setting k to the number of colours used in the resultant solution. During runs, k is then decremented by 1 each time a feasible k-colouring is found, with the algorithms being restarted.

---

[2]These parameters can be altered in our source code, however.

In all trials, a computation limit of 5×10^{11} constraint checks is imposed. This value is chosen to be deliberately high to provide some notion of excess time in the trials. Example run times (in seconds) using this computation limit are given in Table 5.6 later.

5.7.1 Random Graphs

As we saw in Definition 3.15, random graphs are generated such that each pair of vertices is made adjacent with probability p. This gives an average of $\binom{n}{2}p$ edges per graph. It also means that the distribution of vertex degrees is characterised by the binomial distribution $B(n - 1, p)$. For the following experiments, we used values of p ranging from 0.05 (sparse) to 0.95 (dense), incrementing in steps of 0.05, with $n \in \{250, 500, 1000\}$. Twenty-five instances were generated in each case.

Table 5.1 shows the number of colours used in solutions produced by the six algorithms for random graphs with edge probability $p = 0.5$ and varying numbers of vertices. The results indicate that for the smaller graphs ($n = 250$), the TABUCOL, PARTIALCOL, and HEA algorithms produce solutions with fewer colours than the remaining algorithms.[3] However, no statistical difference between these three algorithms is apparent. For larger graphs, however, the HEA produces the best results, allowing us to conclude that, for $n = 500$ and $n = 1000$, the HEA algorithm produces the best solutions across the set of all graphs and their isomorphisms under this particular computation limit.

Considering other graph densities, the charts shown in Fig. 5.6 summarise the mean solution quality achieved by the six algorithms on all random graphs generated. In each figure, the bars show the number of colours used in solutions produced by DSATUR and the lines then give the proportion of this number used in the solutions returned by each of the six algorithms. Note that all algorithms achieve a reduction in the number of colours realised by DSATUR, though in all but the smallest, sparsest graphs, the backtracking algorithm exhibits the smallest margins of improvement.

It is clear from Fig. 5.6 that TABUCOL, PARTIALCOL, and the HEA, in particular, produce the best results for the random graphs. For $n = 250$, these algorithms produce mean results that, across the range of values for p, show no significant difference among one another, perhaps indicating that the achieved solutions are consistently close to being optimal. For larger graphs, however, the HEA's solutions are seen to be significantly better, though its rates of improvement are slightly slower than those of TABUCOL and PARTIALCOL, as illustrated in Fig. 5.7. Similar behaviour during runs was also witnessed with the smaller random instances.

Overall, the patterns shown in Fig. 5.6 indicate that the HEA's strategy of exploring the space of infeasible solutions using both global and local search operators is

---

[3] As in Chap. 3, statistical significance is claimed here according to the nonparametric related samples Wilcoxon signed-rank test (for pairwise comparisons), and the related samples Friedman's two-way analysis of variance by ranks (for group comparisons). For the remainder of this chapter, statistical significance is claimed at the 1% level.

Table 5.1 Summary of results produced at the computation limit using random graphs $G_{n,0.5}$

	Algorithm[a]					
n	TABUCOL	PARTIALCOL	HEA	ANTCOL	HC	Bktr
250	**28.04±0.20**	**28.08 ±0.28**	**28.04±0.33**	28.56 ± 0.51	29.28 ± 0.46	34.24 ± 0.78
500	49.08 ± 0.28	49.24 ± 0.44	**47.88 ±0.51**	49.76 ± 0.44	54.52 ± 0.77	62.24 ± 0.72
1000	88.92 ± 0.40	89.08 ± 0.28	**85.48 ±0.46**	89.44 ± 0.58	101.44 ± 0.82	112.88 ± 0.97

[a]Mean plus/minus standard deviation in number of colours, taken from runs across 25 graphs

the most beneficial of those considered here. Indeed, although the HC algorithm also uses both global and local search operators, here its insistence on preserving feasibility implies a lower level of connectivity in its underlying solution space, making navigation more restricted and resulting in noticeably inferior solutions.

Figure 5.6 also reveals that ANTCOL does not perform well with large sparse instances, though it does become more competitive with denser instances. The reasons for this are twofold. First, the degrees of vertices in sparse graphs are naturally lower, reducing the heuristic bias provided by η and perhaps implying an over-dominant role of τ during applications of BUILDSOLUTION (see Eq. (5.4)). Secondly, sparse graphs also feature greater numbers of vertices per colour—thus, even if very promising independent sets *are* identified by ANTCOL, their reconstruction by later ants will naturally depend on a longer sequence of random trials, making them less likely to reoccur. To back these assertions, we also repeated the trials of ANTCOL using the same local search iteration limit as the HEA, namely, $I = 16n$. However, though this brought slight improvements for denser graphs, the results were still observed to be significantly worse than the HEA's, suggesting the difference in performance indeed lies with the global search element of ANTCOL in these cases.

5.7.2   Flat Graphs

Our second set of experiments concerns flat graphs. Flat graphs are produced by starting with an empty graph $G = (V, E = \emptyset)$ and then partitioning the n vertices into q almost equal-sized independent sets (i.e., each set contains either $\lfloor n/q \rfloor$ or $\lceil n/q \rceil$ vertices). Edges are then added between pairs of vertices in different independent sets in such a way that the variance in vertex degrees is kept to a minimum. This is continued until a user-specified density of p is reached.

It is well known that q-coloured solutions to flat graphs are quite easy to achieve for most values of p. This is because, for lower values for p, problems will be under-constrained, perhaps giving $\chi(G) < q$, and making q-coloured solutions easily identifiable. On the other hand, high values for p can result in over-constrained problems with prominent global optima that are also easily discovered. Hard-to-solve q-colourable graphs are known to occur for a region of p's at the boundary of these extremes, commonly termed the *phase transition region* [11, 12]. Flat graphs, in particular, are known to have rather pronounced phase transition regions because

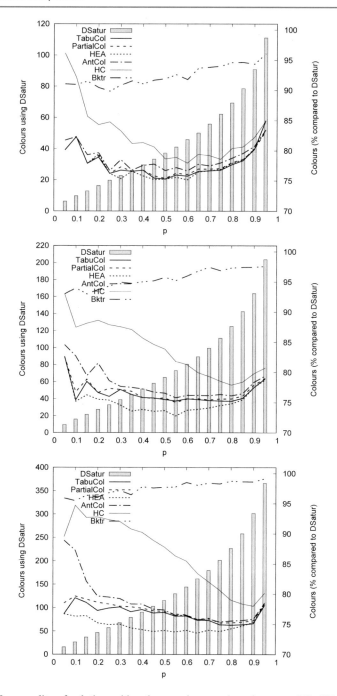

Fig. 5.6 Mean quality of solution achieved on random graphs using $n = 250$, 500, and 1000 (respectively) for various edge probabilities p. All points are the mean of 25 runs on 25 different instances

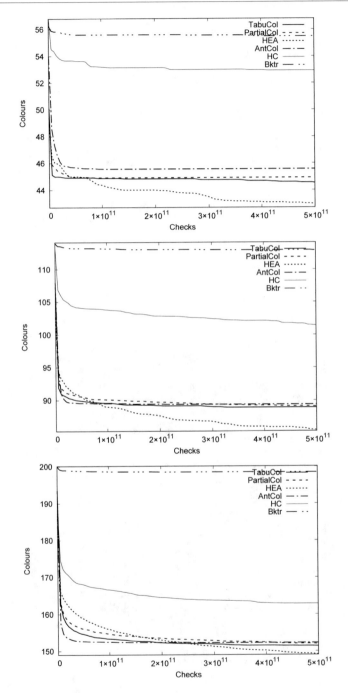

Fig. 5.7 Run profiles on random graphs of $n = 1000$ with edge probabilities $p = 0.25$, 0.5, and 0.75, respectively. Each line represents a mean of 25 runs on 25 different instances

each colour class and vertex degree is deliberately similar, implying a lack of heuristic information for algorithms to exploit (i.e., vertices tend to "look the same").

For our experiments, flat graphs were generated using publicly available software designed by Joseph Culberson which can be downloaded at http://webdocs.cs. ualberta.ca/~joe/Coloring/. Graphs were produced for $q \in \{10, 50, 100\}$ using various settings of p in and around the phase transition regions. In each case, we used $n = 500$, implying approximately 50, 10, and 5 vertices per colour, respectively. Twenty instances were generated in each case.

The relative performance of the six graph colouring algorithms on these flat graphs is shown in Fig. 5.8. Similarly, to random graphs, we see that the HEA, TABUCOL, and PARTIALCOL generally exhibit the best performance on instances within the phase transition regions, with the HC and backtracking algorithms proving the least favourable. One pattern to note is that, for all three values of q, the HEA tends to produce the best quality results on the left side of the phase transition region, while PARTIALCOL produces better results for a small range of p's on the right side. However, this difference is not due to the "FOO" tabu tenure mechanism of PARTIALCOL, because no significant difference was observed when we repeated our experiments using PARTIALCOL under TABUCOL's tabu tenure scheme. Thus, it seems that PARTIALCOL's strategy of only allowing solutions to be built from independent sets is favourable in these cases, presumably because this restriction facilitates the formation of independent sets of size n/q—structures that will be less abundant in denser graphs, but which also serve as the underlying building blocks in these cases.

Another striking feature of Fig. 5.8 is the poor performance of ANTCOL on the right side of the phase transition regions. This again seems to be due to the diminished effect of heuristic value η which, in this case, occurs because of the very low variance in vertex degrees. Furthermore, in denser graphs, fewer combinations comprising n/q vertices will form independent sets, decreasing the chances of an ant constructing one. This reasoning is also backed by the fact that ANTCOL's poor performance lessens with larger values of q where, due to there being fewer vertices per colour, the reproduction of independent sets is dependent on shorter sequences of random trials.

5.7.3 Planar Graphs

Our next set of experiments concerns planar graphs. These provide some contrasting results to those of the previous two subsections because, in many cases, the backtracking algorithm is often able to quickly solve these problems to optimality. However, there are still places where difficulties are encountered.

As we saw in Sect. 1.2, planar graphs are structured so that they can be drawn on a two-dimensional plane such that no edges cross. Because of this, they are quite sparse; indeed, the maximum possible number of edges in a planar graph with n vertices is just $3n - 6$ (see Theorem 6.2). In Sect. 6.1, we will see that all planar graphs are actually four-colourable. In practice, this means that planar graphs can be optimally coloured in polynomial time using, for example, the $\mathcal{O}(n^2)$ algorithm

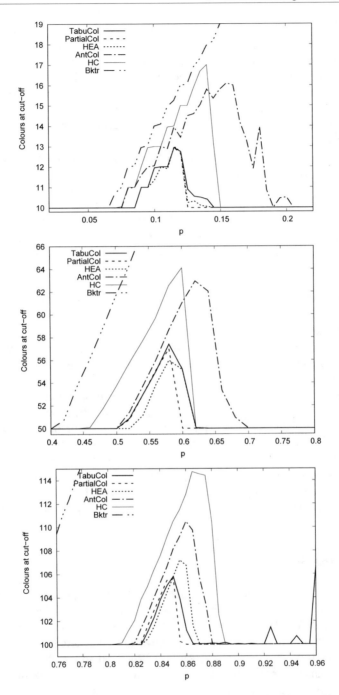

Fig. 5.8 Mean solution quality achieved with flat graphs of $n = 500$ with $q = 10$, 50, and 100 (respectively) for various edge probabilities p. All points are the mean of 20 runs on 20 different instances

of Robertson et al. [13]. However, it is still interesting to see how our six case-study algorithms can perform with these problem instances.

Example Python code for generating planar graphs is given in Appendix A.4. This operates by first randomly placing n points into the unit square. A Delaunay triangulation is then generated from these points to give a planar graph using approximately (but not exceeding) $3n - 6$ edges. A subset of these edges is then taken to give a planar graph with the required number of edges m, ensuring that the resultant graph is also connected. An illustration of this process is shown in Fig. A.3. In our trials, planar graphs were generated using $n \in \{100, 1000, 2000\}$. In each case, 40 different values of m between n and $3n - 6$ were considered and 25 different graphs were then generated for each (n, m) pair.

Figure 5.9 summarises the performance of the backtracking algorithm on this set of planar graphs by considering its success rate and computational requirements. The success rate gives the percentage of instances in which the algorithm has navigated its way back to the root of the search tree (and therefore produced an optimal solution) within the computational limit. The computational requirements are then calculated by taking the mean number of checks performed across these successful runs. For $n = 100$, we see that the algorithm solves all instances to optimality using very small amounts of computation. Factors contributing to this success are the relatively small number of vertices and colours (contributing to smaller search trees), and the fact that $\chi(G)$-colourings are established quickly, allowing much of the remaining search tree to be pruned.

For the larger values of n shown in Fig. 5.9, other patterns start to emerge, with dips in the success rate occurring in two areas. These are somewhat reminiscent of the phase transition regions experienced with flat graphs, seen in the previous subsection. For the lowest values of m used in these figures, the chromatic number of the graphs is nearly always three. Because there are fewer edges, there are also many different three-colourings. The algorithm is therefore able to quickly identify one of these and prune much of the remaining search tree. As m is increased from this point, the number of feasible three colourings diminishes, causing a drop in success rates and increases to the required computational effort. Next, at around $m = 2n$, the increased number of edges means that the chromatic number is now usually four. As before, there are now many different four-colourings, allowing the backtracking algorithm to terminate with an optimal solution quite quickly. Then, as m is raised further, the number of feasible four-colourings drops, once again causing a decrease in the success rate and an increase in the required computational effort.

In Fig. 5.10, we compare the quality of solutions returned by the backtracking algorithm to those of the other five algorithms. In these cases, the other five algorithms have produced identical results, so just one line is used for them in the charts. As shown, for $n = 100$, all six algorithms produce the same results. This indicates that the remaining five algorithms are also producing optimal solutions. For larger instances, the five algorithms give better solutions in the areas corresponding to the phase transition regions of the backtracking algorithm. This is particularly so for the densest graphs, where they always produce four-colourings, whereas backtracking often only produces five-colourings.

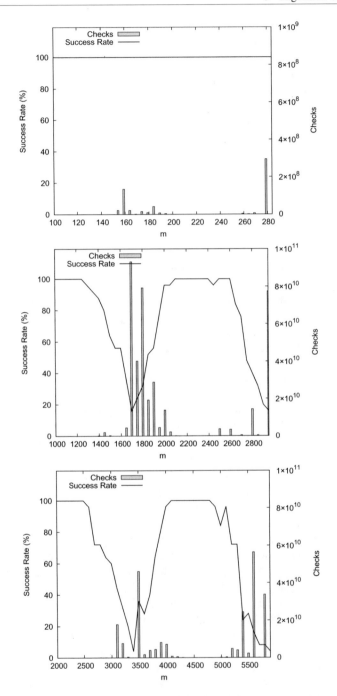

Fig. 5.9 Performance of the backtracking algorithm on planar graphs of differing densities for $n = 100$, 1000, and 2000, respectively. All points in the figures are means taken across 25 different problem instances

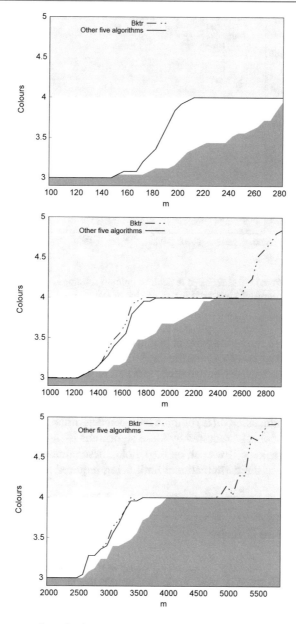

Fig. 5.10 Average number of colours used in solutions returned by the backtracking algorithm and the remaining five heuristics, for $n = 100$, 1000, and 2000, respectively. The lower shaded areas indicate lower bounds on the chromatic number. These were determined using the NetworkX command `nx.graph_clique_number(G)` to calculate the clique number $\omega(G)$ for each graph G. Although the algorithm used by this command has an exponential complexity, we found that these operations were able to complete quickly with these planar graphs. The upper shaded area indicates four colours, which is the maximum number of colours required by any planar graph. All points are means taken across 25 instances

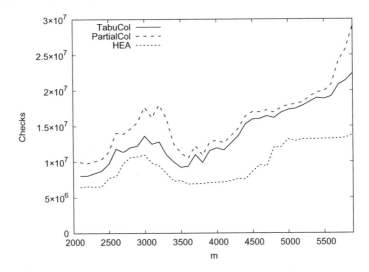

Fig. 5.11 Computational effort required to achieve optimal solutions for planar graphs with $n = 2000$ vertices (mean across 25 instances for each value of m)

Finally, Fig. 5.11 shows the computational effort required to find optimal solutions using TABUCOL, PARTIALCOL, and the HEA algorithm (these are chosen here as they seem to require the least effort overall). Note that these algorithms do not prove the optimality of a solution by themselves; in these cases, optimality is therefore claimed either because $\chi(G)$ has been previously determined by the backtracking algorithm, or because a solution using $\omega(G)$ colours has been determined. In this figure, we see that the number of checks required by these algorithms is slightly higher within the phase transition regions. However, on the whole, these numbers are quite low, with less than 0.003% of the computational limit being required to produce the optimal on average.

5.7.4 Scale-Free Graphs

We now consider another type of graph topology in which the backtracking algorithm is observed to perform very well. Scale-free graphs are known to model many real-world applications of networks, including the World Wide Web, citation networks of academic papers, and flight connections between airports [14]. In essence, these graphs are based around the idea of "preferential attachment" in that, when a vertex is added to a graph, it is more likely to be made adjacent to existing vertices that have high degrees. As a result, the degree distributions of these graphs follow a power law, in which a small number of "hub" vertices feature very high degrees compared to the remaining vertices.

Scale-free graphs can be artificially generated using the Barabási–Albert method [15]. For these trials, we employed the NetworkX implementation of this algorithm.

This uses the command $\texttt{G = nx.barabasi\_albert\_graph(n,q)}$, where n is the number of vertices required in the final graph, and q is the number of edges that are to be added for each newly inserted vertex.

The NetworkX method starts with an empty graph G on q vertices, that is, G initially comprises a vertex set $V = \{v_1, v_2, \ldots, v_q\}$ and an edge set $E = \emptyset$. In each step, a new vertex v is then added to G, together with q edges that connect v to vertices already present in G. This is done via a series of q roulette-wheel trials where, in each case, the probability $\mathrm{P}(u, v)$ of adding the edge $\{u, v\}$ to E is calculated by

$$\mathrm{P}(u, v) = \begin{cases} \dfrac{\deg(u)}{\sum_{w \in (V - \Gamma(v))} \deg(w)} & \text{if } \{u, v\} \notin E \\ 0 & \text{otherwise.} \end{cases} \tag{5.7}$$

This means that the edge $\{u, v\}$ is more likely to be added to G when the degree of u is already large in comparison to the other vertices in the graph.

These steps continue until a graph with n vertices—and thus $m = q(n - q)$ edges—has been produced. According to this method, a setting of $q = 0$ or $q = n$ therefore gives an empty graph on n vertices, $q = 1$ gives a tree, and $q = n - 1$ gives a star graph.

To show the effects of this construction method, Fig. 5.12 contrasts the degree distribution of a scale-free graph to that of a random graph of the same density. Due to the binomial characteristics of random graphs, we see that the degrees are symmetrically distributed around the average. On the other hand, the scale-free graph contains a high number of vertices with degree q, but also a long tail to the right, indicating the presence of high-degree "hub" vertices.

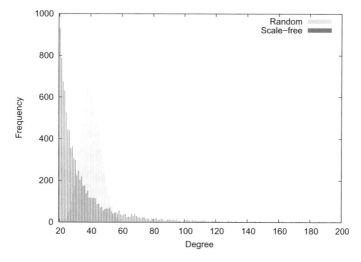

Fig. 5.12 Degree distributions for a random graph with $n = 10{,}000$ and $p = 0.0004$, and a scale-free graph with $n = 10{,}000$ and $q = 20$. The maximum degree for the random graph is 65; for the scale-free graph the maximum is 721

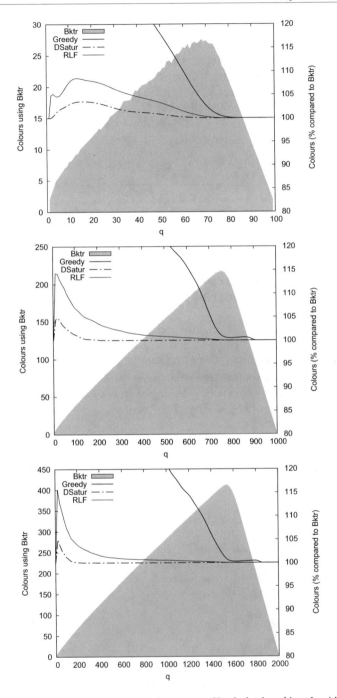

Fig. 5.13 Number of colours achieved in solutions returned by the backtracking algorithm on scale-free graphs for $n = 100$, 1000, and 2000 respectively. The corresponding results for the GREEDY, DSATUR, and RLF heuristics are also shown. All points are the mean twenty problem instances

In our trials, scale-free graphs for $n = \{100, 1000, 2000\}$ were considered using values of $q \in \{1, 2, \ldots, n - 1\}$. Twenty graphs were then generated for each (n, q) pair, giving 61,940 problem instances in total. Although this resultant graph set contains a wide range of sizes and densities, ultimately the backtracking algorithm was seen to perform very well in these cases, with just 14 of the 61,940 instances not being solved to optimality within the computation limit (specifically, 3 for $n = 1000$ and 11 for $n = 2000$). In addition, the time to find the optimal solutions for these 61,926 instances was less than one-quarter of a second per instance on average.[4]

The results of the backtracking algorithm with these instances are summarised in Fig. 5.13. As expected, $\chi(G) = 2$ when $q = 1$ and $q = n - 1$, because the corresponding graphs are trees. More generally, low and high values for q result in sparser graphs contributing to lower chromatic numbers. On the other hand, the highest chromatic numbers are seen when q is set to around three-quarters of n. For comparative purposes, results for the GREEDY, DSATUR and RLF heuristics are also shown in the figure. For the densest graphs, DSATUR and RLF are producing solutions of similar quality; however, for sparser graphs, their solutions are suboptimal except for where $q = 1$.

The success of the backtracking algorithm seen here is due to the special structure of these scale-free graphs. In these cases, the high-degree "hub" vertices tend to belong to a maximum clique C. Because of the selection rules used by the backtracking algorithm, the vertices of C are coloured first, giving a partial $|C|$-colouring. Usually, the remaining vertices are then easily coloured using the available colours, giving a full, feasible $|C|$-colouring. If this is not possible, then only minor adjustments near the leaves of the search tree are usually required. At this point, the number of permitted colours is decremented by one and the algorithm backtracks to level $|C|$ of the search tree; however, because of the presence of the clique C, solutions using fewer than $|C|$ colours cannot be achieved. This allows the algorithm to quickly backtrack to the root of the search tree, therefore providing an optimal solution.

5.7.5 Exam Timetabling Graphs

Our next set of problem instances concerns graphs representing real-world university timetabling problems. As we saw in Sect. 1.1.2, timetabling problems involve assigning a set of "events" (exams, lectures, etc.) to a fixed number of "timeslots". A pair of events then "conflict" when they require the same single resource, e.g., there may be a student or lecturer who needs to attend both events, or the events may require the use of the same room. As a result, conflicting events should be assigned to different timeslots. Under this constraint, a timetabling problem can be modelled using graph colouring by considering each event as a vertex, with edges occurring between pairs of conflicting events. Each colour then represents a timeslot, and a feasible colouring corresponds to a complete timetable with no conflict violations.

---

[4]Using a 3.0 GHz Windows 7 machine with 3.87 GB RAM.

Table 5.2 Details of the 13 timetabling instances of Carter et al. [16]. The degree coefficient of variation (CV) is defined as the ratio of the degree standard deviation to the degree mean

Instance	n	Density	Degree Min; Med; Max	Mean μ	CV (%)
hec-s-92	81	0.415	9; 33; 62	33.7	36.3
sta-f-83	139	0.143	7; 16; 61	19.9	67.4
yor-f-83	181	0.287	7; 51; 117	52.0	35.2
ute-s-92	184	0.084	2; 13; 58	15.5	69.1
ear-f-83	190	0.266	4; 45; 134	50.5	56.1
tre-s-92	261	0.180	0; 45; 145	47.0	59.6
lse-f-91	381	0.062	0; 16; 134	23.8	93.2
kfu-s-93	461	0.055	0; 18; 247	25.6	120.0
rye-s-93	486	0.075	0; 24; 274	36.5	111.8
car-f-92	543	0.138	0; 64; 381	74.8	75.3
uta-s-92	622	0.125	1; 65; 303	78.0	73.7
car-s-91	682	0.128	0; 77; 472	87.4	70.9
pur-s-93	2419	0.029	0; 47; 857	71.3	129.5

In practice, universities will often have a predefined number of timeslots in their timetable and the task will be to determine a feasible solution using an equal number (or fewer) of timeslots than this. In many cases, however, it might be difficult to ascertain whether a timetable with a given number of timeslots is achievable for a particular problem, or it may be desirable to use as few timeslots as possible, particularly if it provides extra time for marking, or allows for a shorter teaching day. Here we concern ourselves with the latter problem and use a set of timetabling problems compiled by Carter et al. [16]. This contains 13 exam timetabling problems encountered at different universities from around the world during the 1980s and 1990s.

A summary of these problem instances is provided in Table 5.2. The names of the graphs start with a three-letter code denoting the name of the university. This is followed by an "s" or "f" (specifying whether the problem occurred in the summer or fall semester) and the year. We see that the set contains problems ranging in size from $n = 81$ to 2419 vertices, and densities of 2.9% up to 41.5%.

It is also known that many of these problem instances feature high numbers of rather large cliques. As Ross et al. [17] have noted, for example:

Table 5.3 Summary of algorithm performance on the 13 timetabling instances of Carter et al. [16]. All statistics are collected from 50 runs on each instance. Asterisks in the rightmost column indicate where the backtracking algorithm was able to produce a provably optimal solution. In these cases, the square brackets indicate the percentage of runs where this occurred, and the average percentage of the computation limit that this took

Instance	Colours at cut-off: mean (best)					
	TABUCOL	PARTIALCOL	HEA	ANTCOL	HC	Bktr
hec-s-92	17.22 (17)	17.00 (17)	17.00 (17)	17.04 (17)	17.00 (17)	19.00 (19)
sta-f-83	13.35 (13)	13.00 (13)	13.00 (13)	13.13 (13)	13.00 (13)	*13.00 (13) [100%, <0.1%]
yor-f-83	19.74 (19)	19.00 (19)	19.06 (19)	19.87 (19)	19.00 (19)	20.00 (20)
ute-s-92	10.00 (10)	10.00 (10)	10.00 (10)	11.09 (10)	10.00 (10)	10.00 (10)
ear-f-83	26.21 (24)	22.46 (22)	22.02 (22)	22.48 (22)	22.00 (22)	*22.00 (22) [100%, 0.7%]
tre-s-92	20.58 (20)	20.00 (20)	20.00 (20)	20.04 (20)	20.00 (20)	23.00 (23)
lse-f-91	19.42 (18)	17.02 (17)	17.00 (17)	17.00 (17)	17.00 (17)	*17.00 (17) [100%, 1.3%]
kfu-s-93	20.76 (19)	19.00 (19)	19.00 (19)	19.00 (19)	19.00 (19)	19.00 (19)
rye-s-93	22.40 (21)	21.06 (21)	21.04 (21)	21.55 (21)	**21.00** (21)	22.00 (22)
car-f-92	39.92 (36)	32.48 (31)	28.50 (28)	30.04 (29)	27.96 (27)	***27.00** (27) [100%, 8.2%]
uta-s-92	41.65 (39)	35.66 (34)	30.80 (30)	32.89 (32)	30.27 (30)	**29.00 (29)**
car-s-91	39.10 (32)	30.20 (29)	29.04 (28)	29.23 (29)	29.10 (28)	**28.00** (28)
pur-s-93	50.70 (47)	45.48 (42)	33.70 (33)	33.47 (33)	33.87 (33)	**33.00** (33)
Total	341.05 (315)	302.36 (294)	280.16 (277)	286.84 (281)	**279.20 (276)**	282.00 (282)
Rank	(6)	(5)	(2)	(4)	(1)	(3)

Consider the instance kfu-s-93, by no means the hardest or largest in this set. It involves 5349 students sitting 461 exams, ideally fitted into 20 timeslots. The problem contains two cliques of size 19 and huge numbers of smaller ones. There are 16 exams that clash with over 100 others.

Table 5.3 summarises the results achieved at the computation limit with the six graph colouring algorithms. In contrast to many of our previous results, the worst overall performance now occurs with the methods relying solely on local search, that is, TABUCOL and to a lesser extent PARTIALCOL. Indeed, we find that these methods are often incapable of achieving feasible solutions even using the initial setting for k determined by DSATUR.[5] The cause of this poor performance seems

---

[5]Consequently, the reported results for TABUCOL and PARTIALCOL in Table 5.3 are produced using an initial k generated by executing the GREEDY algorithm with a random permutation of the vertices.

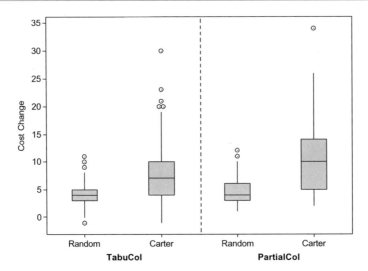

Fig. 5.14 Cost-change distributions for a random graph ($n = 500$, $p = 0.15$, CV= 10.7%, using $k = 16$) and timetable graph car-f-92 ($n = 543$, $p = 0.138$, CV = 75.3%, using $k = 27$). In all cases, samples are taken from candidate solutions with costs of 8

due to the large degree variances seen in these graphs, particularly in comparison to the variances seen in random, flat, and planar graphs seen earlier. The effects of this are demonstrated in Fig. 5.14 where, compared to a random graph of a similar size and density, the differences in cost between neighbouring solutions vary much more widely. This suggests a more "spiky" cost landscape in which the use of local search mechanisms in isolation is insufficient, exhibiting a susceptibility to becoming trapped at local optima.

Table 5.3 also shows that the most consistent performance with these graphs is achieved by the HC and HEA algorithms (no significant difference between the two methods across the instances is apparent). This demonstrates that the issues of using local search in isolation are alleviated by the addition of a global search-based operator. On individual instances, the relative performances of HC and HEA do seem to vary, however. With the problem instances car-f-92 and car-s-91, for example, the HEA's best observed solutions are determined within approximately 1% of the computation limit, while HC's progress is much slower. On the other hand, with instances such as rye-s-93, HC consistently produces the best observed results, using less than 0.3% of the computation limit. This suggests that its operators are somehow suited to this instance (this issue is considered further in Sect. 5.8.1).

We also observe that the backtracking algorithm is again quite competitive with these instances. For four of the problem instances, the algorithm has managed to find and prove the optimal solutions in all runs using a small fraction of the computation limit. Also, the algorithm has produced the best average performance out of all algorithms with the four *largest* problem instances. It seems in these cases

that the abundance of large cliques in the graphs together with their large degree
CVs characterise an abundance of heuristic information that can be successfully
exploited by the algorithm. Indeed, for the four largest instances, all of the solutions
reported in Table 5.3 were found in less than 2% of the computation limit, implying
that the algorithm quickly identifies the correct regions of the search tree. However, counterexamples in which the backtracking algorithm consistently produces the
worst performance can also be seen in Table 5.3, such as with the smallest instance,
hec-s-92.

Finally, we also note the sporadic performance of ANTCOL with these instances.
For all but the four largest problems, ANTCOL's best solutions equal those of the
other algorithms; however, its averages are less favourable, particularly compared to
the HEA and HC algorithms. Consider, for example, the results of ute-s-92 in the
table. This problem is consistently solved using ten colours by all methods except
ANTCOL, which often requires 11 or 12 colours. We find that for instances such
as these, ANTCOL's performance depends very much on the quality of solutions
produced in the first cycle of the algorithm. Due to the low vertex degrees (and
reduced influence of η that results), Equation (5.4) is predominantly influenced by
the pheromone values τ; however, if an 11- or 12-colour solution is produced during
the first cycle, features of these suboptimal solutions are still used to update the
pheromone matrix t, making their reoccurrence in later cycles more likely. The
upshot is that ANTCOL is rarely seen to improve upon solutions found in the initial
cycle of the algorithm with these instances.

5.7.6   Social Network Graphs

Our final set of experiments in this chapter involves graphs representing social networks. Here we consider the social networks of school friends, compiled as part of
the USA-based National Longitudinal Study of Adolescent Health project [18]. The
colouring of such networks might be required when we wish to partition the students
into groups such that individuals are kept separate from their friends, e.g., for group
assignments and team-building exercises (see also Sect. 1.1.1).

To construct these networks, surveys were conducted in various schools, with
each student being asked to list all of his or her friends. In some cases, students were
only allowed to nominate friends attending the same school, while in others they
could include friends attending a "sister school" (e.g., middle-school students could
include friends in the local high school), leading to single-cluster and double-cluster
networks, respectively. In the resultant graphs, each student is represented by a vertex,
with edges signifying a claimed friendship between the associated individuals (see
Fig. 5.15). Note that in the original data, edges signifying friendships are both directed
and weighted; however, in our case, directions and weights have been removed to
form a simple graph.

For these trials, we took a random sample of ten single-cluster networks and ten
double-cluster networks from the Adolescent Health dataset. Summary statistics of

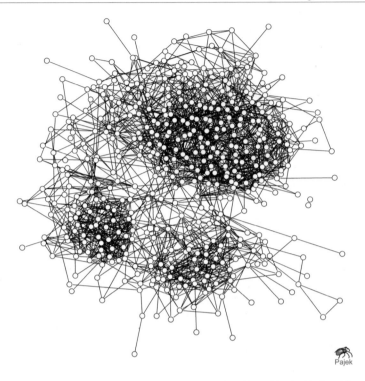

Fig. 5.15 Visualisation of a double-cluster social network collected in the National Longitudinal Study of Adolescent Health project [18]

these graphs are given in Table 5.4. These figures indicate that the vertex degrees are far lower than the timetabling graphs from the previous section, with the highest degree across the whole set being just 29. Consequently, the densities of the graphs are also much lower.

As before, each algorithm was executed 50 times on each instance. The relatively straightforward outcomes of these trials are summarised in Table 5.5. Here, we see that the number of colours needed for these problems ranges from five to ten, though no obvious correlations exist to suggest any links with instance size, density, or the presence of clusters. We also see that the HEA, HC, TABUCOL, and PARTIALCOL methods have all produced the best observed (or optimal) solutions for all instances in all runs. It seems, therefore, that the underlying structures and relative sparsity of these graphs make them relatively "easy" to solve with these algorithms.

In addition, for six of the instances, the backtracking algorithm has managed to find provably optimal solutions, though this does not occur in all runs. Indeed, when this does happen, it seems to occur early in the process (<5% of the computation limit), suggesting that the random elements of the algorithm can have a large effect on the structure of the search tree in these cases. We also observe the poor performance of ANTCOL, which seems to be due to the negative performance features noted in the

Table 5.4 Details of the 20 social networks used. The degree coefficient of variation (CV) is defined as the ratio of the degree standard deviation to the degree mean

Instance	n	Density	Degree		
			Min; Med; Max	Mean μ	CV (%)
Single cluster					
#1	380	0.021	0; 8; 23	8.1	50.5
#2	542	0.013	0; 7; 35	7.1	61.7
#3	563	0.013	0; 7; 23	7.3	55.4
#4	578	0.015	0; 8; 24	8.8	52.7
#5	626	0.013	0; 7; 30	7.8	58.7
#6	746	0.010	0; 7; 28	7.3	58.6
#7	828	0.008	0; 6; 23	6.2	59.3
#8	877	0.009	0; 7; 29	7.8	58.2
#9	1229	0.003	0; 4; 17	4.1	54.6
#10	2250	0.002	0; 4; 25	4.3	78.0
Double cluster					
#11	291	0.027	0; 8; 21	7.8	54.6
#12	426	0.018	0; 7; 26	7.5	56.2
#13	457	0.016	0; 7; 23	7.4	58.8
#14	495	0.017	0; 8; 22	8.5	46.8
#15	569	0.017	0; 9; 34	9.4	50.9
#16	586	0.016	0; 9; 30	9.6	48.4
#17	689	0.010	0; 6; 22	6.8	62.0
#18	795	0.011	0; 9; 24	8.7	53.7
#19	1089	0.007	0; 8; 29	8.1	57.9
#20	1246	0.007	0; 9; 33	8.6	54.4

previous subsection, with a high-quality solution either being produced very quickly (in the first cycle), or not at all.

5.7.7 Comparison Discussion

The results of the above comparison reveal a complicated picture, with different algorithms outperforming others on different occasions. This suggests that the underlying structures of graphs are often critical in an algorithm's resultant performance. In terms of overall patterns, we offer the following observations:

- Algorithms that rely solely on local search (in this case TABUCOL and PARTIAL-COL) often struggle with instances whose cost landscapes are "spiky", commonly

Table 5.5 Summary of algorithm performance on the 20 social networks. All statistics are collected from 50 runs on each instance. Asterisks in the rightmost column indicate where the backtracking algorithm was able to produce a provably optimal solution. In these cases, the square brackets indicate the percentage of runs where this occurred, and the average percentage of the computation limit that this took

Instance	Colours at cut-off: mean (best)					
	TABUCOL	PARTIALCOL	HEA	ANTCOL	HC	Bktr
Single cluster						
#1	8 (8)	8 (8)	8 (8)	8.15 (8)	8 (8)	8 (8)
#2	6 (6)	6 (6)	6 (6)	6.76 (6)	6 (6)	*6 (6) [100%, <1%]
#3	7 (7)	7 (7)	7 (7)	7.45 (7)	7 (7)	7.02 (7)
#4	8 (8)	8 (8)	8 (8)	8.75 (8)	8 (8)	8 (8)
#5	8 (8)	8 (8)	8 (8)	8.41 (8)	8 (8)	8 (8)
#6	6 (6)	6 (6)	6 (6)	6 (6)	6 (6)	*6 (6) [90%, <1%]
#7	6 (6)	6 (6)	6 (6)	6.38 (6)	6 (6)	6 (6)
#8	8 (8)	8 (8)	8 (8)	8.23 (8)	8 (8)	8 (8)
#9	6 (6)	6 (6)	6 (6)	6.10 (6)	6 (6)	6 (6
#10	5 (5)	5 (5)	5 (5)	5 (5)	5 (5)	*5.38 (5) [52%, <1%]
Double cluster						
#11	6 (6)	6 (6)	6 (6)	6.70 (6)	6 (6)	6.02 (6)
#12	5 (5)	5 (5)	5 (5)	5 (5)	5 (5)	*5 (5) [96%, 4%]
#13	6 (6)	6 (6)	6 (6)	6 (6)	6 (6)	*6.32 (6) [46%, 1%]
#14	7 (7)	7 (7)	7 (7)	7.46 (7)	7 (7)	*7 (7) [42%, <1%]
#15	7 (7)	7 (7)	7 (7)	7 (7)	7 (7)	*7 (7) [100%, <1%]
#16	10 (10)	10 (10)	10 (10)	10.13 (10)	10 (10)	10 (10)
#17	7 (7)	7 (7)	7 (7)	7.28 (7)	7 (7)	7 (7)
#18	6 (6)	6 (6)	6 (6)	6 (6)	6 (6)	*6.14 (6) [86%, 1%]
#19	7 (7)	7 (7)	7 (7)	7.65 (7)	7 (7)	7.13 (7)
#20	7 (7)	7 (7)	7 (7)	7.69 (7)	7 (7)	7.02 (7)
Total	136 (136)	136 (136)	136 (136)	142.14 (136)	136 (136)	137.03 (136)
Rank	(1)	(1)	(1)	(6)	(1)	(5)

characterised by a high coefficient of variation (CV) in vertex degrees. On the other hand, these methods do show more promise when the degree CV is quite low, such as with random and flat graphs, suggesting that they have a natural aptitude for navigating spaces in which neighbouring solutions feature costs that are often close or equal to that of the incumbent.

- One obvious advantage of the backtracking algorithm is its ability to produce provably optimal solutions. In our trials, this has been observed with nearly all scale-free graphs, and also some timetabling, planar, and social network graphs. On the other hand, for graphs that are more "regular" in structure such as the random and flat instances, the performance of the backtracking algorithm is significantly worse than the other approaches.

- Across the trials, HEA has proved to be by far the most consistent of the six approaches. We suggest that this is due to a combination of the following attributes:

  - *The HEA operates in the space of infeasible solutions.* Unlike the HC algorithm, which only permits changes to a solution that maintains feasibility, the strategy of allowing infeasible solutions seems to offer higher levels of connectivity (and thus less restriction of movement) within the solution space, helping the algorithm to navigate its way towards high-quality solutions more effectively.

  - *The HEA makes use of global as well as local search operators.* On many occasions, TABUCOL performs poorly when used in isolation; however, the HEA's use of global search operators in conjunction with TABUCOL seems to alleviate these problems by allowing the algorithm to regularly escape from local optima.

  - *The HEA's global search operators are robust.* Unlike ANTCOL's global search operator, which sometimes hinders performance, the HEA's use of recombination in conjunction with a small population of candidate solutions seems beneficial across the instances. This is despite the fact that across all of our tests, recombination was never seen to consume more than 2% of the available run time. Note, in particular, that the GPX operator does not consider any problem-specific information in its operations (such as the connectivity or degree of vertices), yet it still seems to strike a useful balance between (a) altering the solution sufficiently, while (b) propagating useful substructures within the population.

5.8 Further Improvements to the HEA

Before concluding this chapter, we now take a look at some of the individual elements of the HEA and give some ideas as to how the performance of this algorithm can be improved in some cases. These ideas concern maintaining diversity in the population, using alternative recombination operators, and modifying the HEA's local search procedure. They are considered in turn in the following subsections.

5.8.1 Maintaining Diversity

In general, an important factor in the behaviour of an evolutionary algorithm (EA) is the level of diversity that is maintained within its population during a run. Typically, in early iterations of an EA, the diversity of a population will be high, allowing the algorithm to consider many different parts of the solution space. This is often known as the *exploration* phase of the algorithm. As the population evolves, this diversity then generally falls as the algorithm homes in on promising regions of the solution space and seeks to search within these areas more thoroughly. This is often called the *exploitation* phase.

When applying an EA to any computational problem, a suitable balance will need to be established between exploration and exploitation. A fall in diversity that is too slow is undesirable because the algorithm will devote too much energy towards broadly scanning the whole solution space, as opposed to intensively searching specific regions within it. On the other hand, a fall in diversity that is too rapid can also be problematic because the EA will spend too much time focussing on limited regions of the solution space. This latter issue is often called *premature convergence*.

To examine the issue of diversity with the HEA for graph colouring, let us first define a metric for measuring the distance between two candidate solutions.

Definition 5.1 Given a solution S, let P_S be the set of all vertex pairs that are assigned to the same colour in S. That is, $P_S = \{\{u, v\} : u, v \in V \wedge u \neq v \wedge c(u) = c(v)\}$. The *distance* between two solutions S_1 and S_2 can then be calculated using the Jaccard distance measure on the sets P_{S_1} and P_{S_2}. That is:

$$D(S_1, S_2) = \frac{|P_{S_1} \cup P_{S_2}| - |P_{S_1} \cap P_{S_2}|}{|P_{S_1} \cup P_{S_2}|}. \tag{5.8}$$

This distance measure gives the proportion of vertex pairs (assigned to the same colour) that exist in just one of the two solutions. Consequently, if the solutions S_1 and S_2 are identical, then $P_{S_1} \cup P_{S_2} = P_{S_1} \cap P_{S_2}$, giving $D(S_1, S_2) = 0$. Conversely, if no vertex pair is assigned to the same colour in both solutions, then $P_{S_1} \cap P_{S_2} = \emptyset$, implying that $D(S_1, S_2) = 1$. An example of this calculation is shown in Fig. 5.16.

Given this distance measure, we are also able to define a population diversity metric. This is calculated by taking the mean distance between all pairs of solutions in the population.

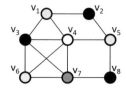 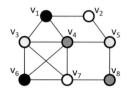

Fig. 5.16 Demonstration of how to measure the distance between two colourings according to Definition 5.1. Here, the left solution $S_1 = \{\{v_1, v_5, v_6\}, \{v_2, v_3, v_8\}, \{v_4\}, \{v_7\}\}$ and the right solution $S_2 = \{\{v_1, v_6\}, \{v_2, v_7\}, \{v_3, v_5\}, \{v_4, v_8\}\}$. This gives two sets, $P_{S_1} = \{\{v_1, v_5\}, \{v_1, v_6\}, \{v_2, v_3\}, \{v_2, v_8\}, \{v_3, v_8\}, \{v_5, v_6\}\}$ and $P_{S_2} = \{\{v_1, v_6\}, \{v_2, v_7\}, \{v_3, v_5\}, \{v_4, v_8\}\}$, leading to $D(S_1, S_2) = (9 - 1)/9 = 8/9$

Definition 5.2 Given a population of solutions defined as a multiset $\mathbf{S} = \{S_1, S_2, \ldots, S_l\}$, the *diversity* of $\mathbf{S}$ is calculated as follows:

$$\text{Diversity}(\mathbf{S}) = \frac{1}{\binom{l}{2}} \sum_{\forall S_i, S_j \in \mathbf{S}: i < j} D(S_i, S_j). \tag{5.9}$$

When applying the HEA to the graphs considered in this chapter, we found that satisfactory levels of diversity were maintained in most cases. However, for some graphs such as the timetabling problem instances, we also observed that large colour classes of low-degree vertices were often formed in the early stages of the algorithm and that these quickly came to dominate the population, causing premature convergence. Indeed, as we saw in Table 5.3, the HEA can sometimes produce inferior results with these problems.

One method by which population diversity might be prolonged in EAs is to make larger changes (mutations) to an offspring to increase its distance from its parents. However, this must be used with care, particularly because changes that are too large might significantly worsen a solution, undoing much of the work carried out in previous iterations of the algorithm. For the HEA, one obvious way of decreasing the distance between parent and offspring is to increase the iteration limit of the local search procedure. However, although this might allow further improvements to be made to a solution, it could also slow the algorithm unnecessarily.

An alternative method for maintaining diversity in this case is to alter the HEA's recombination operator so that it works exclusively with proper colourings. As noted in Sect. 5.3, the GPX operator considers candidate solutions in which clashes are permitted. In practice, however, this could allow large colour classes containing clashes to be unduly promoted in the population, when perhaps the real emphasis should be on the promotion of large *independent sets*. Consequently, we might refine the GPX operator so that it first removes all clashing vertices from each parent before performing recombination. This implies that, before the assignment of missing vertices to random colours, the partial offspring will always be proper. A further effect is that a greater number of vertices will usually need to be recoloured because the

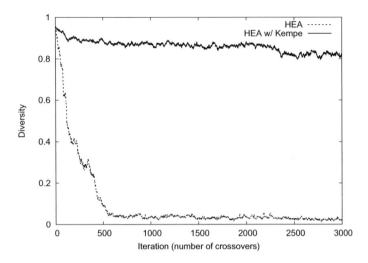

Fig. 5.17 Population diversity using a population of size 10 with the timetabling problem instance car-s-91, using $k = 28$

vertices originally removed from the parents may also be missing in the resultant offspring. Hence, the resultant offspring will tend to be less similar to its parents.

If the above option is chosen, then before randomly reassigning missing vertices to colours, we also have the opportunity to alter the partial proper solution using Kempe chain interchanges. Recall from Theorem 4.1 that this operator, when applied to a proper solution, does not introduce any clashes. Thus, it provides a mechanism by which we can make changes to a solution without compromising its quality in any way.

To illustrate the potential effects of this latter scheme, Fig. 5.17 shows the levels of diversity that exist in the HEA's population for the first 3000 iterations of a run using the timetabling graph car-s-91, which has a chromatic number of 28 (see Table 5.3). When using the original HEA, the population has converged at around 500 iterations and, as we saw in Table 5.3, the algorithm produces solutions using more than 29 colours on average. On the other hand, by applying a series of random Kempe chain moves ($2k$ moves per each application of recombination in this case), population diversity is maintained. In our tests, this modification enabled the algorithm to quickly determine optimal 28-colourings in all runs.

Using Kempe chain interchanges in this way is not always beneficial, however. For instance, similar tests to the above were also carried using random and flat graphs. When using a suitably low value for k in these cases, we found that the Kempe chain interchange operator was usually unable to alter the underlying structures of offspring solutions because its application nearly always resulted in colour relabellings (or, in other words, the bipartite graphs induced by each pair of colour classes in these solutions were nearly always connected, giving total Kempe chains).

Note that within this book's suite of graph colouring algorithms, the HEA contains run-time options for outputting the population diversity and for applying Kempe chain interchanges in the manner described above. (Refer to the algorithm user guide in Appendix A.1 for further information.)

5.8.2 Recombination

Since the proposal of the GPX recombination operator by Galinier and Hao [3], further recombination operators based on this scheme have been suggested, differing primarily on the criteria used for deciding which colour classes to copy from parent to offspring. Porumbel et al. [19], for example, suggest that instead of choosing the largest available colour class at each stage of the recombination process, colour classes with the *least number of clashes* should be prioritised, with class size and information regarding the degrees of the vertices then being used to break ties. Lü and Hao [20], on the other hand, have proposed extending the GPX operator to allow more than two parents to play a part in producing a single offspring. In their operator, the offspring are constructed in the same manner as the GPX, except that at each stage the largest colour class from *multiple parents* is chosen to be copied into the offspring. The intention behind this increased choice is that larger colour classes will be identified, resulting in fewer uncoloured vertices once the k colour classes have been constructed. To prohibit too many colours from being inherited from one particular parent, the authors make use of a parameter q, specifying that if the ith colour class in an offspring is copied from a particular parent, then this parent should not be considered for a further q colours. Note, then, that GPX is simply an application of this operator using two parents with $q = 1$.

Another method of recombination with the graph colouring problem involves considering the individual assignments of vertices to colours as opposed to their partitions. Here, a natural way of representing a solution is to use a vector $(c(v_1), c(v_2), \ldots, c(v_n))$, where $c(v_i)$ gives the colour of vertex v_i. However, it has long been argued that this sort of approach has disadvantages, not least because it leads to a solution space that is far larger than it needs to be (since any solution using k colours can be represented in $k!$ ways—refer to Sect. 2.2). Furthermore, authors such as Falkenauer [21] and Coll et al. [22] have also argued that "traditional" recombination schemes such as 1-, 2-, and n-point crossover with this method of representation tend to recklessly break up building blocks that we might want to be promoted in a population.

In recognition of the perceived disadvantages of the assignment-based representation, Coll et al. [22] have proposed a procedure for relabelling the colours of one of the parents before applying one of these "traditional" crossover operators. Consider two (not necessarily feasible) parent solutions represented as partitions: $\mathcal{S}_1 = \{S_{1,1}, \ldots, S_{1,k}\}$ and $\mathcal{S}_2 = \{S_{2,1}, \ldots, S_{2,k}\}$. Now, using $\mathcal{S}_1$ and $\mathcal{S}_2$, a complete bipartite graph $K_{k,k}$ is formed. This bipartite graph has k vertices in each partition, and the weight between two vertices from different partitions is defined as $W_{i,j} = |S_{1,i} \cap S_{2,j}|$. Given $K_{k,k}$, a maximum weighted matching can then be

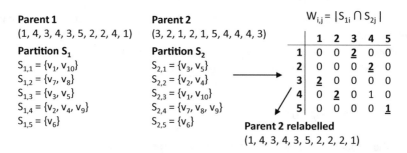

Fig. 5.18 Example of the relabelling procedure proposed by Coll et al. [22]. Here, parent 2 is relabelled using $1 \rightarrow 3, 2 \rightarrow 4, 3 \rightarrow 1, 4 \rightarrow 2$, and $5 \rightarrow 5$

Parent S_1	**Parent** S_2	**Offspring** S'
$S_{1,1} = \{v_1, v_{10}\}$	$S_{2,1} = \{v_1, v_9\}$	$S'_1 = \{v_1, v_{10}\}$
$S_{1,2} = \{v_7, v_8\}$	$S_{2,2} = \{v_7\}$	$S'_2 = \{v_7, \cancel{v_8}\}$
$S_{1,3} = \{v_3, v_5\}$	$\blacktriangleright S_{2,3} = \{v_3, v_5\}$	$S'_3 = \{v_3, v_5\}$
$S_{1,4} = \{v_2, v_4, v_9\}$	$\blacktriangleright S_{2,4} = \{v_2, v_4, v_8\}$	$S'_4 = \{v_2, v_4, v_8\}$
$S_{1,5} = \{v_6\}$	$S_{2,5} = \{v_6, v_{10}\}$	$S'_5 = \{v_6\}$ **Uncoloured** $= \{v_9\}$

Fig. 5.19 Demonstration of the GGA recombination operator. Here, the colour classes in Parent 2 have first been labelled to maximally match those of Parent 1

determined using any suitable algorithm (such as the Hungarian algorithm [23] or Auction algorithm [24]), and this matching can be used to relabel the colours in one of the parents. Figure 5.18 gives an example of this procedure and shows how the second parent can be altered so that its colour labellings maximally match those of the first parent. In this example, we see that the colour classes $\{v_1, v_{10}\}, \{v_3, v_5\}$, and $\{v_6\}$ occur in both parents and will be preserved in any offspring produced via a traditional operator such as uniform crossover. However, this will not always be the case and will depend very much on the best matching available in each case.

An interesting point regarding the structure of solutions and the resultant effects of recombination have also been raised by Porumbel et al. [19]. Specifically, they propose that when solutions to graph colouring problems involve a small number of large colour classes, good quality solutions will tend to occur through the identification of large independent sets, perhaps suggesting that the GPX and its multi-parent variant are naturally suited in these cases. On the other hand, if a solution involves many small colour classes, quality seems to be determined more through the identification of good *combinations* of independent sets.

To these ends, a further recombination operator for graph colouring is also proposed by Lewis [25] which, unlike GPX, shows no bias towards offspring inheriting larger colour classes, or towards offspring inheriting half of its colour classes from each parent. An example of this operator is given in Fig. 5.19. Given two parents, the colour classes in the second parent are first relabelled using procedure of Coll et al. from above. Using the partition-based representations of these solutions, a subset of colour classes from the second parent is then selected randomly, and these replace the corresponding colours in a copy of the first parent. Duplicate vertices are then removed from colour classes originating from the first parent, and any uncoloured vertices are assigned to random colour classes. Tests by Lewis [25] indicate that this recombination operator can produce marginally better solutions than the GPX operator when colour classes are small (approximately five vertices per colour), though worse results can occur in other cases.

Note that the recombination operators listed in this subsection are also included as run-time options within this book's suite of graph colouring algorithms (see Appendix A.1).

5.8.3   Local Search

Finally, from the analyses in this chapter, it is apparent that graph colouring algorithms such as the HEA benefit greatly when used in conjunction with an appropriate local search procedure. For algorithms operating in the space of complete improper solutions, this is usually provided by the TABUCOL algorithm. The tabu search metaheuristic seems very suitable for this purpose because, by extending the steepest descent algorithm, it allows rapid improvements to be made to a solution.

To contrast this, consider the rates of improvement achieved by an analogous simulated annealing algorithm that uses the same neighbourhood operator as TABUCOL but which follows the pseudocode given in Fig. 4.12. For this algorithm, values need to be determined for the initial temperature t, the cooling rate α, and the frequency of temperature updates z. Figure 5.20 compares the run profile of TABUCOL to this simulated annealing algorithm on an example random graph. It can be seen that

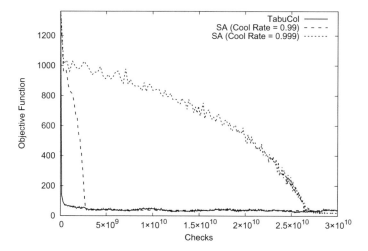

Fig. 5.20 Example run profiles of TABUCOL and an analogous SA algorithm using a random graph $G_{1000,0.5}$ with $k = 86$ colours. Here the SA algorithm uses an initial value for $t = 0.7$, with $z = 500,000$

TABUCOL quickly reduces the objective function value, while the SA approach takes much longer. In addition, the SA algorithm seems quite sensitive to adjustments in its parameters, with inappropriate values potentially hindering performance. On the other hand, it is well known that when the temperature is reduced more slowly, runs of SA tend to produce better quality solutions [26]. Hence, with extended run times, SA may have the potential to produce superior solutions to TABUCOL in some cases.

5.9 Chapter Summary and Further Reading

This chapter has described six different high-performance algorithms for the graph colouring problem. These algorithms have been compared and contrasted over a wide range of problem instances including random, flat, planar, scale-free, and timetabling graphs. Implementations of these methods can be found online (see Appendix A.1).

As with earlier chapters, this chapter's comparison has been carried out using a platform-independent measure of computational effort. In terms of CPU time, Table 5.6 shows the relative run times of the algorithms using a small sample of random graphs. Perhaps the most striking feature is that the HEA is among one of the quickest to execute, a fact that further endorses the method. On the other hand, the ANTCOL and the HC algorithms seem to require significantly more time, apparently due to the computational overheads associated with their BUILDSOLUTION and Kempe chain operators, respectively.

One of the intentions in this chapter has been to test the robustness of our six case-study algorithms by executing them blindly on different problem instances. As we have seen, this has involved using the same parameter values (or methods for calculating them) across all trials. However, different settings may lead to better results in some cases. A broader issue concerns how we might go about *predicting* the performance of a particular algorithm on a previously unseen problem instance. Accurate predictions are useful here because, given a particular graph, we could then apply the most appropriate method from our available portfolio of algorithms.

Table 5.6 Time (in seconds) to complete runs of 5×10^{11} constraint checks with random graphs $G_{n,0.5}$ using a 3.0 GHz Windows 7 PC with 3.87 GB RAM

	$n = 250$	500	1000
TABUCOL	1346	1622	1250
PARTIALCOL	1435	1372	1356
HEA	1469	1400	1337
ANTCOL	4152	3840	4349
HC	5829	5473	5320
Bktr	6328	4794	3930

Research in this area has been carried out by Smith-Miles et al. [27]. In their work, the authors consider 18 different graph metrics. These are then used to help predict which graph colouring algorithm will perform best on any given problem instance. Metrics considered by the authors include the following:

1. The size and density of the graph.
2. The mean and standard deviation of the degrees.
3. The average path length, diameter, and girth of the graph.
4. The *betweenness centrality* of a graph. (This considers the number of shortest paths in a graph that pass through each vertex. If a vertex has many shortest paths passing through it, it is considered to be more "central". Having calculated this metric for each vertex, summary measures can then be taken.)
5. The *energy* of a graph. (This is calculated as the mean of the absolute values of the eigenvalues of the adjacency matrix.)

To achieve their aims, Smith-Miles et al. [27] executed this chapter's algorithms on a wide range of different problem instances. Machine learning methods were then used to classify the types of graph that the different algorithms were seen to perform well with. This information can then be used for predicting algorithm performance on new problem instances. One observation from this work is that, of the six algorithms, only three seem to consistently show regions of the instance space where they are uniquely best, namely, HEA, HC, and ANTCOL. As we have seen, each of these methods combines local search strategies with global operators.

Subsequent work in this area is also due to Neis and Lewis [28]. Here, the authors again use the above algorithms but they also consider different values of the algorithms' control parameters (local search iteration limits, tabu tenures, population sizes, and so on). Similar graph metrics to the list above are used and, after performing a multidimensional database analysis on their results, the authors propose that the most useful metrics for predicting algorithm performance are the standard deviation of betweenness centrality, together with the density and energy of the graph.

References

1. Avanthay C, Hertz A, Zufferey N (2003) A variable neighborhood search for graph coloring. Eur J Oper Res 151:379–388
2. Dorne R, Hao J-K (1998) A new genetic local search algorithm for graph coloring. In: Eiben A, Back T, Schoenauer M, Schwefel H (eds) Parallel problem solving from nature (PPSN) V. LNCS, vol 1498. Springer, pp 745–754
3. Galinier P, Hao J-K (1999) Hybrid evolutionary algorithms for graph coloring. J Comb Optim 3:379–397
4. Thompson J, Dowsland K (2008) An improved ant colony optimisation heuristic for graph colouring. Discret Appl Math 156:313–324

5. Galinier P, Hertz A (2006) A survey of local search algorithms for graph coloring. Comput Oper Res 33:2547–2562
6. Blöchliger I, Zufferey N (2008) A graph coloring heuristic using partial solutions and a reactive tabu scheme. Comput Oper Res 35:960–975
7. Dorigo M, Maniezzo V, Colorni A (1996) The ant system: optimisation by a colony of cooperating agents. IEEE Trans Syst Man Cybern 26(1):29–41
8. Rizzoli A, Montemanni R, Lucibello E, Gambardella L (2007) Ant colony optimization for real-world vehicle routing problems. Swarm Intell 1(2):135–151
9. Lewis R (2009) A general-purpose hill-climbing method for order independent minimum grouping problems: a case study in graph colouring and bin packing. Comput Oper Res 36(7):2295–2310
10. Korman S (1979) Combinatorial optimization. In: The graph-coloring problem. Wiley, New York, pp 211–235
11. Cheeseman P, Kanefsky B, Taylor W (1991) Where the really hard problems are. In: Proceedings of IJCAI-91, pp 331–337
12. Turner J (1988) Almost all k-colorable graphs are easy to color. J Algorithms 9:63–82
13. Robertson N, Sanders D, Seymour P, Thomas R (1997) The four color theorem. J Comb Theory Ser B 70:2–44
14. Barabási A, Bonabeau E (2003) Scale-free networks. Scientific American, May 2003
15. Barabási A (2016) Network science. Cambridge University Press
16. Carter M, Laporte G, Lee SY (1996) Examination timetabling: algorithmic strategies and applications. J Oper Res Soc 47:373–383
17. Ross P, Hart E, Corne D (2003) Genetic algorithms and timetabling. In: Ghosh A, Tsutsui K (eds) Advances in evolutionary computing: theory and applications. Natural computing. Springer, pp 755–771
18. Moody J, White D (2003) Structural cohesion and embeddedness: a hierarchical concept of social groups. Am Sociol Rev 68(1):103–127
19. Porumbel D, Hao J-K, Kuntz P (2010) An evolutionary approach with diversity guarantee and well-informed grouping recombination for graph coloring. Comput Oper Res 37:1822–1832
20. Lü Z, Hao J-K (2010) A memetic algorithm for graph coloring. Eur J Oper Res 203(1):241–250
21. Falkenauer E (1998) Genetic algorithms and grouping problems. Wiley
22. Coll E, Duran G, Moscato P (1995) A discussion on some design principles for efficient crossover operators for graph coloring problems. Anais do XXVII Simposio Brasileiro de Pesquisa Operacional, Vitoria-Brazil
23. Munkres J (1957) Algorithms for the assignment and transportation problems. J Soc Ind Appl Math 5(1):32–38
24. Bertsekas D (1992) Auction algorithms for network flow problems: a tutorial introduction. Comput Optim Appl 1:7–66
25. Lewis R (2015) Springer handbook of computational intelligence. In: Graph coloring and recombination. Studies in computational intelligence. Springer, pp 1239–1254
26. van Laarhoven P, Aarts E (1987) Simulated annealing: theory and applications. Kluwer Academic Publishers
27. Smith-Miles K, Baatar D, Wreford B, Lewis R (2014) Towards objective measures of algorithm performance across instance space. Comput Oper Res 45:12–24
28. Neis P, Lewis R (2020) Evaluating the influence of parameter setup on the performance of heuristics for the graph colouring problem. Int J Metaheurist 7(4):352–378

Applications and Extensions

We are now at a point in this book where we have seen several different algorithms for the graph colouring problem and have noted many of their relative strengths and weaknesses. This chapter now presents a range of problems, both theoretical and practical based, for which such algorithms might be applied. These include face colouring, edge colouring, precolouring, constructing Latin squares, solving Sudoku puzzles, and testing for short circuits in circuit boards. Note that these problems are either equivalent to, or represent special cases of, the general graph colouring problem.

This chapter also considers variants of the graph colouring problem where not all of the graph is visible to an algorithm, or where the graph's structure is subject to change over time. Such problems can arise when setting up wireless networks and also in some timetabling applications. We then go on to consider problems that *extend* and therefore generalise the graph colouring problem, specifically list colouring, equitable colouring, weighted graph colouring, and chromatic polynomials. Detailed real-world applications of graph colouring are also the subject of Chaps. 7, 8, and 9.

Note that, in contrast to the rest of this book, the first two sections of this chapter are concerned with colouring the *faces* of graphs and the *edges* of graphs. As we will see, these two problems can be converted into equivalent formulations of the vertex colouring problem using the concepts of *dual graphs* and *line graphs*, respectively. However, it is often useful for face and edge colouring problems to be considered as separate problems; hence, we will often use the term "vertex colouring" instead of "graph colouring" to avoid any ambiguities.

© The Author(s), under exclusive license to Springer Nature Switzerland AG 2021   155
R. M. R. Lewis, *Guide to Graph Colouring*, Texts in Computer Science,
https://doi.org/10.1007/978-3-030-81054-2_6

6.1 Face Colouring

In the face colouring problem, we want to colour the *spaces* between vertices and edges, as opposed to the vertices themselves. Face colouring is specifically concerned with planar graphs which, as we saw in Chap. 1, are graphs that can be drawn on a plane so that no edges cross one another. When drawn in this way planar graphs can be divided into faces, including one unbounded face that surrounds the graph. Figure 6.1, for example, shows a planar graph comprising ten faces: nine bounded faces and one unbounded face (numbered 10 in the figure). The *boundary* of a face is the set of edges that surrounds it. When a face is bounded, its boundary forms a cycle.

It is evident by inspecting Fig. 6.1 that the number of faces seems to be related to the number of vertices and edges of the graph. In fact, this relationship can be stated explicitly due to an elegant theorem that was first noted by Leonhard Euler (1707–1783):

> **Theorem 6.1** (Euler's Characteristic) *Let G be a planar graph with n vertices, m edges, and f faces. Then*
>
> $$n - m + f = 2.$$

Proof The proof is via induction on the number of faces f. If $f = 1$, then the graph contains no cycles and must therefore be a tree. Since the number of edges in a tree $m = n - 1$, the theorem holds because $n - (n - 1) + 1 = 2$.

Now assume $f \geq 2$, meaning that G must contain at least one cycle. Let $\{u, v\}$ be an edge in one of these cycles. Since this cycle divides two faces, say F_1 and F_2, removing $\{u, v\}$ from G to form a subgraph G' will have the effect of joining F_1 and F_2, with all other faces remaining unchanged. Hence, G' has $f - 1$ faces.

Let n', m', and f' be the number of vertices, edges, and faces in G'. Thus, $n' = n$, $m' = m - 1$, $f' = f - 1$, and $n - m + f = n' - m' + f' = 2$. $\square$

We see that Euler's characteristic does indeed hold for the example graph in Fig. 6.1 since $n - m + f = 15 - 23 + 10 = 2$ as expected.

When considering the face colouring problem it is necessary to restrict ourselves to planar graphs that contain no *bridges*. A bridge is defined as an edge in a graph G whose removal increases the number of components. When a graph contains a

Fig. 6.1 Planar graph with $n = 15$ vertices, $m = 23$ edges, and $f = 10$ faces

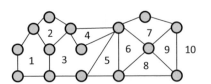

bridge $\{u, v\}$, the unbounded face will surround the graph, but will also feature $\{u, v\}$ on its boundary twice, making it impossible to colour feasibly. Hence, planar graphs containing bridges are not considered further in this section.

Let us now consider the maximum number of edges that a graph can feature while retaining the property of planarity. Consider a connected planar graph G with n vertices, m edges, and f faces. Also write f_i for the number of faces in G that contain exactly i edges in their boundaries. Clearly $\sum_i f_i = f$ and, assuming that G does not contain a bridge,

$$\sum_i = i f_i = 2m \tag{6.1}$$

since every edge is on the boundary of exactly two faces. We can use this relationship in conjunction with Euler's characteristic to give an upper bound on the number of edges in a planar graph. This result also involves knowledge of the *girth* of a graph, defined as follows.

Definition 6.1 The *girth* of a graph G is the length of the shortest cycle in G. If G is acyclic (i.e., contains no cycles), then its girth equals infinity.

Theorem 6.2 *Let G be a planar graph with $n \geq 3$ vertices, m edges, f faces, and no bridges. Then G has at most $3n - 6$ edges. Furthermore, if G has a girth g (where $3 \leq g \leq \infty$), then:*

$$m \leq \max \left\{ \frac{g}{g-2}(n-2), n-1 \right\}.$$

Proof For $g = 3$, we get $\max\{\frac{3}{3-2}(n-2), n-1\} = 3(n-2)$, giving $m \leq 3n - 6$ as required. Hence, we only need to prove the second assertion above.

If $g > n$, then this implies $g = \infty$ meaning that G has no cycles and is therefore a tree. Hence, $m = (n-1) \leq n$. Now assume that $g \leq n$ and that the assertion holds for smaller values of n. Also, assume without loss of generality that G is connected. From earlier, we know that

$$2m = \sum_i i f_i = \sum_{i \geq g} i f_i \geq g \sum_i f_i = gf.$$

Hence, by Euler's characteristic (Theorem 6.1), we get

$$m + 2 = n + f \leq n + \frac{2}{g}m$$

and so

$$m \leq \frac{g}{g-2}(n-2)$$

as required.   □

Theorem 6.2 can sometimes be used to decide whether a graph is planar or not. For example, the complete graph with five vertices K_5 cannot possibly be planar because it has $n = 5$ vertices and $m = 10$ edges, meaning $m \leq 3n - 6$ is not satisfied. As another example, the complete bipartite graph with six vertices $G = (V_1, V_2, E)$, where $E = \{\{u, v\} : u \in V_1, v \in V_2\}$ and $|V_1| = |V_2| = 3$, is also not bipartite since it has $m = 9$ edges, $n = 6$ vertices, and a girth of four, meaning that $m = 9 > \frac{4}{4-2}(6 - 2) = 8$. Less obvious, but profoundly more useful, however, is the amazing fact that a graph is planar *if and only if* it does not contain a subgraph that is a subdivision of either K_5 or $K_{3,3}$. This result, due to Kuratowski [1], has been used alongside similar results to help construct several efficient (polynomial time) algorithms for determining whether a graph is planar or not, including the Path Addition method of Hopcroft and Tarjan [2] and the more recent Edge Addition method of Boyer and Myrvold [3].

6.1.1 Dual Graphs, Colouring Maps, and the Four Colour Theorem

The close relationship between the problems of vertex colouring and face colouring becomes apparent when we consider the concept of *dual graphs*. Given a planar graph G, the *dual* of G, denoted by G^*, is constructed according to the following steps. First, draw a single vertex v_i^* inside each face F_i of G. Second, for each edge e in G, draw a line e^* that crosses e but no other edge in G, and that links the two vertices in G^* corresponding to the two faces in G that e is separating.

This procedure is demonstrated in Fig. 6.2. Here, the vertices in G are shown in grey, and the vertices in G^* are shown in black. G has six faces in total: five bounded faces and one unbounded face. The unbounded face is represented by the top vertex of G^* in the example and is made adjacent to all vertices in G^* whose corresponding faces in G have an edge on the exterior of the graph. Note that G^* may also have multiple edges between a pair of vertices, as it occurs on the right-hand side of the example graph.

It is clear from the figure that the process of forming duals is reversible, that is, we can use the same process to form G from G^*. It is also clear that because G is planar, its dual G^* must also be planar. We can now state relationships between the number of vertices, faces, and edges in G and G^* such as the following.

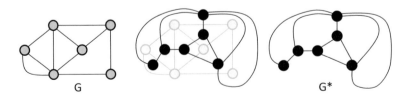

Fig. 6.2 Illustration of how to convert a planar graph G to its dual G^*

Fig. 6.3 The states of mainland Australia (left), and the corresponding planar graph (right)

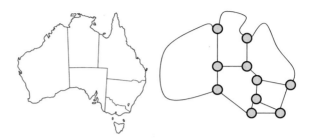

> **Theorem 6.3** *Let G be a connected planar graph with n vertices, m edges, and f faces. Also, let G^* be the dual of G, comprising n^* vertices, m^* edges, and f^* faces. Then $n^* = f$, $m^* = m$, and $f^* = n$.*

Proof It is clear that $n^* = f$ due to the method by which duals are constructed. Similarly, $m^* = m$ because all edges in G^* intersect exactly one edge each in G (and vice versa). The third relation follows by substituting the previous two relationships into Euler's characteristic applied to both G and G^*.   ☐

Recall from Chap. 1 that the four colour theorem (or "conjecture" as it was at the time) was originally stated in 1852 by Francis Guthrie, who hypothesised that four colours are sufficient for colouring the faces of any map such that neighbouring faces have different colours. In the context of graph theory, a map can be represented by a bridge-free planar graph G, with the faces of G representing the various regions of the map, edges representing borders between regions, and vertices representing points where the borders intersect. An illustration using a map of Australia is given in Fig. 6.3.

The following theorem now reveals the close relationship between the vertex colouring and face colouring problems.

> **Theorem 6.4** *Let G be a connected planar graph without loops, and let G^* be its dual. Then the vertices of G are k-colourable if and only if the faces of G^* are k-colourable.*

Proof Since G is connected, planar, and without loops, its dual G^* is a planar graph with no bridges. If we have a k-colouring of the vertices of G, then each face of G^* can now be assigned to the same colour as its corresponding vertex in G. Because no adjacent vertices in G have the same colour, it follows that no adjacent faces in G^* have the same colour. Thus, the faces of G^* are k-colourable.

Now suppose that we have a k-colouring of the faces of G^*. Since every vertex of G is contained in a face of G^*, each vertex in G can assume the colour of its corresponding face in G^*. Again, since no adjacent faces in G^* are allocated the same colour, this implies no adjacent vertices in G are given the same colour.   □

This result is important because it tells us that the faces of any map (represented as a planar graph G^* with no bridges) can be k-coloured by simply determining a vertex k-colouring of its dual graph G. The result also tells us that we can take any theorem concerning the vertex colouring of a planar graph and then state a corresponding theorem on the face colouring of its dual, and vice versa.

One elegant theorem that arises from this relationship demonstrates a link between Eulerian graphs and graphs that are bipartite.

Definition 6.2 A graph G is *Eulerian* if and only if it is connected and the degrees of all its vertices are even.

This gives rise to the following theorem.

Theorem 6.5 *The faces of a planar graph with no bridges G are two-colourable if and only if G is Eulerian.*

Proof Recall that a graph's vertices are two-colourable if and only if it is bipartite. Hence, we need to show that the dual of any planar Eulerian graph is bipartite, and vice versa.

Let G be an Eulerian planar graph. By definition, all vertices in G are even in degree. Since the degree of a vertex in G corresponds to the number of edges surrounding a face in the dual G^*, the edges surrounding each face in G^* constitute cycles of even length. Hence, according to Theorem 3.10, G^* is bipartite.

Conversely, let G^* be bipartite. This means G^* contains no odd cycles and, since G is planar, all faces are surrounded by an even number of edges. Hence, all vertex degrees in G are even, making G Eulerian.   □

Practical examples of Theorem 6.5 arise in the tiling industry where we are often interested in laying tiles of two different colours such that adjacent tiles do not have the same colour. Example titling patterns are shown in Fig. 6.4a, b. Close examination of these patterns reveals that the underlying graphs are Eulerian as expected. Two-colourings also arise when a picture is drawn using a single line that is joined at either end, such as with the geometric drawing device "Spirograph". Figure 6.4c shows an example of this. We see that each time the line crosses itself, the degree of the "vertex" existing at this intersection increases by two; hence, the vertex degrees will always be even.

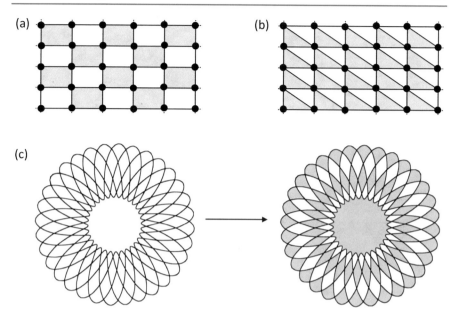

Fig. 6.4 Examples of face colourings using two colours

The connection between face k-colourings of maps and vertex k-colourings of their planar duals allows us to conclude that the four colour theorem for maps is equivalent to the statement that the vertices of all loop-free planar graphs are four-colourable.[1] This concept was hinted upon in Sect. 1.2 where, in Fig. 1.7, we took a map, constructed its (planar) dual graph, four-coloured the vertices of this dual, and then converted this solution back to a four-colouring of the faces of the original map. An example four-colouring of a larger planar graph is shown in Fig. 6.5.

The task of proving that four colours are sufficient for the vertices of *any* planar graph (and therefore the faces of any map) was formally one of the most famous unsolved problems in the whole of mathematics. It was eventually solved in controversial circumstances by Kenneth Appel and Wolfgang Haken in 1976. Their proof is very long and required many months of computation time to test and classify a large number of different graph configurations. Consequently, we end this section by restricting ourselves to proving the weaker six and five colour theorems, before giving a more general history of the four colour theorem itself.

Theorem 6.6 *The vertices of any loop-free planar graph are six-colourable.*

---

[1] Recall that loops (i.e., edges of the form $\{v, v\}$) are disallowed in the vertex colouring problem.

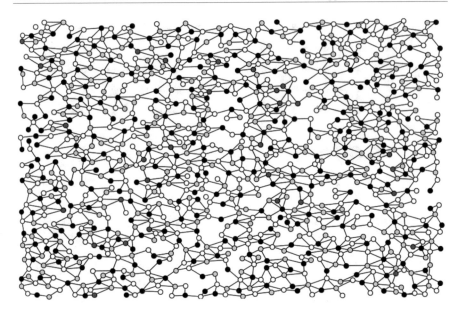

Fig. 6.5 A four-colouring of a 1000-vertex planar graph

Proof Let G be a planar graph with $n \geq 3$. According to Theorem 6.2, G has at most $3n - 6$ edges. This means that the minimal degree of G cannot exceed 5. Thus, in every subgraph G' of G, there is a vertex with degree of at most $\delta = 5$. Therefore, according to Theorem 3.7, we get $\chi(G) \leq 5 + 1$. $\qquad\qquad\square$

With some additional reasoning we can improve this result to get the following.

Theorem 6.7 (Heawood [4]) *The vertices of any loop-free planar graph are five-colourable.*

Proof For contradictory purposes, suppose this statement to be false, and let G be a planar graph with chromatic number $\chi(G) = 6$ and a minimal number of vertices n. Because of Theorem 6.2, G must have a vertex v with $\deg(v) \leq 5$. Now let $G' = G - \{v\}$. We know that G' can be five-coloured using, say, colours labelled 1–5. Each of these colours must also be used to colour at least one neighbour of v (otherwise G would also be five-colourable). We can now assume that v has five neighbours, say $u_1, u_2, \ldots, u_5$, arranged in a clockwise fashion around v, with colours $c(u_i) = i$.

Now denote by $G'(i, j)$ the subgraph of G' spanned by vertices with colours i and j. Suppose that u_1 and u_3 belong to separate components of $G'(1, 3)$. Interchanging the colours 1 and 3 in the component of $G'(1, 3)$ containing u_1 will give us another

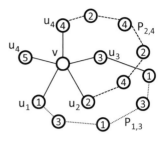

Fig. 6.6 Depicting paths $P_{1,3}$ and $P_{2,4}$ used in the proof of Theorem 6.7. Colour labels are written inside the vertices

feasible five-colouring of G'. However, in this five-colouring, both u_1 and u_3 will have the same colour, meaning that a spare colour now exists for v. This implies that G is five-colourable.

Since u_1 and u_3 must belong to the same component $G'(1, 3)$, we now deduce the existence of a path $P_{1,3}$ in G' whose vertices are coloured using colours 1 and 3 only. Similarly, G' must also contain a path $P_{2,4}$ using colours 2 and 4. However, this is impossible in a planar graph since the cycle u_1, $P_{1,3}$, u_3 separates u_2 from u_4, meaning that $P_{2,4}$ cannot be drawn without edges crossing (see Fig. 6.6). Hence, G cannot be planar. □

6.1.2 Four Colours Suffice

In the proof of Theorem 6.7, the notation $G(i, j)$ denotes the subgraph induced by taking the vertices coloured with colours i and j in G. Individual components of $G(i, j)$ are Kempe chains (see Definition 4.3), which are named after the mathematician Alfred Kempe (1849–1922), who used them in an infamous incorrect proof for the four colour theorem in 1879.

As we saw in Chap. 1, the conjecture that all maps can be coloured using at most four colours was first pointed out in 1852 by Francis Guthrie (1831–1899) who, at the time, was a student at University College London. Guthrie passed these observations on to his brother Frederick who, in turn, passed them on to his mathematics tutor Augustus De Morgan (1807–1871). De Morgan was not able to provide a conclusive proof for this conjecture, but the problem, being both easy to state and tantalisingly difficult to solve, captured the interest of many notable mathematicians of the era, including William Hamilton (1805–1865), Arthur Cayley (1821–1895), and Charles Pierce (1839–1914).

Indeed, over time the four colour conjecture was to become one of the most famous unsolved problems in all of mathematics.

In 1879, a student of Arthur Cayley, Alfred Kempe, announced in *Nature* magazine that he had proved the four colour theorem, publishing his result in the *American Journal of Mathematics* [5]. In his arguments, Kempe made use of his eponymous Kempe chains in the following way. Suppose we have a map in which all faces except one are coloured using colours 1, 2, 3, or 4. If the uncoloured face, which we shall call F, is not surrounded by faces featuring all four colours, then obviously we can colour F using the missing colour. Therefore, suppose now that F is surrounded by faces F_1, F_2, F_3, and F_4 (in that order), which are coloured using colours 1, 2, 3, and 4, respectively. There are now two cases to consider:

Case 1: There exists no chain of adjacent faces from F_1 to F_3 that are alternately coloured with colours 1 and 3.

Case 2: There *is* a chain of adjacent faces from F_1 to F_3 that are alternately coloured with colours 1 and 3.

If Case 1 holds then F_1 can be switched to colour 3, and any remaining faces in the chain can also have their colours interchanged. This operation retains the feasibility of the solution (no adjacent faces will have the same colour) and also means that no face adjacent to F will have colour 1. Consequently, F can be assigned to this colour.

If Case 2 holds then there cannot exist a chain of faces from F_2 to F_4 using only colours 2 and 4. This is because, for such a chain to exist, it would need to cross the chain from F_1 to F_3, which is impossible on a map. Thus, Case 1 holds for F_2 and F_4, allowing us to switch colours as with Case 1.

The arguments of Kempe were widely accepted among mathematicians of the day. He was later elected a Fellow of the Royal Society and also went on to be knighted in 1912. The four colour *conjecture* was now considered to be the four colour *theorem*.

This all changed 11 years later when, in 1890, English mathematician Percy Heawood (1861–1955) shocked the mathematics fraternity by publishing an example map that exposed a flaw in Kempe's arguments [4]. Though he failed to supply his own proof, Heawood had shown that the four colour theorem was indeed still a conjecture. In the same publication, Heawood did show, however, that arguments analogous to Kempe's could be used to prove that all maps are five-colourable, as we saw in Theorem 6.7. In later work, Heawood also proved that if the number of edges around each region of a map is divisible by 3 then the map can be four-coloured.

As the decades passed, the problem that had first been pointed out by Guthrie in 1852 remained unproven. Some piecemeal progress towards a solution was made with one proof showing that four colours were sufficient for colouring maps of up to 27 faces. This was followed by proofs for up to 31 faces, and then 35 faces. However, it would turn out that methods used by Kempe and his contemporaries in early papers would ultimately pave the way.

To start, the focus of research turned towards proofs concerning the vertices of loop-free planar graphs (i.e., the dual graphs of maps). In the first half of the twentieth century, researchers also concentrated their efforts on reducing these graphs to special cases that could be identified and classified. The idea was to produce a minimal set of configurations that could each be tested. Initially, this set was thought to contain nearly 9000 members, which was considered far too large for mathematicians to study individually. This compelled some to turn towards using computers to design specialised algorithms for testing them.

Ultimately, the first conclusive proof of the four colour theorem was produced in 1976 by mathematicians Kenneth Appel (1932–2013) and Wolfgang Haken (b. 1928). Their proof is based on the idea that if the four colour conjecture were false, then there would exist at least one planar graph G with the smallest possible number of vertices such that $\chi(G) = 5$. They then showed that G cannot exist. To do this, they used the notions of unavoidable sets and reducibility.

1. An *unavoidable set* is a set of configurations such that any planar graph has at least one member of this set as a subgraph.
2. A *reducible configuration* is a planar graph that cannot occur in a minimal counterexample G. If a planar graph contains a reducible configuration, then it can be reduced to a smaller planar graph. This smaller graph also has the condition that if it can be four-coloured, then so can the original. Also, if the original graph cannot be four-coloured then neither can the smaller graph, so the original graph is not minimal.

Appel and Haken's proof involved constructing an unavoidable set and therefore proving that G cannot exist. The number of members in this set was found to be 1936, which were then checked one by one by hand and by computer [6–9]. As was later stated in Appel's obituary in *The Economist* on 4 May 2013:

> Both he and Dr. Haken hugely exceeded their time allocation on the computer, which belonged to the university administration department. ...Their proof depended on both hand-checking by family members and then brute-force computer power; the result was published in over 140 pages in the *Illinois Journal of Mathematics* and 400 pages of further diagrams on microfiche. They also, in the old fashioned way, chalked the message on a blackboard in the mathematics department: FOUR COLOURS SUFFICE.

At the time, this work was controversial, with some mathematicians questioning the legitimacy of a proof in which much of the work had been carried out by computer. (How might we guarantee the reliability of the algorithms and hardware?) However, despite these concerns, independent verification soon convinced the community that the four colour theorem had indeed finally been proved. Hence, we are now able to state:

Theorem 6.8 (The Four Colour Theorem) *The vertices of any loop-free planar graph are four-colourable. Equivalently, the faces of any map are four-colourable.*

In more recent years, Robertson et al. [10] have proposed an algorithm for four-colouring planar graphs that operates in $\mathcal{O}(n^2)$ time. They have also shown how to construct an unavoidable set containing just 633 reducible configurations. However, a proof along more "traditional" lines remains elusive and, to this day, the four colour theorem remains an excellent example, along with Fermat's last theorem, of a problem that is very easy to state, but exceptionally difficult to solve.

Readers interested in finding out more about the fascinating history of the four colour theorem are invited to consult the very accessible book *Four Colors Suffice: How the Map Problem Was Solved* by Wilson [11].

6.2 Edge Colouring

Another way in which graphs can be coloured is to assign colours to their *edges*, as opposed to their vertices or faces. This gives rise to the *edge colouring problem* where we seek to colour the edges of a graph so that no pair of edges sharing an endpoint (i.e., incident edges) have the same colour, and so that the number of colours used is minimal.

The edge colouring problem has applications in scheduling round-robin tournaments and also transferring files in computer networks [12,13]. The minimum number of colours needed to edge colour a graph G is called the *chromatic index*, denoted by $\chi'(G)$. This should not be confused with the *chromatic number* $\chi(G)$, which is the minimum number of colours needed to colour the *vertices* of a graph G.

As mentioned earlier, the edge colouring and vertex colouring problems are closely related because we can colour the edges of a graph by simply colouring the vertices of its corresponding *line graph*.

Definition 6.3 Given a graph G, the *line graph* of G, denoted by $L(G)$, is constructed by using each edge in G as a vertex in $L(G)$, and then connecting pairs of vertices in $L(G)$ if and only if the corresponding edges in G share a common vertex as an endpoint.

An example conversion between a graph G and its line graph $L(G)$ is shown in Fig. 6.7a. From this process, it is natural that the number of vertices and edges in $L(G)$ is related to the number of vertices and edges in G.

Theorem 6.9 *Let $G = (V, E)$ be a graph with n vertices and m edges. Then its line graph $L(G)$ has m vertices and*

$$\frac{1}{2} \sum_{v \in V} \deg(v)^2 - m$$

edges.

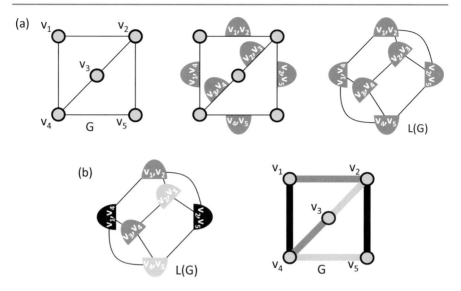

Fig. 6.7 Illustration of **a** how to convert a graph G into its line graph $L(G)$, and **b** how a vertex k-colouring of $L(G)$ corresponds to an edge k-colouring of G

Proof Since each edge in G corresponds to a vertex in $L(G)$ it is obvious that $L(G)$ has m vertices. Now let $\{u, v\}$ be an edge in G. This means that $\{u, v\}$ is a vertex in $L(G)$ with degree $\deg(u) + \deg(v) - 2$. Hence, the total number of edges in $L(G)$ is

$$\frac{1}{2} \sum_{\{u,v\} \in E} (\deg(u) + \deg(v) - 2) = \frac{1}{2} \sum_{\{u,v\} \in E} (\deg(u) + \deg(v)) - m.$$

Note that the degree of each vertex v appears exactly $\deg(v)$ times in this sum. Hence, we can simplify the expression to that stated in Theorem 6.9 as required. □

Figure 6.7b also demonstrates how a vertex k-colouring of the line graph $L(G)$ corresponds to an edge colouring of G. Consequently, rather like the way in which a face colouring problem can be tackled by colouring the vertices of a graph's dual, any edge colouring problem stated on a graph G can be tackled by colouring the vertices its line graph $L(G)$.

We now discuss some important results concerning the chromatic index of a graph.

Theorem 6.10 *Let K_n be the complete graph with $n > 1$ vertices. Then its chromatic index $\chi'(K_n) = n - 1$ if n is even; otherwise $\chi'(K_n) = n$.*

Proof When n is odd, the edges of K_n can be coloured using n colours by the following process. First, draw the vertices of K_n in the form of a regular n-sided

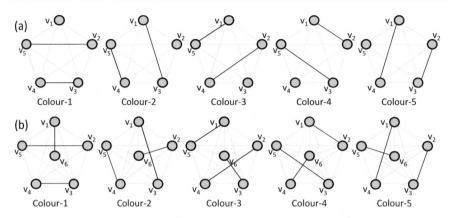

Fig. 6.8 Illustrating how optimal edge colourings can be constructed for complete graphs with **a** K_5 and **b** K_6 using the circle method

polygon. Next, select an arbitrary edge on the boundary of this polygon and colour it, together with all edges parallel to it, using colour 1. Now moving in a clockwise direction, select the next edge on the boundary and colour it, together with its parallel edges, with colour 2. Continue this process until all edges have been coloured.

It is easy to demonstrate that the edges of K_n are not $(n-1)$-colourable by the fact that the largest number of edges that can be assigned the same colour is $(n-1)/2$; it then follows, because the number of edges in K_n is $\frac{n(n-1)}{2}$, that n colours are required.

When n is even, a similar process can be followed, where a regular $(n-1)$-sided polygon is constructed, with the remaining vertex being placed in the centre. The same method for the $(n-1)$ case is then followed, with edges perpendicular to the edges currently being coloured also being assigned to the same colour. As in the previous case, it is easily shown that no feasible edge colouring of K_n exists using fewer than n colours.   □

The method used in the proof of Theorem 6.10 is often referred to as the circle (or polygon) method and was originally proposed by mathematician and Church of England Minister Thomas Kirkman (1806–1895) [14]. An important practical use of this method is for constructing round-robin sports leagues, where we have a set of n teams that are required to play each other once across a sequence of rounds. Figure 6.8 provides examples of this method for $n = 5$ and $n = 6$. Here, the vertices can be thought of as "teams", with edges representing "matches" between these teams. Each colour then represents a round in the schedule. Considering Fig. 6.8a, where $n = 5$, the first round involves matches between team-v_2 and team-v_5 and between team-v_3 and team-v_4, with team-v_1 receiving a bye. The next round then involves matches between team-v_1 and team-v_3 and between team-v_4 and team-v_5, with team-v_2 receiving a bye, and so on. The pattern is similar when n is even, as shown in Fig. 6.8b, except that no team receives a bye. Applications of graph colouring to sports scheduling problems are considered in more detail in Chap. 8.

A further result, due to König [15], concerns the chromatic index of bipartite graphs.

> **Theorem 6.11** (König's Line Colouring Theorem) *Let $G = (V_1, V_2, E)$ be a bipartite graph with maximal degree $\Delta(G)$. Then $\chi'(G) = \Delta(G)$.*

Proof The proof is via induction on the number of edges m in G. It is sufficient to prove that if $m - 1$ edges have been coloured using at most $\Delta(G)$ colours, then the remaining edge can be coloured using one of the $\Delta(G)$ available colours.

Suppose all edges except $\{u, v\}$ have been coloured. Then there exists at least one colour not incident to u, and one colour not incident to v. If the same colour is not incident to both u and v, then edge $\{u, v\}$ can be assigned to this colour. If this is not the case, without loss of generality we can say that: (a) $u \in V_1$ is incident to a grey edge but not a black edge and (b) $v \in V_2$ is incident to a black edge but not a grey edge.

Now consider a grey-black Kempe chain starting from u (that is, a component of G containing u that comprises only those vertices incident to a grey or black edge). Travelling along this chain from u involves alternating between vertex sets V_1 and V_2. However, we will never reach v because each time we arrive at a vertex in V_2 we do so via a grey edge, which v is not incident to.

Since v is not in the Kempe chain, we can now interchange the colours of the edges in this chain without affecting v (or indeed any other vertices outside of the chain). Hence, the edge $\{u, v\}$ can be coloured grey, completing the edge colouring. □

The previous two theorems demonstrate that the edge colouring problem is solvable in polynomial time for both complete and bipartite graphs. We have also seen that, for both topologies, their chromatic indices $\chi'(G)$ are either $\Delta(G)$ or $\Delta(G)+1$. Somewhat surprisingly, it turns out that this feature applies to *any* graph G, as proved by Vizing [16].

> **Theorem 6.12** (Vizing's Theorem) *Let G be a simple graph with maximal degree $\Delta(G)$; then $\Delta(G) \leq \chi'(G) \leq \Delta(G) + 1$.*

Proof When $\Delta(G)$ edges are incident to a vertex, these edges all require a different colour. Hence, the lower bound is proved: $\Delta(G) \leq \chi'(G)$.

The upper bound can be proved via induction on the number of edges. Suppose that, using $\Delta(G) + 1$ colours, we have coloured all edges in G except for the single edge $\{u, v_0\}$. Since $\Delta(G)$ gives the maximal degree, at least one colour will be unused at each of these two vertices. Now construct a series of edges, $\{u, v_0\}, \{u, v_1\}, \ldots,$

and a sequence of colours, $c_0, c_1, \ldots$, as follows: Select a colour c_i that is an unused colour at v_i. Now, let $\{u, v_{i+1}\}$ be an edge with colour c_i. We stop (with $i = k$) when either c_k is an unused colour at u, or c_k is already used on an edge $\{u, v_{j<k}\}$.

Case 1: If c_k is an unused colour at u, then we can recolour $\{u, v_i\}$ with c_i for $0 \leq i \leq k$. We now need to simply recolour edges incident to u to complete the proof.

Case 2: Otherwise, we recolour $\{u, v_i\}$ with c_i for $0 \leq i < j$ and remove the colour from $\{u, v_j\}$. Observe that c_k (say, "grey") is missing at both v_j and v_k. Now let "black" be an used colour at u. If grey is unused at u then we can colour $\{u, v_j\}$ grey. If black is unused at v_j then we can colour $\{u, v_j\}$ black. However, if black is unused at v_k then we colour $\{u, v_i\}$ with c_i for $j \leq i < k$ and colour $\{u, v_k\}$ black, because none of the edges $\{u, v_i\}$ for $j \leq i < k$ will be coloured grey or black.

If neither case above holds, then we consider the subgraph of grey and black edges. The components of this subgraph will be paths and/or cycles. The vertices u, v_i, and v_k are the terminal vertices of paths; hence, they cannot all belong to the same component. In this case, select a component containing just one of these vertices and interchange the colours of its edges. This means that one of the cases above now applies.   □

In essence, Vizing's theorem tells us that the set of all graphs can be partitioned into two classes: "class one" graphs, for which $\chi'(G) = \Delta(G)$, and "class two" graphs, where $\chi'(G) = \Delta(G) + 1$. Holyer [17] has shown that the decision problem of testing determining whether a graph belongs to class one is $\mathcal{NP}$-complete. On the other hand, several polynomially bounded algorithms are available for colouring the edges of any graph using exactly $\Delta(G) + 1$ colours, such as the $\mathcal{O}(nm)$ algorithm of Misra and Gries [18]. The existence of such algorithms tells us that we can colour the edges of any graph using a maximum of one extra colour beyond its chromatic index.

We might now ask whether the existence of such tight bounds for the edge colouring problem helps us to garner further information about the vertex colouring problem. It is clear that if we were given the task of vertex colouring a line graph $L(G)$, one approach would be to convert $L(G)$ into its "original" graph G, and then try to solve the corresponding edge colouring problem on G. Since $\chi(L(G)) = \chi'(G)$, then according to Vizing's theorem this would immediately tell us that we need to use either $\Delta(G)$ or $\Delta(G) + 1$ colours to feasibly colour the vertices of $L(G)$. Indeed, if G were a type two graph, then algorithms such as Mistra and Gries's could be used to quickly find the optimal edge colouring for G and therefore the optimal vertex colouring for $L(G)$. However, it should be remembered that this very attractive sounding proposal is only applicable when we wish to colour the vertices of a line graph that therefore has an "original" graph into which it can be converted. Unfortunately, we cannot convert all graphs into an "original" graph in this way.

6.3 Precolouring

In the precolouring problem, we are given a graph G in which some subset of the vertices $V' \subseteq V$ has already been assigned colours. Our task is to then colour the remaining vertices in the set $V - V'$ so that the resultant solution is feasible and uses a minimal number of colours.

Applications of precolouring arise in register allocation problems (see Sect. 1.1.4) where certain variables must be assigned to specific registers, perhaps due to calling conventions or communication between modules. They also occur in areas such as timetabling and sports scheduling where we might be given a problem instance in which some of the events have already been assigned to particular timeslots.

Precolouring problems can easily be converted into a standard graph colouring problem using graph contraction operations.

> **Definition 6.4** The *contraction* of a pair of vertices $v_i, v_j \in V$ in a graph G produces a new graph in which v_i and v_j are removed and replaced by a single vertex v such that v is adjacent to the union of the neighbourhoods of v_i and v_j, that is, $\Gamma(v) = \Gamma(v_i) \cup \Gamma(v_j)$.

The following steps can now be taken. Given a precolouring problem instance defined on a graph G, let $V'(i)$ define the set of vertices precoloured with colour i. Assuming there are k different colours used in the precolouring, this means that $\bigcup_{i=1}^{k} V'(i) = V'$ and $V'(i) \cap V'(j) = \emptyset$, for $1 \leq i \neq j \leq k$. Now, for each set $V'(i)$, merge all vertices into a single vertex using a series of contraction operations. This has the effect of reducing the number of precoloured vertices to k. Next, add edges between each pair of the k contracted vertices to form a clique. Finally, remove all colours from the vertices of this graph, and apply any arbitrary graph colouring algorithm to produce a feasible solution. A colouring of the original can then be obtained by simply reversing the above process. An example is provided in Fig. 6.9.

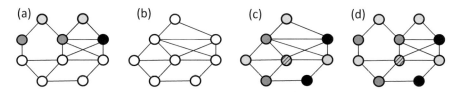

Fig. 6.9 Part **a** shows an example precolouring problem. Part **b** then shows how this can be converted into a new graph by contracting the precoloured vertices and forming a k-clique. A feasible colouring of this graph (shown in **c**) can then be converted back into a solution to the original problem, as shown in (**d**)

6.4   Latin Squares and Sudoku Puzzles

Another prominent area of mathematics for which graph colouring is naturally suited is the field of Latin squares. Latin squares are $l \times l$ grids that are filled with l different symbols, each occurring exactly once per row and once per column. They were originally considered in detail by Leonhard Euler, who filled his grids with symbols from the Latin alphabet, though nowadays it is common to use the integers 1 through to l to fill the grids. Example Latin squares of different sizes are shown in Fig. 6.10.

Latin squares have practical applications in several areas, including scheduling and experimental design. For an application in scheduling, imagine that we have two groups of l people and we want to schedule meetings between all pairs of people belonging to different groups. Clearly l^2 meetings are needed here. Also, since only l meetings can take place simultaneously, at least l timeslots are required. Latin squares give solutions to such problems that make use of exactly l timeslots. To see this, let us name the members of Team One as $r_1, r_2, \ldots, r_l$, which are represented by the rows in the grid, and the members of Team Two as $c_1, c_2, \ldots, c_l$, represented by the columns. The characters within an $l \times l$ Latin square then represent the various timeslots to which the meetings are assigned. For example, the Latin square shown in Fig. 6.10a schedules meetings between r_1 and c_1, r_2 and c_2, and r_3 and c_3 into timeslot 1; meetings between r_1 and c_3, r_2 and c_1, and r_3 and c_2 into timeslot 2; and meetings between r_1 and c_2, r_2 and c_3, and r_3 and c_1 into timeslot 3. Any $l \times l$ Latin square will provide a suitable meeting schedule fitting these criteria.

For an example application of Latin squares in experimental design, imagine that we want to test the effects of l different drugs on a particular illness. Suppose further that the trials are to take place over l weeks using l different patients, with each patient receiving a single drug each week. An $l \times l$ Latin square can be used to allocate treatments in this case, with rows representing patients, and columns representing weeks. This means that over the course of the l weeks each patient receives each of the l drugs once, and in each week all of the l drugs are tested. Looking at the 3×3 Latin square from Fig. 6.10a, for example, we see that Patients 1, 2, and 3 are administered Drugs 1, 2, and 3 (respectively) in Week 1; Drugs 3, 1, and 2 in Week 2; and Drugs 2, 3, and 1 in Week 3, as required.

Note that we can permute the rows and columns of a Latin square and still retain the property of each character occurring exactly one per column and once per row. It

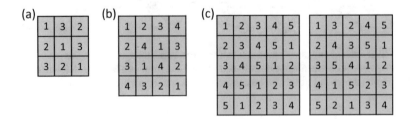

Fig. 6.10 Example Latin squares for **a** $l = 3$, **b** $l = 4$, and **c** $l = 5$

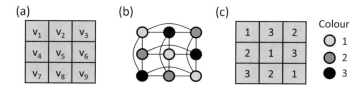

Fig. 6.11 Demonstration of the relationship between graph colouring and Latin squares. Part **a** associates each grid cell with a vertex; Part **b** shows the corresponding graph together with a feasible colouring; and Part **c** gives a valid Latin square corresponding to this colouring

is therefore common to write Latin squares in their *standardised form*, whereby the rows and columns are arranged so that the top row and leftmost column of the grid have the characters in their natural order $1, 2, \ldots, l$. The other $l!(l-1)! - 1$ Latin squares that can be formed by permuting the rows and columns are then considered to be equivalent to this. The Latin square in Fig. 6.10b is in standardised form, while the one in Fig. 6.10a is not.

It is also known that as l is increased, then so does the number of different $l \times l$ Latin squares. For $l = 1, \ldots, 4$, these numbers are 1, 1, 1, and 4, respectively; however, the growth rate is rapid: for $l = 11$ there are more than 5.36×10^{33} different Latin squares. Further information on this growth rate can be found at the Online Encyclopedia of Integer Sequences (https://oeis.org/A000315).

6.4.1 Relationship to Graph Colouring

Figure 6.11 shows how the production of a Latin square can be expressed as a graph colouring problem. As illustrated, the symbols used within the grid represent the colours. Each cell of the grid is then associated with a vertex, and edges are added between all pairs of vertices in the same row and all pairs of vertices in the same column. This results in a graph $G = (V, E)$ with $n = l^2$ vertices and $m = l^2(l-1)$ edges, for which $\deg(v) = 2(l-1) \; \forall v \in V$. (This graph is equivalent to the *Cartesian product* of the complete graphs K_l and K_l.) Note that the set of vertices in each row forms a clique of size l, as do vertices in each column. This implies that solutions using fewer than l colours are not possible.

Note that it is simple to produce a Latin square in standardised form for any value of l by simply using values $(1, 2, \ldots, l)$ for the first row, $(2, 3, \ldots, l, 1)$ for the second row, $(3, 4, \ldots, l, 1, 2)$ for the third row, and so on (see, for example, the left Latin square in Fig. 6.10c). This tells us that Latin squares are a particular topology for which the associated graph colouring problem can be easily solved in polynomial time for any value of l, without a need for resorting to heuristics or approximation algorithms. Graph colouring algorithms can, however, be used for producing different Latin squares to this.

Graph colouring algorithms arguably become more useful in this area when we consider the *partial* Latin square problem. This is the problem of taking a partially

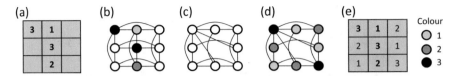

Fig. 6.12 Part **a** shows a partial 3 × 3 Latin square with four filled-in cells. Part **b** shows the corresponding precolouring problem. In Part **c**, the graph has been modified using the steps described in Sect. 6.3 and, in Part **d**, a three-colouring of this graph has been established. Part **e** shows the final Latin square

filled $l \times l$ grid and deciding whether or not it can be completed to form a Latin square. This problem has been proven $\mathcal{NP}$-complete by Colbourn [19].

Figure 6.12 demonstrates how the partial Latin square problem can be tackled using graph colouring principles. It follows the same method as the previous example given in Fig. 6.11, except that certain vertices are now also precoloured. This means that the same steps as those used with the precolouring problem (Sect. 6.3) can now be followed, with an l-colouring of this graph corresponding to a completed $l \times l$ Latin square. Of course, depending on the values of the filled-in cells in the original problem, there could be zero, one, or multiple feasible l-colourings available.

6.4.2 Solving Sudoku Puzzles

The partial Latin square problem has become very popular in recent decades in the form of Sudoku puzzles. In Sudoku, we are given a partially filled Latin square and the objective is to complete the remaining cells so that each column and row contains the characters $1, \ldots, l$ exactly once. In addition, Sudoku grids are also divided into l "boxes" (usually marked by bold lines) which are also required to contain the characters $1, \ldots, l$ exactly once; thus, Sudoku can be considered a special case of the partial Latin square problem in which the constraint of appropriately filling out the "boxes" must also be satisfied. An example 9×9 Sudoku puzzle and a corresponding solution is shown in Fig. 6.13.

Because Sudoku is intended to be an enjoyable puzzle, problems posed in books and newspapers will nearly always be *logic solvable*.

> **Definition 6.5** A Sudoku puzzle is *logic solvable* if and only if it features exactly one solution, which is achievable via forward chaining logic only.

Puzzles that are not logic solvable require random choices to be made. In general, these should be avoided because players will have to go through the tedious process of backtracking and re-guessing if their original guesses turn out to be wrong.

As an example of how a player might deduce the contents of cells, consider the puzzle given in Fig. 6.13. Here, we see that the cell in the seventh row and sixth column (shaded) must contain a 6 because all numbers 1–5 and 7–9 appear either in the same column, the same row, or the same box as this cell. If the problem instance is logic solvable (as indeed this one is), the filling-in of this cell will present further clues, allowing the user to eventually complete the puzzle.

Many algorithms for solving Sudoku puzzles are available online, such as those at http://www.sudokuwiki.org and http://www.sudoku-solutions.com. Such algorithms typically mimic the logical processes that a human might follow, with popular deductive techniques, such as the so-called X-wing and Swordfish rules, also being commonplace. In other areas of Sudoku research, Russell and Jarvis [20] have shown that the number of essentially different Sudoku solutions (when symmetries such as rotation, reflection, permutation, and relabelling are taken into account) is 5,472,730,538 for the popular 9×9 grids. McGuire et al. [21] have also shown that 9×9 Sudoku puzzles must contain at least 17 filled-in cells to be logic solvable and that 9×9 puzzles with 16 or fewer filled-in cells will always admit more than one solution. Similar results for larger grids are unknown, however. Herzberg and Murty [22] have also shown that at least $l - 1$ of the l characters must be present in the filled cells of a Sudoku puzzle for it to be logic solvable.

Although Sudoku is a special case of the partial Latin squares problem, Yato and Seta [23] have demonstrated that the problem of deciding whether or not a Sudoku puzzle features a valid solution is still $\mathcal{NP}$-complete. Graph colouring algorithms can therefore be useful for solving instances of Sudoku, particularly those that are not necessarily logic solvable.

Sudoku puzzles can be transformed into a corresponding graph colouring problem in the same fashion as partial Latin square problems (see Fig. 6.12), with additional edges also being imposed to enforce the extra constraint concerning the "boxes" of the grid. We now present two sets of experiments that illustrate the capabilities of the HEA and backtracking algorithms from Chap. 5 for solving Sudoku puzzles. In the first set of experiments, we focus on Sudoku problems that are not necessarily logic solvable (random puzzles), while in the second set we focus on 9×9 grids that are logic solvable.

	2	4			7			
6								
		3	6	8		4	1	5
4	3	1			5			
5							3	2
7	9						6	
2		9	7	1		8		
	4			9	3			
3	1				4	7	5	

1	2	4	9	5	7	3	8	6
6	8	5	3	4	1	2	9	7
7	9	3	6	8	2	4	1	5
4	3	1	2	6	5	9	7	8
5	6	8	4	7	9	1	3	2
7	9	2	1	3	8	5	6	4
2	5	9	7	1	6	8	4	3
8	4	7	5	9	3	6	2	1
3	1	6	8	2	4	7	5	9

Fig. 6.13 A 9×9 Sudoku puzzle and corresponding solution

6.4.2.1 Solving Random Sudoku Puzzles

To generate problem instances that are not necessarily logic solvable, we can start by taking a completed Sudoku solution of a given size. Such solutions can be obtained from a variety of places such as the solution pages of a Sudoku book or newspaper, or by simply executing a suitable graph colouring algorithm on a graph representing a blank puzzle. In the next step of the procedure, this completed grid can then be randomly shuffled using the following five operators:

- Transpose the grid;
- Permute columns of boxes within the grid;
- Permute rows of boxes within the grid;
- Permute columns of cells within columns of boxes; and
- Permute rows of cells within rows of boxes.

Each of these shuffle operators preserves the validity of a Sudoku solution. Finally, some cells in the grid should be made blank by going through each cell in turn and deleting its contents with probability $1 - p$, where p is a parameter to be defined by the user. This means that instances generated with a low value for p have a lower proportion of filled-in cells.

Figure 6.14 illustrates the performance of the HEA and backtracking algorithms on 9×9, 16×16, and 25×25 Sudoku grids, respectively. In each case, 100 instances for each value of p have been generated and, as in Chap. 5, a computation limit of 5×10^{11} constraint checks has been imposed. For each algorithm, two statistics are displayed. The *success rate* (SR) indicates the percentage of runs for which the algorithms have found a valid Sudoku solution (a feasible l-colouring) within the computation limit. The *solution time* then indicates the mean number of constraint checks that it took to achieve these solutions. Note that only successful runs are considered in the latter statistic.

Looking at the results for 9×9 Sudoku puzzles first, we see that both algorithms feature a 100% success rate across all instances with only a very small proportion of the computation limit being required.[2] For 16×16 puzzles, similar patterns occur for the HEA, with all problem instances being solved, and no runs requiring more than one second of computation time. On the other hand, the backtracking algorithm features a dip in its success rate for values of p between 0.1 and 0.55, with a corresponding increase in solution times. With the larger 25×25 puzzles, this pattern becomes more apparent, with both algorithms featuring dips in their success rates and subsequent increases in their solution times. However, these dips are less pronounced with the HEA, indicating its superior performance overall.

The dips in the success rates of these algorithms are analogous to the phase transition regions we saw with the flat graphs in Sect. 5.7. When p is low, although solution spaces will be larger, there will tend to be many optimal solutions within

---

[2]On our equipment (3.0 GHz Windows 7 PC with 3.87 GB RAM) the longest run in the entire set took just 0.02 s.

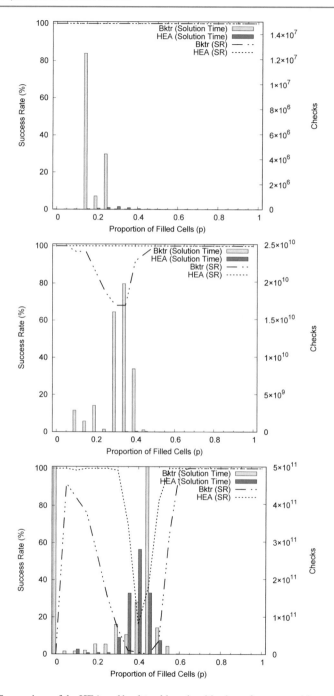

Fig. 6.14 Comparison of the HEA and backtracking algorithm's performance with random Sudoku instances of size 9×9, 16×16, and 25×25, respectively. Note the different scales on the vertical axes in each case

these spaces. Consequently, an effective algorithm should be able to find one of these within a reasonable amount of computation time, as is the case with the HEA. For high vales for p meanwhile, although there will only be a very small number of optimal solutions (and perhaps only one), the solution space will be much smaller. Additionally, solutions to these highly constrained instances will also tend to reside at prominent optima, thus also allowing easy discovery by an effective algorithm. However, instances at the boundary of these two extremes will cause greater difficulties. First, the solution spaces for these instances will still be relatively large, but they will also tend to admit only a small number of optimal solutions. Second, because of their moderate number of constraints, the cost landscapes will also tend to feature more plateaus and local optima, making navigation towards a global optimum more difficult for the algorithm.

6.4.2.2 Logic Solvable Sudoku Puzzles

We now examine the performance of the HEA and backtracking algorithms on logic solvable instances. This will allow us to examine the effect that the size of the solution space has on the difficulty of a puzzle when its solution is known to be unique. As we have seen, it is known that a 9×9 Sudoku puzzle must contain at least 17 filled-in cells to be logic solvable [21].

For our tests, we took a random sample of one hundred 17-clue instances together with their corresponding (unique) solutions. Logic solvable puzzles with more than 17 filled-in cells were then also generated for each of these by randomly selecting an appropriate number of blank cells in the puzzles, and adding the corresponding entries from their solutions. This operation maintains the uniqueness of each puzzle's solution while reducing the size of its solution space. All other experimental details are the same as in the previous subsection.

Figure 6.15 shows the relative performance of the two graph colouring algorithms. In all cases, valid solutions were found within the computation limit. When the solution space size is relatively large (17–20 filled-in cells) we see that the HEA requires up to 2.4×10^{10} constraint checks to find a solution.[3] However, beyond this point the puzzles are solved using very little computational effort—indeed, for more than 35 filled-in cells, solutions are achieved by the initial solutions produced by the DSATUR algorithm.

In contrast to our earlier results on random Sudoku puzzles, Fig. 6.15 also shows that the backtracking algorithm outperforms the HEA with these problem instances. With 17-clue puzzles, for example, the algorithm has identified the unique solutions using just 0.03% of the computational effort required by the HEA. Thus, unlike in the larger puzzles seen in the previous section, here the solution spaces seem suitably sized and structured for the backtracking algorithm to be able to identify the unique Sudoku solutions in very short spaces of time.

---

[3]This equated to approximately 3 min on our computer.

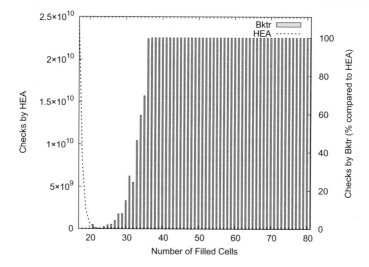

Fig. 6.15 Comparison of the HEA and backtracking algorithm's performance on 9×9 Sudoku grids with unique solutions

6.5 Short Circuit Testing

Another practical application of graph colouring is due to Garey et al. [24], who suggest its use in the process of testing for (undesired) short circuits in printed circuit boards. In their model, a circuit board is represented by a finite lattice of evenly spaced points onto which a set of n cycle-free components has been printed. This set P is referred to as a *net pattern*, with individual components $p \in P$ being called *nets*. Each net connects points that are intended to be electrically common. An example net pattern comprising four components is shown in Fig. 6.16a. Note that connections between points are only permitted in vertical or horizontal directions.

Given a net pattern P, the problem of interest is to determine whether there exists some fault on the circuit board (due to the manufacturing process) whereby an extra conductor path has been introduced between two nets that are not intended to be electrically common. This is the case in Fig. 6.16b. These extra conductor paths are known as "shorts".

An obvious strategy to determine whether a short has occurred is to test each pair of nets $p_i, p_j \in P$ in turn by applying an electrical current to p_i and seeing if this

Fig. 6.16 An example net pattern (**a**), and a net pattern containing a short between two nets (**b**)

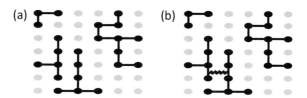

Fig. 6.17 Two further net patterns

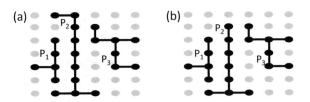

current spreads to p_j. However, Garey et al. [24] suggest that the number of pairwise tests can be reduced significantly by making use of the following two observations.

First, note that many pairs do not need to be tested. Consider, for example, the net pattern in Fig. 6.17a. Here it is unnecessary to test the pair p_1, p_3 because, if there is a short between them, then shorts must also exist between pairs p_1, p_2 and p_2, p_3. Since the objective of the problem is to determine if *any* shorts exist, testing either p_1, p_2 or p_2, p_3 is therefore sufficient. Furthermore, if we consider the net pattern in Fig. 6.17b, it might also be reasonable to assume that shorts cannot occur between p_1 and p_3 without also causing a short involving p_2. Thus, depending on the criteria used for deciding how and where shorts can occur, we have the opportunity to exclude many pairs of nets from the testing procedure. If it is deemed necessary to test a pair of nets, these are called *critical pairs*; otherwise, they are deemed *noncritical*.

The second observation is as follows. Let $G = (V, E)$ be a graph with a set of vertices $V = \{v_1, v_2, \ldots, v_n\}$, where each vertex $v_i \in V$ corresponds to a particular net $p_i \in P$ (for $1 \leq i \leq n$). Also, let each edge $\{v_i, v_j\} \in E$ correspond to a pair of nets p_i, p_j judged to be critical. Now let $\mathcal{S} = \{S_1, \ldots, S_k\}$ be a partition of V such that no pair of vertices v_i, v_j in any subset S_l forms a critical pair. From a graph colouring perspective, $\mathcal{S}$ therefore defines a feasible k-colouring of the vertices of G. Now suppose that, for the printed circuit board in question, external conductor paths are provided so that all nets in any subset $S \in \mathcal{S}$ can be made electrically common during testing. This means that there are k "supernets" that need to be tested. It can now be seen that the printed circuit board contains no short if and only if no pair of "supernets" is seen to be electrically common. Therefore, we only have to perform a maximum of $\binom{k}{2}$ tests as opposed to our original figure of $\binom{n}{2}$ tests. Naturally, it is desirable to reduce k as far as possible to minimise the number of tests needed.

In their paper, Garey et al. [24] propose several criteria for deciding whether a pair of nets should be deemed critical, with associated theorems then being presented. We now review some of these.

Theorem 6.13 (Garey et al. [24]) *Consider a pair of nets $p_i, p_j \in P$ to be critical if and only if a straight vertical line of sight can be drawn that connects p_i and p_j. Then the corresponding graph $G = (V, E)$ is planar and has a chromatic number $\chi(G) \leq 4$.*

Proof Given a net pattern P, for each pair of nets for which a vertical line of sight exists, draw such a line. Since each line is vertical, none can intersect. It is now possible to contract each net into a single point, deforming the lines of sight (which may no longer be straight lines) in such a way that they remain nonintersecting. This structure now corresponds to the graph $G = (V, E)$, with each vertex corresponding to a contracted net, and each edge corresponding to the lines of sight. Since G is planar, $\chi(G) \leq 4$ according to the four colour theorem (Theorem 6.8).   □

Theorem 6.14 (Garey et al. [24]) *Consider a pair of nets $p_i, p_j \in P$ to be critical if and only if a straight vertical line of sight or a straight horizontal line of sight can be drawn that connects p_i and p_j. Then the corresponding graph $G = (V, E)$ has a chromatic number $\chi(G) \leq 12$.*

Proof It is first necessary to show that any graph G formed in this way has a vertex v with $\deg(v) \leq 11$. Let $G_1 = (V, E_1)$ and $G_2 = (V, E_2)$ be subgraphs of G such that E_1 is the set of edges formed from vertical lines and E_2 is the set of edges formed from horizontal lines. Hence $E = E_1 \cup E_2$. By Theorem 6.13, both G_1 and G_2 are planar. We can also assume without loss of generality that the number of vertices $n > 12$. According to Theorem 6.2, the number of edges in a planar graph with n vertices is less than or equal to $3n - 6$. Thus:

$$m \leq |E_1| + |E_2| \leq (3n - 6) + (3n - 6) = 6n - 12.$$

Since each edge contributes to the degree of two distinct vertices, this gives

$$\sum_{v \in V} \deg(v) = 2m \leq 12n - 24.$$

Hence, it follows that some vertex in G must have a degree of 11 or less.

Now consider any subset $V' \subseteq V$ with $|V'| > 12$. The induced subgraph of V' must contain a vertex with a degree at most 11. Consequently, according to Theorem 3.7, $\chi(G) \leq 11 + 1$.   □

In their paper, Garey et al. [24] conjecture that the result of Theorem 6.14 might be improved to $\chi(G) \leq 8$ because, in their experiments, they were not able to produce graphs featuring chromatic numbers higher than this. They also go on to consider the maximum length of lines of sight and show that:

- If lines of sight can be both horizontal and vertical but are limited to a maximum length of 1 (where one unit of length corresponds to the distance between a pair of vertically adjacent points or a pair of horizontally adjacent points on the circuit board), then G will be planar, giving $\chi(G) \leq 4$.
- If lines of sight can be both horizontal and vertical but are limited to a maximum length of 2, then G will have a chromatic number $\chi(G) \leq 8$.

Finally, they also note that if arbitrarily long lines of sight travelling in *any* direction are permitted (as opposed to merely horizontal or vertical) then it is possible to form all sorts of different graphs, including complete graphs. Hence, arbitrarily high chromatic numbers can occur.

6.6 Graph Colouring with Incomplete Information

In this section, we now consider graph colouring problems for which information about a graph is incomplete at the beginning of execution. In the following subsections, we discuss three different interpretations, specifically decentralised graph colouring, online graph colouring, and dynamic graph colouring, and give practical examples of each.

6.6.1 Decentralised Graph Colouring

In decentralised graph colouring, we assume that each vertex of a graph is an individual entity responsible for choosing its colour. Moreover, the only information available to each vertex is who its neighbours are, and what their colours are. In other words, vertices are unaware of the structure of the graph beyond their neighbouring vertices.

A practical example of this problem might occur in the setting up of a wireless ad hoc network. Imagine a situation where a network is to be created by randomly dropping a set of wireless devices (equipped with radio transmitters and receivers) into a particular environment. Imagine further that each device in this network will broadcast information at a particular frequency on the radio spectrum. Finally, also consider the fact that if two devices are close together and using the same frequency (or suitably similar frequencies) their transmissions will interfere with one another, inhibiting the ability of other devices to decipher their signals.

The above situation is illustrated in Fig. 6.18, where each wireless device appears at the centre of a grey circle denoting its transmission range. Figure 6.18a shows two devices, u and v, that are situated in each other's transmission ranges. This is sometimes known as a *primary collision* and implies that u and v should not broadcast using the same frequency. Figure 6.18b, meanwhile, denotes a *secondary collision*. Here, although there is no primary collision between devices v_1 and v_2, it is still necessary that they broadcast at different frequencies to allow u to be able to distinguish between them.

The problem of choosing suitable frequencies for each device in a wireless network can be modelled as a graph colouring problem by relating each device to a vertex, with edges then occurring between any pair of vertices subject to a primary and/or secondary collision. Each frequency then corresponds to a colour and, due to the limited number of sufficiently different frequencies that exist in the radio spectrum, we now wish to colour this graph using the minimum number of colours.

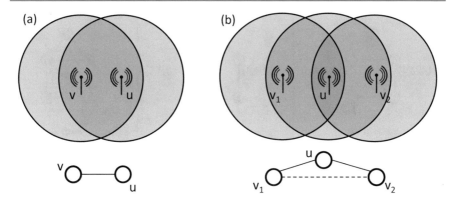

Fig. 6.18 Illustration of a primary collision (**a**), and (**b**) a secondary collision (dotted line) in a wireless network

More precisely, let $G = (V, E)$ be a graph with vertex set V and an edge E. The set of edges due to primary collisions, E_1, contains all pairs of devices that are close enough to be able to receive each other's transmissions (as with Fig. 6.18a). The set E_2 then contains pairs of devices subject to secondary collisions, that is, $\{v_i, v_j\} \in E_2$ if and only if the distance between v_i and v_j in the graph $G_1 = (V, E_1)$ is exactly two (as is the case in Fig. 6.18b). If only primary collisions need to be considered when assigning frequencies, we only need to colour the graph G_1; otherwise, we will need to colour the graph $G = (V, E = E_1 \cup E_2)$. In either case, this task is a type of decentralised graph colouring problem because each vertex (wireless device) is responsible for choosing its colour (frequency) while being aware only of its neighbours and their current colours.

One simple but effective algorithm for the decentralised graph colouring problem is due to Finocchi et al. [25]. This operates as follows. Let $G = (V, E)$ be a graph with maximal degree $\Delta(G)$. To begin, all vertices in G are set as uncoloured. Each vertex is also allocated a set of candidate colours, defined $L_v = \{1, 2, \ldots, \deg(v)+1\} \ \forall v \in V$. A single iteration of the algorithm now involves the following four steps:

1. In parallel, each uncoloured vertex v selects a *tentative* colour $t_v \in L_v$ at random.
2. In parallel, consider each tentatively coloured vertex v. If no neighbour of v is coloured with t_v, then set t_v as the *final* colour of v.
3. In parallel, consider each remaining tentatively coloured vertex v and

   a. Remove its tentative colour.
   b. Update L_v by deleting all colours from L_v that are assigned as final colours to neighbours of v.
   c. If $L_v = \emptyset$ then let l be the largest colour label assigned as a final colour in v's neighbourhood and set $L_v = \{1, 2, \ldots, \min\{l + 1, \Delta(G) + 1\}\}$.

4. If any uncoloured vertices remain, return to Step 1.

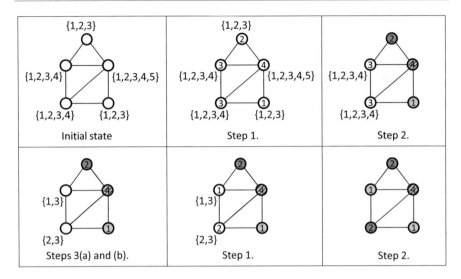

Fig. 6.19 Example run of algorithm of Finocchi et al. Here tentatively coloured vertices are shown in white. Labels within the vertices indicate colours

An example run of this algorithm is shown in Fig. 6.19. In the first iteration, we see that three of the five vertices are allocated final colours; the remaining two vertices are then allocated final colours in the second iteration.

Note that in the above algorithm, each vertex v is initially assigned a set of candidate colours $L_v = \{1, 2, \ldots, \deg(v) + 1\}$. This means that L_v always contains sufficient options to allow each vertex v to be coloured differently from all of its neighbours; hence, Step 3(c) will never actually be used. If, however, we desire a solution using fewer colours, we might choose to introduce a *shrinking factor* $s > 1$, which can be used to limit the initial set of candidate colours for each vertex v to $L_v = \{1, 2, \ldots, \lfloor \frac{\deg(v)+1}{s} \rfloor\}$. In this case, Step 3(c) might now be needed if the original contents of L_v prove insufficient. The algorithm may also need to execute for an increased number of iterations to achieve a feasible solution (if, indeed, one can be found).

Finocchi et al. [25] also suggest an improvement to this algorithm by replacing Step 2 with a more powerful operator. Observe in Step 2 of the first iteration of Fig. 6.19 that there are two vertices tentatively coloured with colour 3. Accordingly, neither of these vertices receives a final colour at this iteration, though it is obvious that one of them could indeed receive colour 3 as a final colour at this point. An improvement to Step 2 therefore operates as follows. Let $G(i) = (V(i), E(i))$ be the subgraph induced by all vertices tentatively coloured with colour i. We now identify a maximal independent set for $G(i)$ and assign all vertices in this set to a final colour i. All other vertices in $G(i)$ should remain uncoloured. To form this independent set, in parallel each vertex $v \in V(i)$ first generates a random number $r_v \in [0, 1]$. The tentative colour of a vertex v is then selected as its final colour if and only if r_v is less than the random numbers chosen by its neighbours in $G(i)$.

This is equivalent to the Greedy process of randomly permuting the vertices in $V(i)$ and then adding each vertex $v \in V(i)$ to the independent set if and only if it appears before its neighbours in the permutation.

In addition to assigning frequencies in wireless networks, decentralised graph colouring problems are known to arise in several other practical situations, including TDMA slot assignment, wake-up scheduling, and data collection [26]. One particularly noteworthy piece of research is due to Kearns et al. [27], who have examined the decentralised colouring of graphs representing social networks. In their case, each vertex in the graph is a human participant, and two vertices are adjacent if these people are judged to know one another. The objective of the problem is for each person to choose a colour for himself or herself only by using information regarding the colours of his or her neighbours. Participants are also able to change their colour as often as necessary until, ultimately, a feasible colouring of the entire graph is formed. This problem has real-world implications in situations where it is desirable to distinguish oneself from one's neighbours, for example, selecting a mobile phone ringtone that differs from one's friends, or choosing to develop professional expertise that differs from one's colleagues. In their research, Kearns et al. [27] carried out experiments on several graph topologies using segregated participants. Under a time limit of 5 min, topologies such as cycle graphs were optimally coloured quite quickly through the collective efforts of the participants. Other more complex graphs modelling more realistic social network topologies were seen to present significantly more difficulties, however.

6.6.2 Online Graph Colouring

In the online graph colouring problem, a graph is gradually revealed to an algorithm by presenting the vertices one at a time. The algorithm must then assign each vertex to a colour before considering the next vertex in the sequence. In other words, an online graph colouring algorithm receives vertices in a given ordering $v_1, v_2, \ldots, v_n$ and the colour for a vertex v_i is determined by only considering the graph induced by the vertices in $\{v_1, \ldots, v_i\}$. Once v_i has been coloured, it cannot usually be changed by the algorithm.

Much of the research relating to online graph colouring algorithms looks at worst-case behaviour with different topologies. Let A be an online graph colouring algorithm and consider the colourings of a graph G produced by A over all orderings of G's vertices. The maximum number of colours used among these colourings is denoted by $\chi_A(G)$. That is, $\chi_A(G)$ denotes the worst possible performance of A on G.

Kierstead and Trotter [28] have described an online graph colouring algorithm A that, for any interval graph G,

$$\chi_A(G) \leq 3\omega(G) - 2. \tag{6.2}$$

This result contrasts with the (offline) graph colouring problem, where $\omega(G)$-colourings can be achieved in polynomial time for all interval graphs (see Sect. 3.2.1.1).

Studies of online graph colouring have also focussed on the behaviour of the GREEDY algorithm, which, we recall, operates by assigning each vertex to the lowest indexed colour seen to be feasible (see Sect. 3.1). Bounds noted by Gyárfás and Lehel [29] include

$$\chi_{\text{GREEDY}}(G) \leq \omega(G) + 1 \tag{6.3}$$

if G is a split graph (i.e., a graph that can be partitioned into one clique and one independent set),

$$\chi_{\text{GREEDY}}(G) \leq \frac{3}{2}\omega(G) + 1 \tag{6.4}$$

if G is the complement of a bipartite graph, and

$$\chi_{\text{GREEDY}}(G) \leq 2\omega(G) - 1 \tag{6.5}$$

if G is the complement of a chordal graph.

Upper bounds on the quality of solutions produced by GREEDY can also be determined by looking at the *Grundy chromatic number*. For a particular graph G, this is defined as the maximum number of colours used by GREEDY over all orderings of G's vertices. Graphs for which the Grundy chromatic number is equal to the chromatic number include complete graphs, empty graphs, odd cycles, and complete k-partite graphs. In general, however, the problem of determining Grundy chromatic numbers is $\mathcal{NP}$-hard, with the best known exact algorithm operating in $\mathcal{O}(2.443^n)$ time [30,31].

Empirical work by Ouerfelli and Bouziri [32] has also suggested that instead of following the GREEDY algorithm's strategy of assigning vertices to the lowest indexed feasible colour, in online colouring it is often beneficial to assign vertices to the feasible colour containing the *most* vertices. This is because such a heuristic will often assist in the formation of larger independent sets, ultimately helping to reduce the number of colours used in the final solution.

A real-world application of online graph colouring is presented by Dupont et al. [33]. Here, a military-based frequency assignment problem is considered in which wireless communication devices are introduced one by one into a battlefield environment. From a graph colouring perspective, given a graph $G = (V, E)$, the problem starts with an initial colouring of the subgraph induced by the subset of vertices $\{v_1, \ldots, v_i\}$. The remaining vertices $v_{i+1}, \ldots, v_n$ are then introduced one at a time, with the colour (frequency) of each vertex having to be determined before the next vertex in the sequence is considered. In this application, the number of available colours is fixed from the outset, so it is possible that a vertex v_j ($i < j \leq n$) might be introduced for which no feasible colour is available. In this case, a repair operator is used that attempts to rearrange the existing colouring so that a feasible colour is created for v_j. Because such rearrangements are considered expensive, the repair operator also attempts to minimise the number of vertices that have their colours changed during this process.

6.6.3   Dynamic Graph Colouring

Dynamic graph colouring differs from decentralised and online graph colouring in that we again possess a global knowledge of the graph we are trying to colour. However, in this case, graphs are also permitted to change over time.

A practical application of dynamic graph colouring might occur in the timetabling of lectures at a university (see Sect. 1.1.2 and Chap. 9). To begin, a general set of requirements and constraints will be specified by the university and an initial timetable will be produced. However, on viewing this draft timetable, circumstances might dictate that some constraints need to be altered, additional lectures need to be introduced, or other lectures need to be cancelled. This will result in a new timetabling problem that needs to be solved, with the process continuing in this fashion until a finalised solution is agreed upon.

The dynamic graph problem is considered by Hardy [34], who considers two general cases.

1. *Edge dynamic problems.* Here, the vertices of a graph are not altered, but edges can be added and removed.
2. *Vertex dynamic problems.* A generalisation of the above in which vertices (and their edges) are added and removed.

For both of these cases, a dynamic graph colouring problem can be modelled using a sequence of graphs $\mathcal{G} = (G_1, G_2, \ldots, G_l)$. A solution to each graph $G_i \in \mathcal{G}$ will then need to be produced within a limited time frame before the next graph G_{i+1} is considered. An important issue in this problem is to decide how and when solutions to the previously observed graphs, $G_1, \ldots, G_i$, can be used to help establish solutions for the future graphs $G_{i+1}, \ldots, G_l$. When changes between successive graphs are made at random, Hardy finds that making use of a solution for G_i can help establish a solution for G_{i+1}, providing that the number of changes made between time steps is fairly small. In the remaining cases, it is sufficient to produce solutions to G_{i+1} using no previous information.

Hardy [34] also considers situations where future changes to graphs are expected to occur with certain probabilities. Solutions can then be sought that are robust to these potential changes. As an example, consider the situation where two vertices u, v are nonadjacent in G_i, but are expected to become adjacent at a later time step. In these circumstances, it might be advantageous to try and assign u and v to different colours, even though this is not currently required. For both edge and vertex dynamic problems, Hardy finds that it is useful to take these expected changes into account. The scheme suggested involves using a local search routine that maintains the feasibility of a solution with regard to the current graph G_i, but that also seeks to optimise a "robustness" measure that takes future change probabilities into account. Further information is also documented by Hardy et al. [35].

6.7 List Colouring

The list colouring problem is an extension to the graph colouring problem that, as usual, involves assigning differing colours to adjacent vertices. In this case, though, individual vertices are also given a list of *permissible* colours to which they can be assigned.

Defined more precisely, the list colouring problem takes a graph $G = (V, E)$ together with a set L_v of permissible colours for each vertex $v \in V$. The sets L_v are usually referred to as "lists", giving the problem its name. The task is to now produce a feasible colouring of G with the added restriction that all vertices should only be assigned to colours appearing in their corresponding lists (that is, $\forall v \in V$, $c(v) \in L_v$). If a k-colouring exists for a particular problem instance of the list colouring problem, we say that the graph G is "k-choosable". The "choice number" $\chi_L(G)$ then refers to the minimum k for which G is k-choosable. Note that the chromatic number of a graph $\chi(G) \leq \chi_L(G)$.

List colouring problems have obvious applications in areas such as timetabling where, in addition to scheduling events into a minimal number of timeslots (as we saw in Sect. 1.1.2), we might also face constraints of the form "event v can only be assigned to timeslots x and y", or "event u cannot be assigned to timeslot z". The problem is also $\mathcal{NP}$-hard because it generalises the graph colouring problem. Specifically, graph colouring problems can be easily converted into an equivalent list colouring decision problem by simply setting $L_v = \{1, 2, \ldots, \Delta(G) + 1\}$, $\forall v \in V$.

In practice, algorithms for the graph colouring problem can often be used for deciding whether a graph is k-choosable. More specifically, graph colouring algorithms can be used to tackle any list colouring problem for which our chosen $k \geq |L|$, where L is defined as the union of all lists: $L = \bigcup_{v \in V} L_v$.

To see this, imagine we have a list colouring problem defined on a graph G for which $k \geq |L|$ is satisfied. First, we create a new graph G' by copying the vertices and edges of G and then adding k additional vertices, which we label $u_1, u_2, \ldots, u_k$. Next, we add edges between all $\binom{k}{2}$ pairs of these additional vertices so that they form a complete graph K_k. This implies that any feasible colouring of G' must use at least k different colours. Without loss of generality, we can assume that $c(u_i) = i$ for $1 \leq i \leq k$. Finally, we then go through each vertex v in G' that came from the original graph G and consider its colour list, adding an edge between v and u_i if colour $i \notin L_v$. This has the effect of disallowing v from being assigned to colour i, as required.

Figure 6.20 demonstrates this process. In this example, $k = |L| = 4$, meaning that four additional vertices $u_1, \ldots u_4$ are added (larger values for k are also permitted). The colouring produced for the extended graph G' also uses four colours, which is the chromatic number in this case. However, we also observe that none of the vertices originating from G are coloured with colour 1 in this example; hence, we deduce that G is 3-choosable, as shown in Fig. 6.20c.

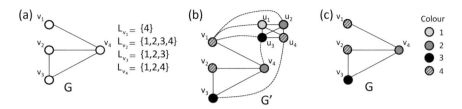

Fig. 6.20 Illustration of how a list colouring problem (**a**) can be converted into an equivalent graph colouring problem (**b**), whose colouring then represents a feasible solution to the original list colouring problem (**c**)

6.8   Equitable Graph Colouring

Another extension to the graph colouring problem is the *equitable* colouring problem, where we seek to establish a feasible colouring of a graph G such that the sizes of the colour classes differ by at most 1. In other words, we desire a feasible k-colouring so that exactly $n \bmod k$ colour classes contain $\lceil n/k \rceil$ vertices, and the remainder contain exactly $\lfloor n/k \rfloor$ vertices.

Examples of equitable graph colouring problems occur quite naturally as extensions to the general graph colouring problem. In university timetabling, for example, it might be desirable to minimise the number of rooms required by balancing the number of events per timeslot (see Sect. 1.1.2). Another application can be found in the creation of table plans for large parties. Imagine, for example, that we have n guests who are to be seated at k equal-sized tables, but that some guests are known to dislike each other and therefore need to be assigned to different tables. In this case, we can model the problem as a graph by using vertices for guests, with edges occurring between pairs of guests who dislike each other. An extension to this application is the subject of Chap. 7.

Let $G = (V, E)$ be a graph with n vertices, a maximal degree $\Delta(G)$, and an independence number $\alpha(G)$.

Definition 6.6 If V can be partitioned into k colour classes $\mathcal{S} = \{S_1, \ldots, S_k\}$ such that each S_i is an independent set and $||S_i| - |S_j|| \le 1 \ \forall i \ne j$, then $\mathcal{S}$ is said to be an *equitable k-colouring* of G.

Definition 6.7 The smallest k for which an equitable k-colouring of G exists is the *equitable chromatic number*, denoted by $\chi_e(G)$.

Like the graph colouring problem, the equitable graph colouring problem is known to be $\mathcal{NP}$-complete. This follows from the fact that the problem of deciding whether

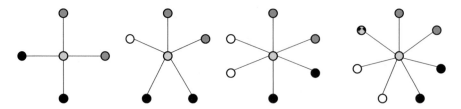

Fig. 6.21 The equitable chromatic numbers for star graphs with $n = 5, 6, 7$, and 8 are 3, 4, 4, and 5, respectively

a graph G is k-colourable can be converted into an equitable k-colouring problem by simply adding an appropriate number of isolated vertices to G. Hence, the equitable graph colouring problem generalises the standard graph colouring problem. An alternative proof is also due to Furmańczyk [36], who shows that the problem of deciding whether $\chi_e(G) \leq 3$ is $\mathcal{NP}$-complete, even when G is the line graph of a cubic graph.

Because a feasible equitable k-colouring of a graph G is also a feasible k-colouring of G, it is obvious that $\chi(G) \leq \chi_e(G)$. In some cases, however, this bound can be very poor. Consider star graphs, for example, which comprise a vertex set $V = \{v_1, \ldots, v_n\}$ and an edge set $E = \{\{v_1, v_i\} : i \in \{2, \ldots, n\}\}$. These are a type of bipartite graph and therefore feature a chromatic number of two. However, v_1 must be assigned a different colour to all other vertices; hence, in an equitable colouring, all other colour classes must contain a maximum of two vertices. Equitable chromatic numbers of star graphs are therefore calculated as $\lceil (n - 1)/2 \rceil + 1$, as illustrated in Fig. 6.21.

A better lower bound for equitable chromatic numbers on general graphs is outlined by Furmańczyk [36]. Suppose that G is equitably coloured, with vertex v assigned to colour 1. The number of vertices coloured with colour 1 is therefore at most $\alpha(G - \Gamma(v) - \{v\}) + 1$. Since this colouring is equitable, the number of vertices coloured with any other colour is then at most $\alpha(G - \Gamma(v) - \{v\}) + 2$. Hence:

$$\left\lceil \frac{n + 1}{\alpha(G - \Gamma(v) - \{v\}) + 2} \right\rceil \leq \chi_e(G). \tag{6.6}$$

For example, using a star graph with $n = 8$ whose internal vertex v_1 is coloured with colour 1, this gives $\left\lceil \frac{8+1}{0+2} \right\rceil = 5$. Note, however, that this bound requires a graph's independence number to be calculated, which is itself an $\mathcal{NP}$-hard problem.

With regard to upper bounds on $\chi_e(G)$, it is known that any graph can be equitably k-coloured when $k \geq \Delta(G) + 1$. Hence:

Theorem 6.15 (Hajnal and Szemerédi [37]) *Let G be a graph with maximal degree $\Delta(G)$. Then $\chi_e(G) \leq \Delta(G) + 1$.*

This fact was initially conjectured by Erdős [38], with a formal proof being published 6 years later by Hajnal and Szemerédi [37]. Shorter proofs of this theorem have also been shown by Kierstead and Kostochka [39] and Kierstead et al. [40]. The latter publication also presents a polynomial-time algorithm for constructing an equitable $(\Delta(G) + 1)$-colouring. The method involves first removing all edges from G and dividing the n vertices arbitrarily into $\Delta(G)$ equal-sized colour classes. In cases where n is not a multiple of $\Delta(G)$, sufficient isolated vertices are added. The vertices are then considered in turn and, in each iteration i, the edges incident to vertex v_i are added to G. If v_i is seen to be adjacent to another vertex in its colour class, it is moved to a different feasible colour class, leading to a feasible colouring with up to $\Delta(G) + 1$ colours. If this colouring is not equitable, then a polynomial-length sequence of adjustments are made to re-establish the balance of the colour classes.

It is notable that Theorem 6.15 is similar to Theorem 3.6 from Chap. 3 which states that for any graph G, the chromatic number $\chi(G) \leq \Delta(G) + 1$. Meyer [41] has gone one step further to even conjecture a form of Brooks' Theorem 3.8 for equitable graph colouring: every graph G has an equitable colouring using $\Delta(G)$ or fewer colours except for complete graphs and odd cycles. Recall, however, that the problem of determining an equitable k-colouring for an arbitrary graph G is still $\mathcal{NP}$-complete, implying the need for approximation algorithms and heuristics in general.

One simple approach for achieving approximate equitable k-colourings can be achieved by making a simple modification of the DSATUR algorithm. Recall from Sect. 3.3 that this algorithm takes vertices one at a time and then colours them using the lowest colour label not assigned to any of their neighbours. Here, we can change this strategy by using k colour classes from the outset and, at each iteration, select the feasible colour class *containing the fewest vertices*, breaking any ties randomly.

Figure 6.22 summarises results achieved by this modification for random graphs $G_{500,p}$, using $p = 0.1, 0.5,$ and 0.9, for a range of suitable k-values. For comparison's sake, the results of a second algorithm are also included here. This operates in the same manner except that vertices are assigned to a randomly chosen feasible colour in each case. The cost here is simply the difference in size between the largest and smaller colour classes in a solution. Hence, a cost of 0 or 1 indicates an equitable k-colouring.

Figure 6.22 demonstrates that, for these random graphs, the policy of assigning vertices to feasible colour classes with the fewest vertices brings about more equitably coloured solutions. We also see that the algorithm consistently achieves equitable colourings for the majority of k-values except for those close to the chromatic number, and those where k is a divisor of n. For the former case, the low number of available colours restricts the choice of feasible colours for each vertex, often leading to inequitable colourings. On the other hand, when k is a divisor of n the algorithm is seeking a solution with a cost of 0, meaning that the last vertex considered by the algorithm must be assigned to the unique colour class containing one fewer vertex than the remaining colour classes. If this colour turns out to be infeasible (which is often the case), this vertex will then need to be assigned to another colour class, resulting in a solution with a cost of 2.

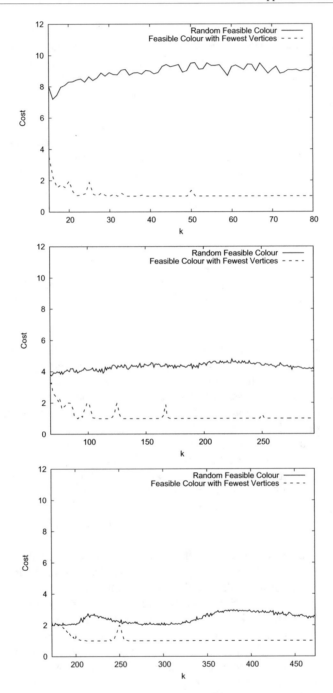

Fig. 6.22 Quality of equitable solutions produced by the modified DSATUR algorithms on random graphs with $n = 500$ for, respectively, $p = 0.1, 0.5$, and 0.9. All figures are the average of 50 instances per k-value

It is also possible to further improve these solutions by, for example, applying a local search algorithm with appropriate neighbourhood operators such as Kempe chain interchanges and pair swaps. An approach along these lines for a related problem is the subject of the case study presented in Chap. 7.

6.9   Weighted Graph Colouring

Further useful extensions of the graph colouring problem can be achieved through the addition of numeric *weights* to a graph. Typically, the term "weighted graph colouring" is used in situations where the vertices of a graph are allocated weights. However, the term is also sometimes used for problems where the edges are weighted, and for the multicolouring problem. These are considered in turn in the following subsections.

6.9.1   Weighted Vertices

A natural formulation of the weighted graph colouring problem is as follows. Let $G(V, E, w)$ be a graph for which each vertex $v \in V$ is given a nonnegative integer weight w_v. Given a fixed number of colours k, our task is to identify a proper (but perhaps partial) solution $\mathcal{S} = \{S_1, \ldots, S_k\}$ that maximises the objective function:

$$f(\mathcal{S}) = \sum_{i=1}^{k} g(S_i), \tag{6.7}$$

where $g(S_i) = \sum_{v \in S_i} w_v$ gives the sum of the vertex weights belonging to colour class S_i.

For this problem, if $\chi(G) \leq k$ then an optimal solution will feature a cost of $\sum_{v \in V} w_v$, corresponding to a feasible graph colouring solution. On the other hand, if $\chi(G) > k$, then the problem involves finding the best subset of vertices V' to colour, where $V' = \bigcup_{i=1}^{k} S_i$.

It is straightforward to show that this problem is $\mathcal{NP}$-hard by noting that any instance of the graph colouring problem can be transformed into this weighted variant by setting $w_v = 1 \, \forall v \in V$. If, in addition to this, $k = 1$, then the problem is also equivalent to the $\mathcal{NP}$-hard maximum independent set problem.

This formulation of the weighted graph colouring problem arises in practical situations where a limited number of colour classes are available and where the colouring of certain vertices is considered more important than others. For example, in an exam timetabling problem, we may be given a fixed number of timeslots (colours), and we might want to prioritise the assignment of larger exams (vertices with higher weightings) to the timetable while also making sure that clashing exams (adjacent vertices) are not assigned to the same timeslots. One simple heuristic for this problem is to employ the GREEDY algorithm using a fixed number of colours k

and an ordering of the vertices $v_1, v_2, \ldots, v_n$ such that $w_{v_1} \geq w_{v_2} \geq \cdots \geq w_{v_n}$. Algorithms that explore the space of partial proper solutions are also very suitable here. For example, we might choose to make use of the PARTIALCOL algorithm while seeking to minimise the objective function $\sum_{v \in U} w_v$, where $U = V - \bigcup_{i=1}^{k} S_i$ is the set of uncoloured vertices (see Sect. 5.2).

Another well-known formulation of the weighted graph colouring problem involves taking a vertex-weighted graph $G(V, E, w)$ as above, and then determining a feasible colouring $S = \{S_1, \ldots, S_k\}$ that minimises Eq. (6.7) using $g(S_i) = \max\{w_v : v \in S_i\}$. A practical example of this occurs in the scheduling of fixed-time jobs to timeslots. Imagine, for example, that we are given a set of jobs V, each with a processing time $w_v \; \forall v \in V$. Imagine further that these jobs are to be scheduled into k timeslots, and that the jobs assigned to a particular timeslot will be carried out simultaneously; hence, the *duration* of a timeslot S_i is set at $\max\{w_v : v \in S_i\}$. Finally, also suppose that some pairs of jobs u, v are *incompatible* and cannot be assigned to the same timeslot. Such pairs correspond to edges $\{u, v\} \in E$.

In this formulation, it is obviously in our interest to try and assign vertices with large weights to the same colour classes. Similarly, it also makes sense in many cases to reduce the number of colours being used by increasing the number of colour classes S_i for which $S_i = \emptyset$. Demange et al. [42] have noted that the optimal number of nonempty colour classes for a particular graph G may well be larger than $\chi(G)$, though it will also always be less than or equal to $\Delta(G) + 1$.

As with the previous example, this current problem formulation can be shown to be $\mathcal{NP}$-hard by observing that any instance for which $w_v = 1$ ($\forall v \in V$) is equivalent to the standard graph colouring problem. Furthermore, the problem remains $\mathcal{NP}$-hard even for interval graphs [43] and bipartite graphs [42]. For the bipartite case, Demange et al. [42] have provided an algorithm with approximation ratio of $4r_w/(3r_w + 2)$ (where $r_w = \frac{\max\{w_v : v \in V\}}{\min\{w_v : v \in V\}}$). They also prove that optimal solutions for bipartite graphs can be found in polynomial time whenever $|\{w_v : v \in V\}| \leq 2$. For general graphs, an approximation algorithm is also suggested that operates as follows. As before, let $G = (V, E, w)$ be a graph with weighted vertices and $g(S_i) = \max\{w_v : v \in S_i\}$.

1. Construct a graph with weighted edges $\bar{G} = (V, E', w')$ where $\bar{G}$ is the complement of G, and for any edge $\{u, v\} \in E'$, $w'_{uv} = w_u + w_v - g(\{w_u, w_v\})$.
2. Compute a maximum weighted matching M^* of $\bar{G}$.
3. For each edge in M^*, colour the end points with a new colour.
4. Colour any remaining vertices with their own new colour.

An example of this process is shown in Fig. 6.23. The matching M^* can be determined in polynomial time using methods such as the $\mathcal{O}(mn^2)$ blossom algorithm [44]. Note that each colour class in the solution is an independent set, but that these are limited to contain a maximum of two vertices. Indeed, in graphs where no independent set contains more than two vertices (such as the complement of a bipartite graph), this algorithm guarantees the optimal. In further work, Hassin and Monnot [45] have

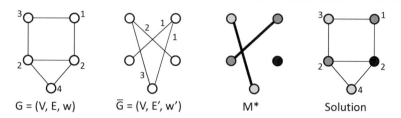

Fig. 6.23 Illustration of the algorithm of Demange et al. [42]

shown that, for any graph, this process produces a solution whose objective function value never exceeds twice the optimum. They also show that the same approximation ratio applies when $g(S_i)$ takes other forms such as $g(S_i) = \min\{w_v : v \in S_i\}$ and $g(S_i) = \frac{1}{|S_i|} \sum_{v \in S_i} w_v$. Malaguti et al. [46] have also proposed several IP-based methods for this problem similar in spirit to those discussed in Sect. 4.1.2. In particular, they propose the use of heuristics for building up a large sample of independent sets and then use an IP model similar to that of Sect. 4.1.3 to select a subset of these. Local search-based methods based on Kempe chain interchanges and pair swaps also seem to be naturally suited to this problem.

6.9.2 Weighted Edges

In many cases, it is more convenient to apply weights to the *edges* of a graph as opposed to the vertices. This allows us to express levels preference for assigning vertices to different (or the same) colours.

One interpretation involves taking a graph $G(V, E, w)$ for which each edge $\{u, v\} \in E$ is allocated an integer weight w_{uv}. Given a fixed number of colours k, our task is to then identify a complete (but perhaps improper) solution $\mathcal{S} = \{S_1, \ldots, S_k\}$ that minimises the objective function:

$$f(\mathcal{S}) = \sum_{i=1}^{k} \sum_{u,v \in S_i : \{u,v\} \in E} w_{uv}. \tag{6.8}$$

Here, if all edge weights in the graph are positive values, then any solution for which $f(\mathcal{S}) = 0$ will correspond to a feasible k-coloured solution.

This sort of formulation is applicable in areas such as exam timetabling and social networking. For the former, imagine that we wish to assign exams (vertices) to timeslots (colours), but that there are insufficient timeslots to feasibly accommodate all exams. To form a complete exam timetable, this means that some clashes will be necessary; however, some types of clashes may be deemed less critical than others. For example, if two clashing exams only have a small number of common students, then we may allow them to both be assigned to the same timeslot, with alternative arrangements then being made for the people affected. On the other hand, if two exams contain a large number of common participants then a clash is far less desirable. Appropriate weights added to the corresponding edges can be used to express such preferences.

Note that due to the nature of this problem's requirements, algorithms that search the space of complete improper solutions will often be naturally suitable here. In Chap. 7, an application along these lines will be made to the problem of partitioning members of social networks, where edge weights are used to express a level of "liking" or "disliking" between pairs of individuals. A simulated annealing-based approach for constructing subject options columns at schools is also described by Lewis et al. [47].

6.9.3   Multicolouring

Another problem that is also sometimes referred to as "weighted graph colouring" is the $\mathcal{NP}$-hard graph *multicolouring* problem. In this problem, we are given a graph $G(V, E, w)$ for which each vertex $v \in V$ is allocated a weight $w_v \in \{1, 2, \ldots\}$. The task is to then assign w_v different colours to each vertex v such that (a) adjacent vertices have no colours in common and (b) the number of colours used is minimal.

Multicolouring has practical applications in areas such as frequency assignment problems where, in some cases, devices should be able to transmit and receive messages on multiple frequencies as opposed to just one [48]. McDiarmid and Reed [49] have shown that this problem is polynomially solvable for bipartite and perfect graphs, but that it remains $\mathcal{NP}$-hard for triangular lattice graphs and their induced subgraphs, which have practical applications in cellular telephone networks. They also suggest a suitable polynomial-time approximation algorithm for the latter topology.

Note that the graph colouring problem is a special case of the multicolouring problem for which $w_v = 1$ $\forall v \in V$. On the other hand, any instance of the multicolouring problem can also be converted into an equivalent graph colouring problem by replacing each vertex $v \in V$ with a clique of size w_v, and then connecting every member of the clique to all neighbours of v in G. This method of conversion allows us to use any graph colouring algorithm (such as those from Chap. 5) with the graph multicolouring problem, though it also increases the number of vertices to colour by a factor of $\sum_{v \in V} w_v / n$. Consequently, graph multicolouring is often studied as a separate computational problem, for which the backtracking algorithm of Caramia and Dell'Olmo [50] and the IP branch-and-price method of Mehrotra and Trick [51] are prominent examples.

6.10   Chromatic Polynomials

The final topic in this chapter concerns counting the number of *different* colourings of a graph G. To do this, let $P(G, k)$ denote the number of distinct feasible colourings of G that are using k or fewer colours. As we will see presently, $P(G, k)$ can always be

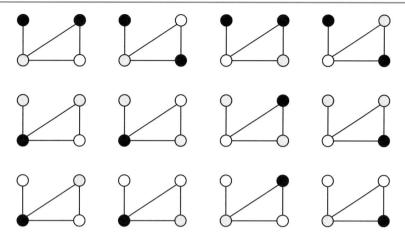

Fig. 6.24 All feasible three-colourings of this particular four-vertex graph

expressed as polynomial; it is therefore usually known as the *chromatic polynomial* of a graph. Chromatic polynomials were first defined by Birkhoff [52]. In this work, he attempted to show that if a graph G is planar, then $P(G, 4) \geq 1$, thereby proving the four colour theorem. These ideas were later expanded to general graphs by Whitney [53].

Consider the four-vertex graph G in Fig. 6.24. As shown, this graph can be three-coloured in 12 different ways but cannot be two-coloured. Consequently, $P(G, 3) = 12$. Now consider the case where $k = 4$ colours are available. Since $k = n$, there are $n! = 4! = 24$ different ways of colouring G using exactly four colours. This gives

$$P(G, 4) = 24 + \binom{4}{3} P(G, 3)$$
$$= 24 + 4 \times 12$$
$$= 72.$$

As shown, the second term in this sum involves multiplying $P(G, 3)$ by $\binom{4}{3}$. This is necessary because, for each of the 12 available three-colourings, there are $\binom{4}{3}$ ways of choosing three colours from the available four. The number of different k-colourings for this example graph are shown in the following table.

k	0	1	2	3	4	...
$P(G, k)$	0	0	0	12	72	...

Note that the lowest k-value for which $P(G, k)$ is non-zero gives the chromatic number of a graph. This fact is also apparent in Fig. 6.24, where none of the presented three-colourings is using fewer than three colours. In this sense, the chromatic polynomial contains at least as much information as its chromatic number.

For $k \geq 3$, the problem of calculating the chromatic polynomial of a graph is #$\mathcal{P}$-complete. This class of problems (pronounced "sharp p complete") defines the set of counting problems associated with decision problems in $\mathcal{NP}$. It also includes examples such as counting the number of shortest uv-paths in a graph, and counting the number of Hamiltonian cycles in a graph. Members of the #$\mathcal{P}$-complete problem class are known to be at least as hard as NP-complete problems, implying that chromatic polynomials cannot be calculated in polynomial time.

Despite this, the chromatic polynomials of some basic graph topologies can be easily expressed as closed formulas [54]. For example:

- If G is a complete graph on n vertices, $P(G, k) = k(k-1)(k-2) \ldots (k-(n-1))$.
- If G is an empty graph on n vertices, $P(G, k) = k^n$.
- If G is a tree containing n vertices, $P(G, k) = k(k-1)^{n-1}$.
- If G is a cycle graph with n vertices, $P(G, k) = (k-1)^n + (-1)^n(k-1)$.

These formulae can be used to help prove the following.

Theorem 6.16 *The number of different k-colourings of a graph G, denoted by $P(G, k)$, is a polynomial.*

Proof Let G_1 be the graph obtained by deleting an edge $\{u, v\}$ from G. Also, let G_2 be the graph obtained by contracting $\{u, v\}$. This means that

$$P(G, k) = P(G_1, k) - P(G_2, k).$$

To see this, observe that the number of k-colourings of G_1 in which u and v have different colours is unchanged if the edge $\{u, v\}$ is added. It is therefore equal to $P(G, k)$. Similarly, the number of k-colourings of G_1 in which u and v have the *same* colour is equal to $P(G_2, k)$. Hence, $P(G_1, k) = P(G, k) + P(G_2, k)$.

The actions of removing and contracting edges can now be repeated on each subsequent subgraph to form a binary tree of successively smaller graphs. The leaves of this tree will be empty graphs. Since the number of ways of k-colouring an empty graph is a polynomial, it follows that the $P(G, k)$ is also a polynomial. □

Figure 6.25 gives an example of the process described in the proof of Theorem 6.16. As demonstrated, rather that continuing these actions until all leaf nodes are empty graphs, it is sufficient for leaves to represent graphs with known chromatic polynomials. In this example, the leaves are path graphs and complete graphs

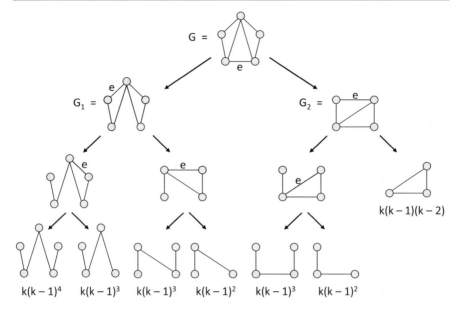

Fig. 6.25 Example binary tree formed using the ideas in the proof of Theorem 6.16. In each step, the edge e is removed and contracted to form the left and right branches, respectively

as indicated. The chromatic polynomial of this example graph is therefore

$$P(G, k) = \left(k(k-1)^4 - k(k-1)^3\right)$$
$$- \left(k(k-1)^3 - k(k-1)^2\right)$$
$$- \left(k(k-1)^3 - k(k-1)^2\right)$$
$$+ (k(k-1)(k-2))$$
$$= k^5 - 7k^4 + 18k^3 + 20k^2 + 8k.$$

The number of different k-colourings for this graph are summarised in the following table:

k	0	1	2	3	4	5	6	...
$P(G, k)$	0	0	0	6	96	540	1920	...

These figures also demonstrate that the chromatic number of this graph is three.

6.11 Chapter Summary

This chapter has reviewed a wide range of computational problems related to graph colouring. Several of our chosen topics, such as face and edge colouring, precolouring, multicolouring, and solving sudoku puzzles, are either equivalent to the graph (vertex) colouring problem or represent special cases of the problem. They can therefore be tackled using the algorithms seen in Chaps. 3–5.

In later sections of this chapter, we have also discussed various extensions to the graph colouring problem. These have included decentralised and online graph colouring, dynamic colouring (where graphs are known to evolve), list colouring, equitable colouring, weighted graph colouring, and chromatic polynomials. Further extensions using real-world operational research problems are considered in the next three chapters.

References

1. Kuratowski K (1930) Sur le probleme des courbes gauches en topologie. Fundam Math 15:271–283
2. Hopcroft J, Tarjan R (1974) Efficient planarity testing. J Assoc Comput Mach 21:549–568
3. Boyer W, Myrvold J (2004) On the cutting edge: simplified $\mathcal{O}(n)$ planarity by edge addition. J Graph Algorithms Appl 8:241–273
4. Heawood P (1890) Map-colour theorems. Q J Math 24:332–338
5. Kempe A (1879) On the geographical problem of the four colours. Am J Math 2:193–200
6. Appel K, Haken W (1977) Solution of the four color map problem. Sci Am 4:108–121
7. Appel K, Haken W (1997) Every planar map is four colorable. Part I. Discharging. Ill J Math 21:429–490
8. Appel K, Haken W (1977) Every planar map is four colorable. Part II. Reducibility. Ill J Math 21:491–567
9. Appel K, Haken W (1989) Every planar map is four colorable. Contemporary Mathematics, AMS. 978-0-8218-5103-6
10. Robertson N, Sanders D, Seymour P, Thomas R (1997) The four color theorem. J Comb Theory, Ser B 70:2–44
11. Wilson R (2003) Four colors suffice: how the map problem was solved. Penguin Books
12. de Werra D (1988) Some models of graphs for scheduling sports competitions. Discret Appl Math 21:47–65
13. Coffman E, Garey M, Johnson D, LaPaugh A (1985) Scheduling file transfers. SIAM J Comput 14(3):744–780
14. Kirkman T (1847) On a problem in combinations. Camb Dublin Math J 2:191–204
15. König D (1916) Gráfok és alkalmazásuk a determinánsok és a halmazok elméletére. Mat Termtud Értesö 34:104–119
16. Vizing V (1964) On an estimate of the chromatic class of a p-graph. Diskret Analiz 3:25–30
17. Holyer I (1981) The NP-completeness of edge-coloring. SIAM J Comput 10:718–720
18. Misra J, Gries D (1992) A constructive proof of Vizing's theorem. Inf Process Lett 41:131–133
19. Colbourn C (1984) The complexity of completing partial Latin squares. Discret Appl Math 8(1):25–30

20. Russell E, Jarvis F (2005) There are 5,472,730,538 essentially different Sudoku grids. http://www.afjarvis.staff.shef.ac.uk/sudoku/sudgroup.html, September 2005
21. McGuire G, Tugemann B, Civario G (2012) There is no 16-clue Sudoku: solving the Sudoku minimum number of clues problem. Comput Res Repos. arXiv:1201.0749
22. Herzberg A, Murty M (2007) Sudoku squares and chromatic polynomials. Not AMS 54(6):708–717
23. Yato T, Seta T (2003) Complexity and completeness of finding another solution and its application to puzzles. IEICE Trans Fundam Electron Commun Comput Sci E86-A:1052–1060
24. Garey M, Johnson D, So H (1976) An application of graph coloring to printed circuit testing. IEEE Trans Circuits Syst CAS-23:591–599
25. Finocchi I, Panconesi A, Silvestri R (2005) An experimental analysis of simple, distributed vertex colouring algorithms. Algorithmica 41:1–23
26. Hernández H, Blum C (2014) FrogSim: distributed graph coloring in wireless ad hoc networks. Telecommun Syst 55:211–223
27. Kearns M, Suri S, Montfort N (2006) An experimental study of the coloring problem on human subject networks. Science 313:824–827
28. Kierstead H, Trotter W (1981) An extremal problem in recursive combinatorics. Congr Numer 33:143–153
29. Gyárfás A, Lehel J (1988) On-line and first fit colourings of graphs. J Graph Theory 12:217–227
30. Zaker M (2006) Results on the Grundy chromatic number of graphs. Discret Math 306(23):3166–3173
31. Bonnet É, Foucaud F, Kim E, Sikora F (2015) Complexity of Grundy coloring and its variants. In: Xu D, Du D, Du D (eds) Computing and combinatorics. Springer International Publishing, pp 109–120. ISBN 978-3-319-21398-9
32. Ouerfelli L, Bouziri H (2011) Greedy algorithms for dynamic graph coloring. In: Proceedings of the international conference on communications, computing and control applications (CCCA), pp 1–5. https://doi.org/10.1109/CCCA.2011.6031437
33. Dupont A, Linhares A, Artigues C, Feillet D, Michelon P, Vasquez M (2009) The dynamic frequency assignment problem. Eur J Oper Res 195:75–88
34. Hardy B (2018) Heuristic methods for colouring dynamic random graphs. PhD thesis, Cardiff University
35. Hardy B, Lewis R, Thompson J (2018) Tackling the edge dynamic graph colouring problem with and without future adjacency information. J Heurist 24(3):321–343
36. Furmańczyk H (2004) Graph colorings. In: Equitable coloring of graphs. American Mathematical Society, pp 35–54
37. Hajnal A, Szemerédi E (1970) Combinatorial theory and it's application. In: Erdős P (ed) Proof of a conjecture. North-Holland, pp 601–623
38. Erdős P (1964) Theory of graphs and its applications. In: Problem 9. Czech Academy of Sciences, p 159.
39. Kierstead H, Kostochka A (2008) A short proof of the Hajnal-Szemerédi theorem on equitable coloring. Comb Probab Comput 17:265–270
40. Kierstead H, Kostochka A, Mydlarz M, Szemerédi E (2010) A fast algorithm for equitable graph coloring. Combinatorica 30:217–224
41. Meyer W (1973) Equitable coloring. Amer Math Monthly 80:920–922
42. Demange M, de Werra D, Monnot J, Paschos V (2007) Time slot scheduling of compatible jobs. J Sched 10:111–127
43. Escoffier B, Monnot J, Paschos V (2006) Weighted coloring: further complexity and approximability results. Inf Process Lett 97(3):98–103
44. Kolmogorov V (2009) Blossom V: aA new implementation of a minimum cost perfect matching algorithm. Math Program Comput 1(1):43–67
45. Hassin R, Monnot J (2005) The maximum saving partition problem. Oper Res Lett 33:242–248

46. Malaguti E, Monaci M, Toth P (2009) Models and heuristic algorithms for a weighted vertex coloring problem. J Heurist 15:503–526
47. Lewis R, Anderson T, Carroll F (2020) Can school enrolment and performance be improved by maximizing students sense of choice in elective subjects? J Learn Anal 7(1):75–87
48. Aardel K, van Hoesel S, Koster A, Mannino C, Sassano A (2002) Models and solution techniques for the frequency assignment problems. 4OR: Q J Belgian, French and Italian Oper Res Soc 1(4):1–40
49. McDiarmid C, Reed B (2000) Channel assignment and weighted coloring. Networks 36(2):114–117
50. Caramia M, Dell'Olmo P (2001) Solving the minimum-weighted coloring problem. Networks 38(2):88–101
51. Mehrotra A, Trick M (2007) Extending the horizons: advances in computing, optimization, and decision technologies. In: A branch-and-price approach for graph multi-coloring. Operations research/computer science interfaces series, vol 37. Springer, pp 15–29
52. Birkhoff G (1912) A determinant formula for the number of ways of coloring a map. Ann Math 14(1/4):42–46
53. Whitney H (1932) The coloring of graphs. Ann Math 33:688–718
54. Zhang J (2018) An introduction to chromatic polynomials. https://math.mit.edu/~apost/courses/18.204_2018/Julie_Zhang_paper.pdf, May 2018

Designing Seating Plans

7

The following three chapters of this book contain detailed case studies showing how graph colouring methods can be used to successfully tackle important real-world problems. The first of these case studies concerns the task of designing table plans for large parties, which, as we will see, combines elements of the $\mathcal{NP}$-hard edge-weighted graph colouring problem, the equitable graph colouring problem and the k-partition problem. A user-friendly implementation of the algorithm proposed in this section can also be found online at http://www.weddingseatplanner.com, which is viewed using Adobe Flash Player.

7.1 Problem Background

Consider an event such as a wedding or gala dinner where, as part of the formalities, the N guests need to be divided on to k dining tables. To ensure that guests are sat at tables with appropriate company, it is often necessary for organisers to specify a seating plan, taking into account the following sorts of factors:

- Guests belonging to groups, such as couples and families, should be sat at the same tables, preferably next to each other.
- If there is any perceived animosity between different guests, these should be sat on different tables. Similarly, if guests are known to like one another, it may be desirable for them to be sat at the same table.
- Some guests might be required to sit at a particular table. Also, some guests might be prohibited from sitting at other tables.
- Since tables may vary in size and shape, each table should be assigned a suitable number of guests, and these guests should be appropriately arranged around the table.

© The Author(s), under exclusive license to Springer Nature Switzerland AG 2021
R. M. R. Lewis, *Guide to Graph Colouring*, Texts in Computer Science,
https://doi.org/10.1007/978-3-030-81054-2_7

A naïve method for producing a seating plan best fitting these sorts of criteria might be to consider all possible plans and then choose the one perceived to be the most suitable. However, for non-trivial values of N or k, the number of possible solutions will be prohibitively large for this to be possible. To illustrate, consider a simple example where we have 48 guests using 6 tables, with exactly 8 guests per table. For the first table, we need to choose 8 people from the 48, for which there are $\binom{48}{8} = 377{,}348{,}994$ possible choices. For the next table, we then choose 8 further people from the remaining 40, giving $\binom{40}{8} = 76{,}904{,}685$ further choices, and so on. Assuming that N is a multiple of k (allowing equal-sized tables), the number of possible plans is thus calculated:

$$
\prod_{i=0}^{k-2} \binom{N - \frac{in}{k}}{N/k}
$$

$$
= \frac{N!}{\left(\frac{N}{k}\right)! \left(N - \frac{N}{k}\right)!} \cdot \frac{\left(N - \frac{N}{k}\right)!}{\left(\frac{N}{k}\right)! \left(N - \frac{2N}{k}\right)!} \cdot \frac{\left(N - \frac{2N}{k}\right)!}{\left(\frac{N}{k}\right)! \left(N - \frac{3N}{k}\right)!} \cdots \cdots \frac{\left(N - \frac{(k-2)N}{k}\right)!}{\left(\frac{N}{k}\right)! \left(N - \frac{(k-1)N}{k}\right)!}
$$

$$
= \frac{N!}{\left(\left(\frac{N}{k}\right)!\right)^{k-1} \left(N - \frac{(k-1)N}{k}\right)!}
$$

$$
= \frac{N!}{\left(\left(\frac{N}{k}\right)!\right)^{k}} \tag{7.1}
$$

This function clearly features a growth rate that is subject to a combinatorial explosion—even for the modestly sized example above, the number of distinct solutions to check is approximately 2.9×10^{33}, which is far beyond the capabilities of any state-of-the-art computing equipment. Furthermore, if we were to relax the problem by allowing tables of any size, the task would now be to partition the N guests into N non-empty subsets, meaning that the number of solutions would be equal to a Stirling number of the second kind $\left\{ {N \atop k} \right\}$ (Eq. (2.1)). These numbers feature even higher growth rates than Eq. (7.1), leading to even larger solution spaces.

Such arguments demonstrate that this sort of naive method for producing desirable seating plans is infeasible in most cases. However, the problem of constructing seating plans is certainly important since (a) it is regularly encountered by event organisers and (b) the quality of the proposed solution could have a significant effect on the success (or failure) of the gathering.

Currently, there is a small amount of commercial software available for constructing seating plans. These include Perfect Table Plan (https://www.perfecttableplan.com/), and Top Table Planner (https://www.toptableplanner.com/). The first of these allows users to input a list of guest names and then specify preferences between these guests (such as whether they need to be sat apart or together). It then allows users to define table shapes, sizes, and locations, before assisting the user in placing the guests at these tables via drag and drop functionality and also an auto-assign tool. The details of the underlying algorithm used with the auto-assign tool are commercially sensitive, though the software's documentation states that an evolutionary algorithm is used, with different penalty costs being applied for different types of constraint violation. The fitness function of the algorithm is simply an aggregate of these penalties.

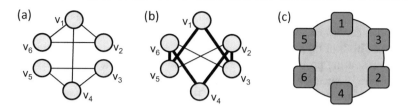

Fig. 7.1 a A graph G in which edges specify pairs of guests who should not be sat in adjacent seats; **b** the compliment graph $\bar{G}$, together with a Hamiltonain cycle (shown in bold); and **c** the corresponding seating arrangement around a circular table

7.1.1   Relation to Graph Problems

In its simplest form, the problem of constructing a seating plan might be defined using an $N \times N$ binary matrix $\mathbf{W}$, where element $W_{ij} = 1$ if guests i and j are required to be sat at different tables and $W_{ij} = 0$ otherwise. We can also assume that $W_{ij} = W_{ji}$. Given this input matrix, our task might then be to partition the N guests into k subsets $\mathcal{S} = \{S_1, \ldots, S_k\}$, such that the objective function:

$$f(\mathcal{S}) = \sum_{t=1}^{k} \sum_{\forall i, j \in S_t : i < j} W_{ij} \qquad (7.2)$$

is minimised.

The problem of confirming the existence of a zero-cost solution to this problem is equivalent to the $\mathcal{NP}$-complete decision variant of the graph colouring problem. Here, the graph $G = (V, E)$ is defined using the vertex set $V = \{v_1, \ldots, v_N\}$ and the edge set $E = \{\{v_i, v_j\} : W_{ij} = 1 \wedge v_i, v_j \in V\}$. That is, each guest corresponds to a vertex, and two vertices v_i and v_j are considered to be adjacent if and only if $W_{ij} = 1$. Colours correspond to tables, and we are now interested in colouring G using k colours.

From an alternative perspective, consider the situation where we are again given the binary matrix $\mathbf{W}$, and are now presented with a subset S of guests that have been assigned to a particular circular-shaped table. Here, we might be interested in arranging the guests onto the table such that, for all pairs of guests $i, j \in S$, if $W_{ij} = 1$ then i and j are not sat in adjacent seats. This problem can also be described by a graph $G = (V, E)$ for which the vertex set $V = \{v_i : i \in S\}$ and the edge set $E = \{\{v_i, v_j\} : W_{ij} = 1 \wedge v_i, v_j \in V\}$. A Hamiltonian cycle of the complement graph $\bar{G}$ defines a seating arrangement satisfying this criterion, as illustrated in the example in Fig. 7.1. However, determining the existence of a Hamiltonian cycle in an arbitrary graph is also an $\mathcal{NP}$-complete problem.

In practical situations, it might be preferable for $\mathbf{W}$ to be an integer or real-valued matrix instead of binary, allowing users to place greater importance on some of their seating preferences compared to others. Assuming that lower values for W_{ij} indicate an increased preference for guests i and j to be sat together, the problem of partitioning the groups on to k tables now becomes equivalent to the edge-weighted graph colouring problem, while the task of arranging people on to circular tables in the manner described above becomes equivalent to the travelling salesman problem. Of course, both of these problems are also $\mathcal{NP}$-complete since they generalise the graph k-colouring problem and the Hamiltonian cycle problem, respectively. Also, the problem of arranging guests around tables can become even more complicated when tables of different shapes are used. For example, with rectangular tables, we might also need to take into account who is sat opposite a guest in addition to their neighbours on either side.

7.1.2 Chapter Outline

In this chapter, we describe a formulation of the above problem that is closely related to graph colouring, but which also involves the additional constraint of grouping together guests who like one another while also maintaining appropriate numbers of guests per table. This problem interpretation is used in conjunction with the commercial website http://www.weddingseatplanner.com, which contains a free tool for inputting and solving instances of the problem. The reader is invited to try out this tool while reading this chapter.

It is stated by Nielsen [1] that users tend to leave a website in less than two minutes if it is not understood or perceived to fulfil their needs. Consequently, the particular problem formulation considered here is intended to strike the right balance between being quickly accessible to users while still being useful and flexible in practice. Since users of the website will typically have little knowledge of optimisation algorithms and the implications of problem intractability, the algorithm is also designed to supply the user with high-quality (though not necessarily optimal) solutions in very short amounts of run time (typically less than 3 seconds). In particular, our approach seeks to exploit the underlying graph-based structures of this problem, encouraging effective navigation of the solution space via specialised neighbourhood operators.

7.2 Problem Definition

In our definition of this problem, we choose to first partition the N guests into $n \leq N$ *guest groups*. Each guest group refers to a subset of guests who are required to sit together (couples, families with young children, etc.) and will usually be known beforehand by the user. In addition to making the problem smaller, specifying guest groups in this way also means that users do not have to subsequently input preferences between pairs of people in the same families, etc., to ensure that they are sat together in a solution.

Having done this, the wedding seating problem (WSP) can now be formally stated as a type of graph partitioning problem. Specifically, we are given a graph $G = (V, E)$ in which each vertex $v \in V$ represents a guest group. The size of each guest group is denoted by s_v. The total number of guests in the problem is thus $N = \sum_{v \in V} s_v$.

In G, each edge $\{u, v\} \in E$ defines the relationship between vertices u and v according to a weighting w_{uv} (where $w_{uv} = w_{vu}$). If $w_{uv} > 0$ this is interpreted to mean that we would prefer the guests associated with vertices u and v to be sat on different tables. Larger values for w_{uv} reflect a strengthening of this requirement. Similarly, negative values for w_{uv} mean that we would rather u and v were assigned to the same table.

A solution to the WSP is now defined as a partition of the vertices into k subsets $S = \{S_1, \ldots, S_k\}$. The requested number of tables k is defined by the user, with each subset S_i defining the guests assigned to a particular table.

7.2.1 Objective Functions

Under this definition of the problem, the quality of a particular candidate solution might be calculated according to various metrics. In our case, we use two objective functions, both that are to be minimised. The first of these is analogous to Eq. (7.2):

$$f_1 = \sum_{i=1}^{k} \sum_{\forall u, v \in S_i : \{u,v\} \in E} (s_v + s_u) w_{uv} \tag{7.3}$$

and reflects the extent to which the rules governing who sits with whom are obeyed. In this case, the weighting w_{uv} is multiplied by the total size of the two guest groups involved $s_v + s_u$. This is done so that violations involving larger numbers of people contribute more to the cost (i.e., it is assumed that s_v people have expressed a seating preference concerning guest group u, and s_u people have expressed a preference concerning guest group v).

The second objective function used in our model is intended to encourage equal numbers of guests being assigned to each table. In practice, some weddings may have varying sized tables, and nearly all weddings will have a special "top table" where the bride, groom, and their associates sit. The top table and its guests can be ignored in this particular formulation because they can easily be added to a table plan once the other guests have been arranged. We also choose to assume that the remaining tables are equal in size, which seems to be a very common option, particularly for large venues. Consequently, the second objective function measures the degree to which the number of guests per table deviates from the required number of either $\lfloor N/k \rfloor$ or $\lceil N/k \rceil$:

$$f_2 = \sum_{i=1}^{k} \left(\min \left(|\tau_i - \lfloor N/k \rfloor|, |\tau_i - \lceil N/k \rceil| \right) \right). \tag{7.4}$$

Here $\tau_i = \left(\sum_{\forall v \in S_i} s_v \right)$ denotes the number of guests assigned to each table i. Obviously, if the number of guests N is a multiple of k, then Eq. (7.4) simplifies to $f_2 = \sum_{i=1}^{k} (|\tau_i - N/k|)$.

7.2.2   Problem Intractability

We now show that this problem is $\mathcal{NP}$-hard. We do this by showing that it generalises two classical $\mathcal{NP}$-hard problems: the k-partition problem, and the equitable graph k-colouring problem. Let us first define the k-partition problem.

Definition 7.1 Let Y be a multiset of n *weights*, represented as integers, and let k be a positive integer. The $\mathcal{NP}$-hard *k-partition problem* involves partitioning Y into k subsets such that the total weight of each subset is equal.

Note that the k-partition problem is also sometimes known as the load balancing problem, the equal piles problem, or the multiprocessor scheduling problem.

Theorem 7.1 *The WSP is $\mathcal{NP}$-hard.*

Proof Let $G = (V, E)$. If $E = \emptyset$ then f_1 (Eq. (7.3)) equals zero for all solutions. Hence, the only goal is to ensure that the number of guests per table is equal (or as close to equal as possible). Consequently, the problem is equivalent to the $\mathcal{NP}$-hard k-partition problem.

From another perspective, let $s_v = 1 \; \forall v \in V$ and let $w_{uv} = 1 \; \forall \{u, v\} \in E$. The number of guests assigned to each table i therefore equals $|S_i|$. This special case is equivalent to the $\mathcal{NP}$-hard optimisation version of the equitable k-colouring problem (see Sect. 6.8). □

7.3   Problem Interpretation and Tabu Search Algorithm

On entering the website, the user is first asked to input (or import) the names of all guests into an embedded interactive table. Guest groups that are to be seated together (families, etc.) are placed on the same rows of the table thus defining the various values for s_v. Guests to be sat at the top table are also specified. At the next step, the user is then asked to define seating preferences between different guest groups. Since guests to be sat at the top table have already been given, constraints only need to be considered between the remaining guest groups (guests at the top table are essentially ignored from this point onwards).

Figure 7.2 shows a small example of this process. Here, nine guest groups ranging in size from 1 to 4 have been input, though one group of four has been allocated to the top table. Consequently, only the remaining eight groups (comprising $N = 20$ guests) are considered. The right-hand grid then shows how the seating preferences (values for w_{uv}) are defined between these. On the website, this is done interactively

(a)

Top table?	Guest name	Companion 1	Companion 2	Companion 3
1 ✓	Cath	Michael	Kurt	Rosie
2	John	Sarah	Jack	Jill
3	Bill	June		
4	Pat	Susan		
5	Una	Tom		
6	Ruth	Kevin	Gareth	
7	Ken	Frank	Bobby	
8	Rod	Dereck	Freddy	
9	Jane			

(b)

	John+3	Bill+1	Pat+1	Una+1	Ruth+2	Ken+2	Rod+2	Jane
John+3	╲	✗			✓			✓
Bill+1		╲			✗	✓		
Pat+1	✗		╲		⊖		✗	
Una+1		✗		╲				
Ruth+2	✓	✓	⊖		╲	⊖		
Ken+2	✓	✓			⊖	╲		
Rod+2		✗					╲	⊖
Jane	✓						⊖	╲

Fig. 7.2   Specification of guest groups (**a**) and seating preferences (**b**)

by clicking on the relevant cells in the grid. In our case, users are limited to three options: (1) "Definitely Apart" (e.g., John and Pat); (2) "Rather Apart" (Pat and Ruth); and (3) "Rather Together" (John and Ken). These are allocated weights of ∞, 1, and -1, respectively, for reasons that will be made clear below. Note that it would have been possible to allow the user to input their own arbitrary weights here; however, while being more flexible, it was felt by the website's interface designers that this ran the risk of bamboozling the user while not improving the effectiveness of the tool [2].

Once the input to the problem has been defined by the user, the overall strategy of our algorithm is to classify the requirements to the problem as either hard (mandatory) constraints or soft (optional) constraints. In our case, we consider just one hard constraint, which we attempt to satisfy in Stage 1—specifically the constraint that all pairs of guest groups required to be "Definitely Apart" are assigned to different tables. In Stage 2, the algorithm then attempts to reduce the number of violations of the remaining constraints via specialised neighbourhood operators that do not allow any of the hard constraints satisfied in Stage 1 to be re-violated. The two stages of the algorithm are now described in more detail.

7.3.1   Stage 1

In Stage 1, the algorithm operates on the subgraph $G' = (V, E')$, where each vertex $v \in V$ represents a guest group, and the edge set $E' = \{\{u, v\} \in E : w_{uv} = \infty\}$. In other words, the graph G' contains only those edges from the original graph G that define the "Definitely Apart" requirement. Using this subgraph, the problem of assigning all guests to k tables (while not violating the "Definitely Apart" constraint) is equivalent to finding a feasible k-colouring of G'.

In our case, an initial solution is produced using the variant of the DSATUR heuristic used with the equitable graph colouring problem in Sect. 6.8. Starting with k empty colour classes (tables), each vertex (guest group) is taken in turn according to the DSATUR heuristic and assigned to the feasible colour class containing the fewest vertices, breaking ties randomly. If no feasible colour exists for a vertex then it is kept to one side and is assigned to a random colour at the end of this process, thereby introducing violations of the hard constraint.

If the solution produced by the above constructive process contains hard constraint violations, an attempt is then made to eliminate them using TABUCOL (see Sect. 5.1). As we saw in Chap. 5, this algorithm can often be outperformed by other approaches in terms of the quality of solution it produces, but it does have the advantage of being very fast, which is an important requirement in this application. Consequently, TABUCOL is only run for a fixed number of iterations, specifically $20n$.

If at the end of this process a feasible k-colouring for G' has not been achieved, k is incremented by 1, and Stage 1 of the algorithm is repeated. Of course, this might occur because the user has specified a k-value for which no k-colouring exists (that is, $k < \chi(G')$) or it might simply be that a solution *does* exist, but that the algorithm has been unable to find it in the given computation limit. The process of incrementing k and reapplying DSATUR and TABUCOL continues until all of the hard constraints have been satisfied, resulting in a feasible colouring of G'.

7.3.2 Stage 2

Having achieved, a feasible k-colouring of $G' = (V, E')$ in Stage 1, Stage 2 is now concerned with eliminating violations of the soft constraints by exploring the space of feasible solutions. That is, the algorithm will make alterations to the seating plan in such a way that no violations of the "Definitely Apart" constraint are reintroduced.

Note that movements in this solution space might be restricted—indeed the space might even be disconnected—and so it is necessary to use neighbourhood operators that provide as much solution space connectivity as possible. In this case, ideal candidates are the Kempe chain interchange and the pair swap operators seen in Chap. 4 (see Definitions 4.3 and 4.4). With seating plans, applications of these operators have the effect of either moving one guest group from one table to another or interchanging two subsets of guest groups between a pair of tables.

In this case, in each iteration of Stage 2, all neighbouring solutions are evaluated, and the same acceptance criteria as TABUCOL are applied. Once a move is performed, all relevant parts of the tabu list **T** are then updated to reflect the changes made to the solution. That is, all of the vertices and colours involved in the move are marked as tabu in **T**. For speed's sake, in our application a fixed-size tabu tenure of 10 is used along with an iteration limit of $10n$.

7.3.2.1 Evaluating All Neighbours

When evaluating the cost of all neighbouring solutions in each iteration of Stage 2, considerable speedups can be achieved by avoiding situations where a particular move is evaluated more than once. Note that a Kempe chain comprising l vertices can actually be generated via l different vertex/colour combinations. For example, the Kempe chain $\{v_4, v_7, v_8, v_9\}$ depicted in Fig. 7.3 corresponds to KEMPE(v_4, 1, 2),

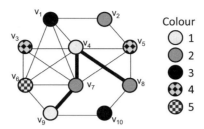

Fig. 7.3 Example Kempe chain, KEMPE(v_4, 1, 2) = $\{v_4, v_7, v_8, v_9\}$

EVALUATE-ALL-KEMPE-CHAIN-INTERCHANGES ($\mathcal{S} = \{S_1, \ldots, S_k\}$)
(1) $K_{vi} \leftarrow 0 \ \forall v \in V, \ \forall i \in \{1, \ldots, k\}$
(2) **forall** $S_i \in \mathcal{S}$ **do**
(3)   **forall** $v \in S_i$ **do**
(4)   **forall** $S_j \in (\mathcal{S} - \{S_i\})$ **do**
(5)   **if** $K_{vj} = 0$ **then**
(6)   Evaluate the cost of $\mathcal{S}$ if a KEMPE(v, i, j) interchange were to be applied.
(7)   **forall** $u \in$ KEMPE(v, i, j) **do**
(8)   **if** $u \in S_i$ **then** $K_{uj} \leftarrow |$KEMPE(v, i, j)$|$
(9)   **else** $K_{ui} \leftarrow |$KEMPE(v, i, j)$|$

Fig. 7.4 Procedure for efficiently evaluating all possible Kempe chain interchanges in a solution $\mathcal{S}$

KEMPE(v_7, 2, 1), KEMPE(v_8, 2, 1), and KEMPE(v_9, 1, 2). Of course, only one of these combinations needs to be considered at each iteration.

To achieve these speed-ups an additional matrix $\mathbf{K}_{n \times k}$ can be used where, given a vertex $v \in S_i$, each element K_{vj} is used to indicate the size of the Kempe chain formed via KEMPE(v, i, j). This matrix is populated in each iteration of tabu search according to the steps shown in Fig. 7.4. As can be seen here, initially all elements of $\mathbf{K}$ are set to zero. The algorithm then considers each vertex $v \in S_i$ in turn (for $1 \leq i \leq k$) and, according to Step (5), only evaluates an interchange involving the chain KEMPE(v, i, j) if the same set of vertices has not previously been considered. If a new Kempe chain is identified, the cost of performing this interchange is then evaluated (Step (6)), and the matrix $\mathbf{K}$ is updated to make sure that this interchange is not evaluated again in this iteration (Steps (7–9)).

Finally, after the evaluation of all possible Kempe chain interchange moves, the information in $\mathbf{K}$ can also be used to quickly identify all possible moves achievable via the pair swap operator. Specifically, for each $v \in S_i$ (for $1 \leq i \leq k-1$) and each $u \in S_j$ (for $i+1 \leq j \leq k$), pair swaps will only occur where both $K_{vj} = 1$ and $K_{ui} = 1$.

7.3.2.2   Cost Function

The objective function used in Stage 2 of this algorithm is simply ($f_1 + f_2$) (see Eqs. (7.3) and (7.4)). Note that this will always evaluate to a value less than ∞

since violations of the hard constraints cannot occur. Although such an aggregate function is not wholly ideal (because it involves adding together two different forms of measurement) it is acceptable in our case because, in some sense, both metrics relate to the number of people affected by the violations—that is, a table that is considered to have x too many (or too few) people will garner the same penalty cost as x violations of the "Rather Apart" constraint.

Finally, the speed of this algorithm can also be further increased by observing that (a) the cost functions f_1 and f_2 only involve the addition of terms relating to the quality of each separate colour class (table), and (b) neighbourhood moves with this algorithm only affect two colours. These features imply that if a move involving colours i and j is made in iteration l of the algorithm, then in iteration $l + 1$, the cost changes involved with moves using any pair of colours from the set $(\{1, \ldots, k\} - \{i, j\})$ will not have changed and do therefore not have to be recalculated by the algorithm.

7.4 Algorithm Performance

In this section, we analyse the performance of our two-stage tabu search algorithm in terms of both computational effort and the costs of its resultant solutions.

The algorithm and interface described above was implemented in ActionScript 3.0 and is executed via a web browser (an installation of Adobe Flash Player is required). The optimisation algorithm is therefore run at the client-side. To ensure run times are kept relatively short, and to also allow the interface to be displayed clearly on the screen, problem size has been limited to $n = 50$, with guest groups of up to 8 people, allowing a maximum of $N = 400$ guests.

To gain an understanding of the performance characteristics of this algorithm, a set of maximum-sized problem instances of $n = 50$ guest groups (vertices) were constructed, with the size of each group chosen uniform randomly in the range 1–8 giving $N \approx 50 \times 4.5 = 225$. These instances were then modified such that each pair of vertices was joined by an ∞-weighted edge with probability p, meaning that a proportion of approximately p guest group pairs would be required to be "Definitely Apart". Tests were then carried out using values of $p = \{0.0, 0.3, 0.6, 0.9\}$ with numbers of tables $k = \{3, 4, \ldots, 40\}$.

Figure 7.5 shows the results of these tests with regard to the costs that were achieved by the algorithm at termination. Note that for $p \geq 0.3$, values are not reported for the lowest k values because feasible k-colourings were not achieved (quite possibly because they do not exist). The figure indicates that, with no hard constraints ($p = 0.0$), balanced table sizes have been achieved for all k-values up to 30. From this point onwards, however, it seems there are simply too many tables (and too few guests per table) to spread the guest groups equally. Higher costs are also often incurred for larger values of p because, in these cases, many guest group combinations (including many of those required for achieving low-cost solutions) will now contain at least one hard constraint violation, meaning that they cannot be

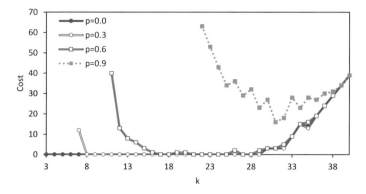

Fig. 7.5 Solution costs for four values of p using various k-values

assigned to the same table. That said, similar solutions are achieved for $p = 0.0$, 0.3, and 0.6 for various values of k, suggesting that the cost of the best solutions found is not unduly affected by the presence of moderate levels of hard constraints. The exception to this pattern, as shown in the figure, is for the smallest achievable values for k. Here, the larger number of guest groups per table makes it more likely that combinations of guest groups will be deemed infeasible, reducing the number of possible feasible solutions and making the presence of a zero-cost solution less likely.

Figure 7.6 now shows the effect that variations in p and k have on the neighbourhood sizes encountered in Stage 2, together with the overall run times of the algorithm. For unconstrained problem instances ($p = 0.0$) all Kempe chains are of size 1, and all pairs of vertices in different colours qualify for a pair swap. Hence, the number of distinct moves available for each operator are $n(k-1)$ and approximately $(n(n - n/k))/2$, respectively. However, for more constrained problems (lower k's and/or larger p's), the number of neighbouring solutions is lower. This means that a smaller number of evaluations need to take place at each iteration of tabu search, resulting in shorter run times. The exception to this pattern is for low values of k using $p = 0.0$, where the larger numbers of guests per table require more overheads in the calculation of Kempe chains and the cost function, resulting in increased run times.

Finally, it is also instructive to consider the proportion of Kempe chains that are seen to be *total* during runs of the tabu search algorithm. Recall from Sect. 5.5 that a Kempe chain KEMPE(v, i, j) is described as total when KEMPE$(v, i, j) = (S_i \cup S_j)$: that is, the graph induced by the set of vertices $S_i \cup S_j$ forms a connected bipartite graph. (Consider, for example, the chain KEMPE$(v_3, 4, 2)$ from Fig. 7.3.) Interchanging the colours of vertices in a total Kempe chain serves no purpose since this only results in the labels of the two colour classes being swapped, leading to no changes in the objective function. Figure 7.7 shows these proportions for the four considered problem instances. We see that total Kempe chains are more likely to occur with higher values of p (due to the greater connectivity of the graphs), and

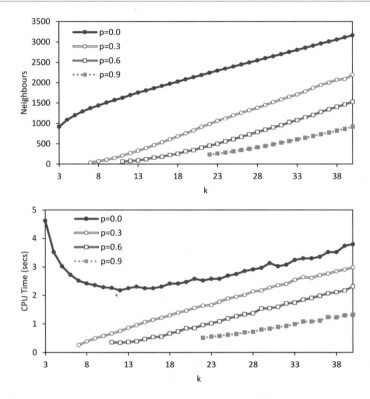

Fig. 7.6 Average number of neighbouring solutions per iteration of the tabu search algorithm (top); and average run times of the algorithm (bottom) for the four problem instances using various k-values (using a 3.0 GHz Windows 7 PC with 3.87 GB RAM)

for lower values of k (because the vertices of the graph are more likely to belong to one of the two colours being considered). Indeed, for $p = 0.9$ and $k = 22$, we see that *all* Kempe chains considered by the algorithm are total, meaning that the neighbourhood operator is ineffective in this case.

7.5 Comparison to an IP Model

In this section, we now compare the results achieved by our two-stage tabu search algorithm to those of a commercial integer programming (IP) solver. As we saw in Sect. 4.1.2, one of the advantages of using an IP approach is that, given excess time, we can determine with certainty the optimal solution to a problem instance (or, indeed, whether a feasible solution exists). In contrast to our tabu search-based method, the IP solver is, therefore, able to provide the user with a certificate of optimality and/or infeasibility, at which point it can be halted. Of course, due to the

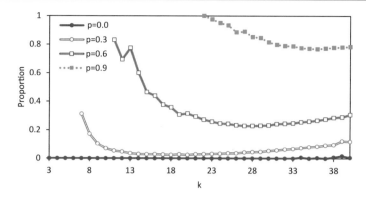

Fig. 7.7 Proportion of Kempe chains seen to be total for the four problem instances using various k-values

underlying intractability of the WSP these certificates will not always be produced in reasonable time, but given the relatively small problem sizes being considered in this chapter, it is still pertinent to ask how often this is the case and to also compare the quality of the IP solver's solutions to our tabu search approach under similar time limits.

The WSP can formulated as an IP problem as follows. Recall that there are n guest groups that we seek to partition on to k tables. Accordingly, the seating preferences of guests can be expressed using a symmetric $n \times n$ matrix $\mathbf{W}$, where:

$$W_{ij} = \begin{cases} \infty & \text{if we require guest groups } i \text{ and } j \text{ to be "definitely apart";} \\ 1 & \text{if we would prefer } i \text{ and } j \text{ to be on different tables ("rather apart");} \\ -1 & \text{if we would prefer } i \text{ and } j \text{ to be on the same table ("rather together");} \\ 0 & \text{otherwise.} \end{cases}$$

$$(7.5)$$

As before, we also let s_i define the size of each guest group $i \in \{1, \ldots, n\}$. A solution to the problem can then be represented by a $n \times n$ binary matrix $\mathbf{X}$, where:

$$X_{it} = \begin{cases} 1 & \text{if guest group } i \text{ is assigned to table } t, \\ 0 & \text{otherwise,} \end{cases} \qquad (7.6)$$

and a binary vector $\mathbf{Y}$ of length n, where:

$$Y_t = \begin{cases} 1 & \text{if at least one guest group is assigned to table } t, \\ 0 & \text{otherwise,} \end{cases} \qquad (7.7)$$

subject to the following constraints:

$$\sum_{t=1}^{n} X_{it} = 1 \qquad \forall i \in \{1, \ldots, n\} \tag{7.8}$$

$$X_{it} + X_{jt} \leq Y_t \qquad \forall i, j : W_{ij} = \infty, \forall t \in \{1, \ldots, n\} \tag{7.9}$$

$$X_{it} = 0 \qquad \forall i \in \{1, \ldots, n\}, \forall t \in \{k+1, \ldots, n\} \tag{7.10}$$

$$Y_t = 0 \qquad \forall t \in \{k+1, \ldots, n\} \tag{7.11}$$

$$X_{it} = 0 \qquad \forall i \in \{1, \ldots, n\}, \forall t \in \{i+1, \ldots, n\} \tag{7.12}$$

$$X_{it} \leq \sum_{j=t-1}^{i-1} X_{jt-1} \qquad \forall i \in \{2, \ldots, n\}, \forall t \in \{2, \ldots, i-1\}. \tag{7.13}$$

Here, Constraints (7.8)–(7.13) stipulate the hard constraints of the problem; hence, a solution satisfying these can be considered feasible. This IP formulation is essentially the same as the graph colouring formulation seen in Sect. 4.1.2.4, except that the required number of colours k is specified as a constraint. Equation (7.8) states that each guest group (vertex) should be assigned to exactly one table (colour), while (7.9) specifies that no pair of guest groups should be assigned to the same table if they are subject to a "Definitely Apart" constraint, with $Y_t = 1$ when at least one guest group has been assigned to table t. Equations (7.10) and (7.11) then ensure that a maximum of k tables are used. Finally, (7.12) and (7.13) impose the anti-symmetry constraints.

As with the tabu search algorithm, the quality of a feasible candidate solution in the IP model is quantified using the sum of the two previously defined objective functions ($f_1 + f_2$). For the IP model, f_1 is rewritten as

$$f_1 = \sum_{t=1}^{k} \sum_{i=1}^{n-1} \sum_{j=i+1}^{n} X_{it} X_{jt} (s_i + s_j) W_{ij} \tag{7.14}$$

in order to cope with the binary matrix method of solution representation; however, it is equivalent in form to Eq. (7.3). Similarly, f_2 in the IP model is defined in the same manner as Eq. (7.4), except that τ_i is now calculated as $\tau_i = \sum_{j=1}^{n} X_{jt} s_j$. Again, this is equivalent to Eq. (7.4).

It is worth noting here that the objective function defined in Eq. (7.14) actually contains a quadratic term, making our proposed mathematical model a binary *quadratic* integer program. Although modern commercial IP solvers such as Xpress and CPLEX can cope with such formulations, the use of quadratic objective functions is sometimes thought to hinder performance. One way to linearise this model is to introduce an additional auxiliary binary variable:

$$Z_{ijt} = \begin{cases} 1 & \text{if guest groups } i \text{ and } j \text{ are both assigned to table } t, \\ 0 & \text{otherwise}, \end{cases} \tag{7.15}$$

together with the following additional constraints:

$$Z_{ijt} = 0 \qquad\qquad \forall i \in \{1, \ldots, n\}, \forall j \in \{1, \ldots, i\}, \forall t \in \{1, \ldots, k\} \qquad\qquad (7.16)$$

$$Z_{ijt} = 0 \qquad\qquad \forall i \in \{1, \ldots, n\}, \forall j \in \{1, \ldots, n\}, \forall t \in \{k+1, \ldots, n\} \qquad (7.17)$$

$$Z_{ijt} \leq \frac{1}{2}(X_{it} + X_{jt}) \qquad \forall i \in \{1, \ldots, n-1\}, \forall j \in \{i+1, \ldots, n\}, \forall t \in \{1, \ldots, k\} \quad (7.18)$$

$$Z_{ijt} \geq (X_{it} + X_{jt}) - 1 \quad \forall i \in \{1, \ldots, n-1\}, \forall j \in \{i+1, \ldots, n\}, \forall t \in \{1, \ldots, k\}. \quad (7.19)$$

These constraints ensure that the correct values are assigned to the variables Z_{ijt}. They also allow us to restate the objective function f_1 in the linear form:

$$f_1 = \sum_{t=1}^{k} \sum_{i=1}^{n-1} \sum_{j=i+1}^{n} Z_{ijt}(s_i + s_j)W_{ij}. \qquad\qquad (7.20)$$

7.5.1   Results

In our experiments, both IP formulations were tested using the commercial software Xpress. We repeated the experiments of Sect. 7.4 using two time limits: 5 seconds, which was approximately the longest time required by our tabu search algorithm (see Fig. 7.6); and 600 s, to gain a broader view of the IP solver's capabilities with these formulations. Across the 152 combinations of p and k, under the five-second limit, the linear model produced feasible solutions for just 11 cases compared to the quadratic model's 112. Similarly, the number of cases where certificates of infeasibility were returned were 24 and 26, respectively. The underperformance of the linear model in these cases may well be due to the much larger number of variables and constraints involved, which seems to present difficulties under this very strict time limit. That said, even though the models' results became more similar under the 600 s time limit, the costs returned by the linear model were still consistently worse than the quadratic model's. Consequently, only the results from the quadratic model are considered for the remainder of this section.

The results of the trials are summarised in Fig. 7.8. The circled lines in the left of the graphs indicate values of k where certificates of infeasibility were produced by the IP solver under the two time limits. As might be expected, these certificates are produced for a larger range of k-values when the longer time limit is used; however, for $p \in \{0.3, 0.6\}$ there remain values of k for which feasible solutions have not been produced (by any algorithm) and where certificates of infeasibility have not been supplied. Thus, we are none the wiser as to whether feasible solutions exist for these particular k-values. Also, note that certificates of *optimality* were not provided by the IP solver in any of the trials conducted.

Figure 7.8 shows that, under the 5 s limit, the IP approach has produced solutions of inferior quality compared to tabu search in all cases. Also, the IP method has failed to achieve feasible solutions in seven of the 121 cases where tabu search has been successful. When the extended run time limit is applied, this performance gap diminishes, but similar patterns still emerge. We see that the tabu search algorithm has produced feasible solutions whenever the IP approach has, plus three further

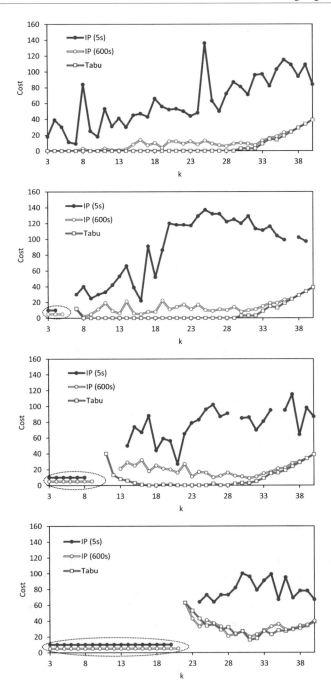

Fig. 7.8 Comparison of solution costs achieved using the IP solver (using 2 different time limits) and the tabu search-based approach for $p = 0.0$, $p = 0.3$, $p = 0.6$, and $p = 0.9$, respectively

cases. Also, in the 119 cases where both algorithms have achieved feasible solutions, tabu search has produced superior quality solutions in 94 cases, compared to the IP method's six. However, we must bear in mind that the IP solver has required more than 400 times the CPU time of tabu search to achieve these particular solutions, making it much less suitable for an online tool.

7.6 Chapter Summary and Discussion

In this chapter, we have examined the interesting combinatorial problem of constructing seating plans for large parties. As we have seen, the underlying graph colouring properties of this problem allow the design of an effective two-stage heuristic algorithm with much better performance than an equivalent IP formulation, even for fairly small problems. This algorithm might also be used for other situations where we are seeking to divide people into groups, such as birthday parties, gala dinners, team-building exercises, and group projects.

The problem formulation considered here (and used with the online tool) is chosen to strike the right balance between being useful to users and being easy to understand. As part of this, we have chosen to allow only three different weights on edges: -1, 1, and ∞, corresponding to the constraints "Rather Together", "Rather Apart", and "Definitely Apart", respectively. In practice, however, this algorithm could be applied to any set of weights. For example, we might choose to define a threshold value c, and then consider any edge $\{u, v\}$ with weight $w_{uv} \geq c$ as a hard constraint, with the remaining edges then being treated as soft constraints.[1] On the other hand, if it is preferable to treat all constraints as soft constraints, we might simply abandon Stage 1 and use the optimisation process of Stage 2 to search through the solution space comprising all k-partitions of the guest groups.

Finally, an additional advantage of using a graph colouring-based model is that it can be easily extended to incorporate "table-specific" constraints that specify which table each guest group can and cannot be assigned to. To impose such constraints, we first need to add k additional vertices to the model, one for each available table. Next, edges of weight ∞ then need to be added between each pair of these "table-vertices", thereby forming a clique of size k and ensuring that each table-vertex is assigned to a different colour in any feasible solution. Having introduced these extra vertices, we can then add other types of constraints into the model:

- If guest group v is not permitted to sit at table i, then an edge of weight ∞ can be imposed between vertex v and the ith table vertex.
- If a guest group v must be assigned to table i, then edges of weight ∞ can be imposed between vertex v and all table vertices except the ith table vertex.

---

[1] That is, the graph G' used in Stages 1 and 2 would comprise edge set $E' = \{\{u, v\} \in E : w_{uv} \geq c\}$.

Note that if our model is to be extended in this way, we will now be associating each subset of guest groups S_i in a solution $\mathcal{S} = \{S_1, \ldots, S_k\}$ with a particular table number i. Hence, we might also permit tables of different sizes and shapes into the model, perhaps incorporating constraints concerning these factors into the objective function. Extensions of this nature will be considered with a different problem in the next chapter.

References

1. Nielsen J (2004) The need for web design standards. https://www.nngroup.com/articles/the-need-for-web-design-standards/, September 2004
2. Carroll F, Lewis R (2013) The "engaged" interaction: important considerations for the HCI design and development of a web application for solving a complex combinatorial optimization problem. World J Comput Appl Technol 1(3):75–82

Designing Sports Leagues 8

In this chapter, our case study considers the applicability of graph colouring methods for producing *round-robin* tournaments. These are particularly common in sports competitions. As we will see, the task of producing valid round-robin tournaments is relatively straightforward, but things become more complicated when additional constraints are added to the problem. The initial sections of this chapter focus on the problem of producing round-robins in general terms and examine the relationship between this problem and graph colouring. A detailed real-world case study that makes use of various graph colouring techniques is then presented in Sect. 8.6.

8.1 Problem Background

Round-robin schedules are used in many sports tournaments and leagues across the globe, including the Six Nations Rugby Championships, various European and South American domestic soccer leagues, and the England and Wales County Cricket Championships. Round-robins are schedules involving t teams, where each team is required to play all other teams exactly l times within a fixed number of *rounds*. The most common types are *single* round-robins, where $l = 1$, and *double* round-robins, where $l = 2$. In the latter, teams are typically scheduled to meet once in each other's home venue.

Usually, the number of teams in a round-robin schedule will be even. In cases where t is odd, an extra "dummy team" can be introduced, and teams assigned to play this dummy team will receive a bye in the appropriate part of the schedule.

Definition 8.1 Round-robin schedules involving t teams are considered *valid* if each team competes at most once per round. They are also described as *compact* if the number of rounds used is minimal at $l(t - 1)$, thus implying $t/2$ matches per round.

We saw in Chap. 6 that compact round-robin schedules can be constructed for any number of teams t by simply making use of Kirkman's circle method (see Theorem 6.10). In addition to this, it is also known that the number of *distinct* round-robin schedules grows exponentially with the number of teams t, since this figure is monotonically related to the number of nonisomorphic one-factorisations of the complete graph K_t.

Definition 8.2 Let K_t be the complete graph with t vertices, where t is even. A *one-factor* of K_t is a perfect matching. A *one-factorisation* of K_t is a partition of the edges into $t - 1$ disjoint one-factors.

An example one-factorisation of K_6, comprising five one-factors, is illustrated in Fig. 6.8. For $t \in \{2, 4, 6\}$ there is just one nonisomorphic one-factorisation available. These numbers then rise to 6, 396, 526,915,620, and 1,132,835,421,602,062,347 for $t = 8$ to 14 respectively [1]. Such a growth rate—combined with the fact that many different round-robins can be generated from a particular one-factorisation by relabelling the teams and reordering the rounds—implies that the enumeration of all round-robin schedules will not be possible in reasonable time for non-trivial values of t.

In addition to Kirkman's circle method, other algorithms of linear complexity are also available for quickly producing valid compact round-robin schedules. The "Greedy round-robin" algorithm, for example, operates by arranging matches into lexicographic order: $\{1, 2\}, \{1, 3\}, \{1, 4\}, \ldots, \{t - 1, t\}$. The first match is then assigned to the first round, and each remaining match is then considered in turn and assigned to the next round where no clash occurs. When the final round is reached, the algorithm loops back to the first round. The solution produced by this method for $t = 8$ teams is given in Fig. 8.1. Note that if a double round-robin is required, the second half of the schedule can be produced by simply copying the first half.

Another linear-complexity method for producing compact round-robin schedules is the "canonical" round-robin algorithm of de Werra [2]. Unlike the circle and Greedy methods, this approach focuses on the issue of deciding whether teams should play at home or away. In this case, a "break" is defined as a situation where a team is required to play two home matches (or away matches) in consecutive rounds, and it is proven that this method always achieves the minimum number of $t - 2$ breaks. Pseudocode for this method is shown in Fig. 8.2. Note that, in this case, matches are denoted by an ordered pair with the first and second elements denoting the home- and away teams, respectively. The canonical schedule for $t = 8$ is also given in Fig. 8.3.

Fig. 8.1 Single round-robin schedule produced by the Greedy round-robin algorithm for $t = 8$ teams

Round Matches

$r_1:$ $\{\{1,2\}, \{3,7\}, \{4,6\}, \{5,8\}\}$
$r_2:$ $\{\{1,3\}, \{2,8\}, \{4,7\}, \{5,6\}\}$
$r_3:$ $\{\{1,4\}, \{2,3\}, \{5,7\}, \{6,8\}\}$
$r_4:$ $\{\{1,5\}, \{2,4\}, \{3,8\}, \{6,7\}\}$
$r_5:$ $\{\{1,6\}, \{2,5\}, \{3,4\}, \{7,8\}\}$
$r_6:$ $\{\{1,7\}, \{2,6\}, \{3,5\}, \{4,8\}\}$
$r_7:$ $\{\{1,8\}, \{2,7\}, \{3,6\}, \{4,5\}\}$

Fig. 8.2 Procedure for producing a canonical single round-robin schedule for t teams, where t is even

MAKE-CANONICAL-SCHEDULE (t)

(1) **for** $i \leftarrow 1$ **to** $n - 1$ **do**
(2)   **if** i is odd **then** assign (i, n) to round r_i
(3)   **else** assign (n, i) to round r_i
(4)   **for** $j \leftarrow 1$ **to** $n/2 - 1$ **do**
(5) $x \leftarrow (i + j) \bmod (n - 1)$
(6) $y \leftarrow (i - j) \bmod (n - 1)$
(7)   **if** $x = 0$ **then** $x \leftarrow n - 1$
(8)   **if** $y = 0$ **then** $y \leftarrow n - 1$
(9)   **if** j is odd **then** assign (x, y) to round r_i
(10)   **else** assign (y, x) to round r_i

The task of minimising breaks has also been explored in other research. Trick [3], Elf et al. [4] and Miyashiro and Matsui [5], for example, have examined the problem of taking an *existing* single round-robin and then assigning home/away values to each of the matches to minimise the number of breaks. Miyashiro and Matsui [6] have also shown that the problem of deciding whether a home/away assignment exists for a particular schedule such that the theoretical minimum of $t - 2$ breaks is achieved is computable in polynomial time. The inverse of this problem—taking a fixed home/away pattern and then assigning matches consistent with this pattern—has also been studied by various other authors [2,7,8].

One interesting feature of the Greedy, circle, and canonical methods is that the solutions they produce are isomorphic. For example, a canonical single round-robin schedule for t teams can be transformed into the schedule produced by the Greedy round-robin algorithm by simply converting the ordered pairs into unordered pairs and then reordering the rounds. Similarly, a circle schedule can be transformed into a Greedy schedule by relabelling the teams using the mapping $t_1 \leftarrow t_t$ and $t_i \leftarrow t_{i-1} \ \forall i \in \{2, \ldots, t\}$, with the rounds then being reordered (see also [9]).

8.1.1 Further Round-Robin Constraints

Although it is straightforward to construct valid, compact round-robin schedules using constructive methods such as the Greedy, circle, and canonical algorithms,

Fig. 8.3 Canonical
round-robin schedule for
$t = 8$ teams

Round Matches
r_1 : $\{(1,8), (2,7), (4,5), (6,3)\}$
r_2 : $\{(3,1), (5,6), (7,4), (8,2)\}$
r_3 : $\{(1,5), (3,8), (4,2), (6,7)\}$
r_4 : $\{(2,6), (5,3), (7,1), (8,4)\}$
r_5 : $\{(1,2), (3,7), (5,8), (6,4)\}$
r_6 : $\{(2,3), (4,1), (7,5), (8,6)\}$
r_7 : $\{(1,6), (3,4), (5,2), (7,8)\}$

in practical circumstances it is often the case that the production of "high-quality" schedules will depend on additional user requirements and constraints. As we have seen, the minimisation of "breaks" is one example of this. Other constraints, however, can include those associated with demands from broadcasters, various economical and logistical factors, inter-team politics, policing, and the perceived fairness of the league. Also, factors such as the type of sport, the level of competition, and the country (or countries) involved will also play a part in determining what is considered "high-quality". An impression of the wide range of such constraints and requirements can be gained by considering the variety of round-robin scheduling problems that have previously been tackled in the literature, including German, Austrian, and Italian soccer leagues [10, 11], New Zealand basketball leagues [12], amateur tennis tournaments [13], English county cricket fixtures [14], and American professional ice hockey [15]. A good survey on the wide range of problems and solution methods for sports scheduling is also provided by Kendall et al. [16].

An oft-quoted example of such requirements is the issue of *carryover* in round-robin schedules. This considers the possibility of a team's performance being influenced by its opponents in previous rounds. For example, if t_i is known to be a very strong team whose opponents are often left injured or demoralised, then a team that plays t_i's opponents in the next round may well be seen to gain an advantage.[1] In such cases the aim is to therefore produce a round-robin schedule in which the overall effects of carryover are minimised. This requirement was first considered by Russell [17], who proposed a constructive algorithm able to produce provably optimal schedules in cases where the number of teams t is a power of 2. More recently, methods for small-sized problems have also been proposed by Trick [3] and Henz et al. [18], both of whom make use of constraint programming techniques.

Another set of schedule requirements is encapsulated in the *travelling tournament problem* (TTP). Originally proposed by Easton et al. [19], in this problem a compact double round-robin schedule is required where teams play each other twice, once in each other's home venue. The overriding aim is to then minimise the *distances travelled* by each team. Geographical constraints like this are particularly relevant in large countries such as Brazil and the USA where the match venues are often

---

[1] As an illustration, in Fig. 8.1 we see that team 2, for instance, is scheduled to play the opponents of team 1 from the previous round on five different occasions. This feature also exists for other teams; thus this schedule contains rather a large amount of carryover.

far apart. Consequently, when a team is scheduled to attend a succession of away matches, instead of returning to their home city after each match, the team travels directly to their next away venue. Note that in addition to the basic round-robin scheduling constraints, this problem also contains elements of the travelling salesman problem, as we are interested in scheduling runs of successive away matches for each team such that they occur in venues that are close to one another. An early solution method proposed for this formulation was proposed by Easton et al. [20], who used integer- and constraint-programming techniques. Subsequent proposals, however, focussed on metaheuristic techniques, and in particular, neighbourhood search-based algorithms. Good examples of these include the simulated annealing approaches of Anagnostopoulos et al. [21] and Lim et al. [22], and the local-search approaches of Di Gaspero and Schaerf [23] and Ribeiro and Urrutia [24]. In the latter study the authors also consider the added restriction that the double round-robin schedule should be "mirrored": that is, if teams t_i and t_j play in t_i's home stadium in round $r \in \{1, \ldots, t - 1\}$, then t_i and t_j should necessarily play each other in t_j's home stadium in round $r + (t - 1)$.

8.1.2 Chapter Outline

The above paragraphs illustrate that the requirements of sports scheduling problems can be complex and idiosyncratic. In the remainder of this chapter, we will examine how graph colouring concepts can be used to help find solutions to such problems. In the next section, we will describe how basic round-robin scheduling problems can be represented as graph colouring problems. In Sect. 8.3 we then assess the "difficulty" of solving such graphs using our suite of graph colouring algorithms from Chap. 5. Following this, in Sect. 8.4 we will then discuss ways in which this model can be extended to incorporate other types of "hard" (i.e., mandatory) constraint, and in Sect. 8.5 we review various neighbourhood operators that can be used with this extended model for exploring the space of *feasible* solutions (that is, round-robin solutions that are compact, valid, and also obey any imposed hard constraints). Finally, in Sect. 8.6 we consider a real-world round-robin scheduling problem from the Welsh Rugby Union and propose two separate algorithms that make use of our proposed algorithmic operators. The performance of these algorithms is then analysed over several different problem instances.

8.2 Representing Round-Robins as Graph Colouring Problems

Round-robin scheduling problems can be represented as graph colouring problems by considering each match as a vertex, with edges then being added between any pair of matches that cannot be scheduled in the same round (i.e., matches featuring a common team). Colours then represent the individual rounds of the schedule, and the task is to colour the graph using k colours, where k represents the number of

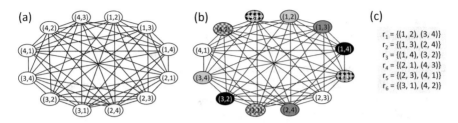

Fig. 8.4 Graph for a double round-robin problem with $t = 4$ teams (**a**), an optimal colouring of this graph (**b**), and the corresponding schedule (**c**)

available rounds. Note that for the remainder of this chapter we only consider the task of producing compact schedules: thus, $k = \chi(G)$ unless otherwise specified.

For a single round-robin, each vertex is associated with an unordered pair $\{t_i, t_j\}$, denoting a match between teams t_i and t_j. The number of vertices n in such graphs is thus $\frac{1}{2}t(t-1)$, with $\deg(v) = 2(t-2)$ $\forall v \in V$. For a compact schedule, the number of available colours $k = t - 1$. For double round-robins the number of vertices $n = t(t-1)$, $\deg(v) = 4(t-2) + 1$ $\forall v \in V$, and $k = 2(t-1)$, since teams will play each other twice. In this case, each vertex is associated with an ordered pair $(t_i, t_{j \neq i})$, with t_i denoting the home-team and t_j the away-team. An example graph for a double round-robin with $t = 4$ teams is provided in Fig. 8.4.

Recall from Sect. 6.2 that the complete graph K_t can also be used to represent a round-robin scheduling problem by associating each vertex with a team and each edge with a match. In such cases, the task is to find a proper edge colouring of K_t, with all edges of a particular colour indicating the matches that occur in a particular round. The graphs generated using our methods above are the corresponding *line graphs* of these complete graphs. Of course, in practice, it is easy to switch between these two representations. However, the main advantage of using our representation is that it allows the exploitation of previously developed vertex-colouring techniques, as the following sections will demonstrate.

8.3 Generating Valid Round-Robin Schedules

Having defined the basic structures of the "round-robin graphs" that we wish to colour, in this section we investigate whether such graphs constitute difficult-to-colour problem instances. Note that by k-colouring such graphs we are doing nothing more than producing valid, compact round-robin schedules which, as we have mentioned, can be easily achieved using the circle, Greedy, and canonical algorithms. However, there are several reasons why solving these problems from the perspective of vertex colouring is worthwhile.

1. Because of the structured, deterministic way in which the circle, Greedy and canonical methods operate, their range of output will only represent a very small part of the space of all valid round-robin schedules.
2. The schedules that are produced by the circle, Greedy and canonical methods also occupy very *particular* parts of the solution space. For example, Lambrechts et al. [25] have shown that, for even numbers of teams, the circle method produces schedules in which the amount of carryover is maximised.
3. As noted earlier, the solutions produced via the Greedy, circle and canonical methods are isomorphic. Moreover, the specific structures present in these isomorphic schedules are often seen to have adverse effects when applying neighbourhood search operators, as we will see in Sect. 8.5.
4. Finally, we are also able to modify the graph colouring model to incorporate additional real-world constraints, as shown in Sect. 8.4.

By using graph colouring methods, particularly those that are stochastic in nature, the hope is that we have a more robust and less biased mechanism for producing round-robin schedules, allowing a larger range of structurally distinct schedules to be sampled. This is especially useful in the application of metaheuristics, where the production of random initial solutions is often desirable.

Figure 8.5 summarises the results of experiments using single and double round-robins of up to $t = 60$ teams. Fifty runs of the backtracking and hybrid evolutionary algorithms were executed in each case using a computation limit of 5×10^{11} constraint checks as before. The success rates in these figures give the percentage of these runs where optimal colourings (compact valid round-robins) were produced. It is obvious from these figures that the HEA is very successful here, featuring 100% success rates across all instances. Indeed, no more than 0.006% of the computation limit on average was required for any of the values of t tested. On the other hand, the backtracking approach experiences more difficulty, with success rates dropping considerably for larger values of t. That said, when the algorithm *does* produce optimal solutions it does so quickly, indicating that solutions are either found early in the search tree or not at all.[2] The success of the HEA with these instances is also reinforced by the fact that its solutions are very diverse, as illustrated in the figure.

8.4   Extending the Graph Colouring Model

Another advantage of transforming the task of round-robin construction into a type of graph colouring problem is that we can easily extend the model to incorporate other

---

[2]On this point, Lewis and Thompson [26] have also found that much better results for the backtracking algorithm on these particular graphs can be achieved by restricting the algorithm to only inspect one additional branch from each node of the search tree. The source code available for this algorithm—see Appendix A.1—can easily be modified to allow this.

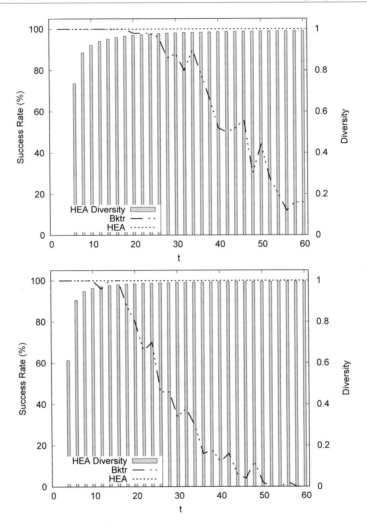

Fig. 8.5 Success rates of the Backtracking and HEA algorithms for finding optimal colourings with, respectively, single round-robin graphs for $t = 2, \ldots, 60$ ($n = 1, \ldots, 1770$), and double round-robin graphs ($n = 2, \ldots, 3540$). All figures are averaged across 50 runs. The bars show the diversity of solutions produced by the HEA across the 50 runs, calculated using Eq. (5.9)

types of sports scheduling constraints. In this section, we specifically consider the imposition of *round-specific* constraints, which specify the rounds that matches can and cannot be assigned to. In many practical cases, round-specific constraints will be a type of hard constraint—that is, they will be mandatory in their satisfaction— and candidate solutions that violate such constraints will be considered infeasible. Encoding such constraints directly into the graph colouring model allows us to attach an importance to these constraints that are equal to the basic round-robin constraints themselves.

Fig. 8.6 Method of
extending the graph
colouring model to
incorporate round-specific
constraints. "Match-vertices"
appear on the left;
"round-vertices" on the right

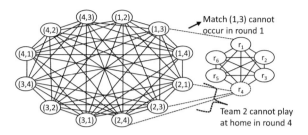

To impose round-specific constraints we follow the method seen in Sect. 6.7 for
the list colouring problem. First, k extra vertices are added to the model, one for
each available round. Next, edges are then added between all pairs of these "round-
vertices" to form a clique of size k, ensuring that each round-vertex will be assigned
to a different colour in any feasible solution. Having introduced these extra vertices,
a variety of different round-specific constraints can then be introduced:

- *Match-Unavailability Constraints.* Often it will be impossible to assign a match
  to a particular round, perhaps because the venue of the match is being used for
  another event, or because league rules state that the associated teams should not
  play each other in specific rounds in the league. Such constraints are introduced
  by adding an edge between the relevant match-vertex and round-vertex.
- *Preassignment Constraints.* In some cases, a match will need to be assigned to
  a specific round r_i, e.g., to increase viewing figures and/or revenue. For such
  constraints, edges are added between the appropriate match-vertex and all round-
  vertices except for the round-vertex corresponding to r_i.
- *Concurrent Match Constraints.* In some cases it might also be undesirable for
  two matches (t_i, t_j) and (t_l, t_m) to occur in the same round. For example, the two
  home teams t_i and t_l may share a stadium, or perhaps it is forbidden for rival fans
  of certain teams to visit the same city on the same day. Such constraints can be
  introduced by assigning an edge between the appropriate pairs of match vertices.

Figure 8.6 gives two examples of how we can impose such constraints. Any feasible k-coloured solution for such graphs will constitute a valid compact round-
robin schedule that obeys the imposed round-specific constraints. As we will see in
Sect. 8.5, incorporating constraints in this fashion also allows us to apply neighbourhood operators stemming from the underlying graph colouring model that ensures
these extra constraints are never re-violated. Note that an alternative strategy for coping with hard constraints such as these is to *allow* their violation within a schedule,
but to then penalise their occurrence via a cost function. Anagnostopoulos et al. [21],
for example, use a strategy whereby the space of all valid compact round-robins is
explored, with a cost function then being used that reflects the number of hard *and*
soft constraint violations. Weights are then used to place a higher penalty on violations of the hard constraints, and it is hoped that by using such weights the search
will eventually move into areas of the solution space where no hard constraint viola-

tions occur. The choice of which strategy to employ will depend largely on practical requirements.

To investigate the effects that the imposition of round-specific constraints has on the difficulty of the underlying graph colouring problem, double round-robin graphs were generated with varying numbers of match-unavailability constraints. Specifically, these constraints were added by considering each match-vertex/round-vertex pair in turn and adding edges between them with probability p. This means, for example, that if $p = 0.5$, each match can only be assigned to approximately half of the available rounds. Graphs were also generated in two ways: one where $k = \chi(G)$ was ensured (by referring to a pre-generated valid round-robin), and one where this matter was ignored, possibly resulting in graphs for which $\chi(G) > k$.

Note that by adding edges in this binomially distributed manner, the expected degrees of each vertex can be calculated in the following way. Let V_1 define the set of match vertices and V_2 the set of round-vertices, and let $v \in V_1$ and $u \in V_2$. Then:

$$\begin{aligned}
\mathbb{E}(\deg(v)) &= 4(t - 2) + 1 + p \times |V_2| \; \forall v \in V_1, \text{ and} \\
\mathbb{E}(\deg(u)) &= 2(t - 1) - 1 + p \times |V_1| \; \forall u \in V_2.
\end{aligned} \tag{8.1}$$

The expected variance in degree across all vertices $V = V_1 \cup V_2$, where $n = |V|$, is thus approximated as

$$\frac{|V_1| \times \mathbb{E}(\deg(v))^2 + |V_2| \times \mathbb{E}(\deg(u))^2}{n} - \left(\frac{|V_1| \times \mathbb{E}(\deg(v)) + |V_2| \times \mathbb{E}(\deg(u))}{n} \right)^2. \tag{8.2}$$

The effect that p has on the overall degree coefficient of variation (CV) of these graphs is demonstrated in Fig. 8.7. As p is increased from zero, $\mathbb{E}(\deg(v))$ and $\mathbb{E}(\deg(u))$ initially become more alike, resulting in a slight drop in the CV. However, as p is increased further, $\mathbb{E}(\deg(u))$ rises more quickly than $\mathbb{E}(\deg(v))$, resulting in large increases to the CV.

The consequences of these specific characteristics help to explain the performance of our six graph colouring algorithms across a large number of instances, as shown in Fig. 8.8. As with the results from Chap. 5, the quality of solution achieved by TABU-COL and PARTIALCOL is observed to be substantially worse than that of the other approaches when p, and therefore the degree CV, is high. In particular, TABUCOL shows very disappointing performance, providing the worst-quality results for both graph sizes where the CV is $\gtrsim 40\%$.

In contrast, some of the best performance across the instances is once again due to the HEA. Surprisingly, ANTCOL also performs well here, with no significant difference being observed in the mean results of the HEA and ANTCOL algorithms across the set. The reasons for the improved performance of ANTCOL, particularly with denser graphs, seems due to two factors: (a) the higher degrees of the vertices in the graphs, and (b) the high variance in degrees. In ANTCOL's BUILDSOLUTION procedure (Sect. 5.4) the first factor naturally increases the influence of the heuristic value η in Eq. (5.4), while the second allows a greater discrimination between vertices. In these cases it seems that a favourable balance between heuristic and pheromone information is being struck, allowing ANTCOL's global operator to effectively contribute to the search.

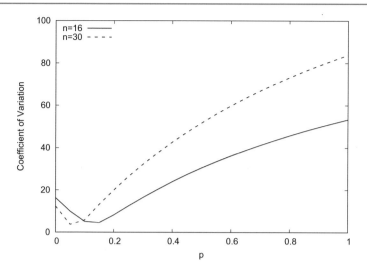

Fig. 8.7 Effect of varying p on the degree coefficient of variation with double round-robin graphs of size $t = 16$ and 30

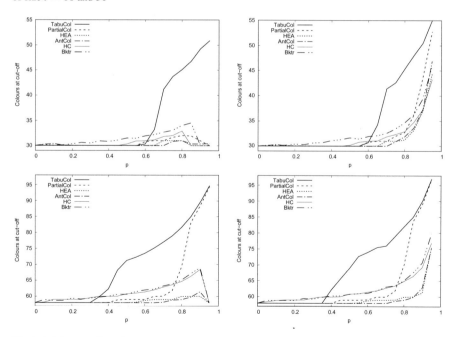

Fig. 8.8 Mean quality of solutions achieved with double round-robin graphs using (respectively): $t = 16$, ($n = 270$, $k = 30$) with $\chi(G) = 30$; $t = 16$ with $\chi(G) \geq 30$; $t = 30$, ($n = 928$, $k = 58$) with $\chi(G) = 58$; and $t = 30$ with $\chi(G) \geq 58$. All points are the average of 25 runs on 25 graphs

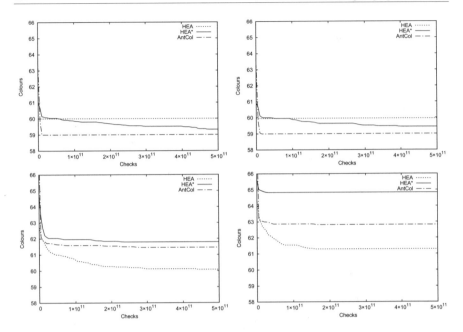

Fig. 8.9 Run profiles for double round-robins with $t = 30$ ($n = 928$) using, respectively: $p = 0.8$, $\chi(G) = 58$; $p = 0.8$, $\chi(G) \geq 58$; $p = 0.9$, $\chi(G) = 58$; and $p = 0.9$, $\chi(G) \geq 58$. HEA* denotes the HEA algorithm with a reduced local-search limit of $I = 2n$

However, if we examine individual values of p, the picture becomes more complicated. Figure 8.9, for example, shows run profiles of AntCol and HEA on four sets of large, highly constrained round-robin graphs. For comparison, the variant HEA* is also included here, which features a local-search iteration limit equal to AntCol's (i.e., I is reduced from $16n$ to $2n$). For $p = 0.8$, we see that AntCol quickly produces the best-observed results, with HEA consistently requiring an additional colour. However, reducing the local-search element of the HEA—thus placing more emphasis on global search—seems to improve performance, though the results of HEA* are still inferior to those of AntCol. On the other hand, for $p = 0.9$ the picture is reversed, with HEA (using $I = 16n$) producing the best results, suggesting that increased amounts of local search are beneficial in this case.

Despite these complications, however, the results of Fig. 8.9 allow us to conclude that methods such as AntCol and the HEA provide useful mechanisms for producing feasible compact round-robin schedules, even in the presence of high levels of additional constraints.

Table 8.1 Description of various neighbourhood operators that preserve the validity and compactness of double round-robin schedules. "Move Size" refers to the number of matches (vertices in the graph colouring model) that are affected by the application of these operators

	Description	Move size
N_1	Select two teams, $t_i \neq t_j$, and swap the rounds of vertices (t_i, t_j) and (t_j, t_i)	2
N_2	Select two teams, $t_i \neq t_j$, and swap their opponents in all rounds	$2(k-2)$
N_3	Select two teams, $t_i \neq t_j$, and swap all occurrences of t_i to t_j and all occurrences of t_j to t_i	$2k-2$
N_4	Select two rounds, $r_i \neq r_j$ and swap their contents	t
N_5	Select a match and move it to a new round. Repair the schedule using an ejection chain repair procedure	Variable

8.5   Exploring the Space of Round-Robins

Upon production of a valid round-robin schedule, we may now choose to apply one or more neighbourhood operators to try and eliminate occurrences of any remaining soft-constraint violations. Table 8.1 lists several neighbourhood operators that have been proposed for round-robin schedules, mostly for use with the travelling tournament problem [21, 24, 27]. Note that the information given in this table applies to double round-robins, so the number of rounds $k = 2(t-1)$; however, simple adjustments can be made for other cases.

A point to note about these operators is that while they all preserve the validity of a round-robin schedule, they will not be useful in all circumstances. For example, applications of N_1, N_2, and N_3 will not affect the amount of carryover in a schedule. Also, while applications of N_2 can change the home/away patterns of individual teams, they cannot alter the *total* number of breaks in a schedule. Finally, perhaps the most salient point from our perspective is that if extra hard constraints are being considered, such as the round-specific constraints listed in Sect. 8.4, then the application of such operators may lead to schedules that, while valid, are not necessarily feasible.

Pursuing the relationship with graph colouring, a promising strategy for exploring the space of round-robin schedules is again presented by the Kempe chain interchange operator (see Definition 4.3). Of course, because this operator is known to preserve the feasibility of a graph colouring solution, it is suitable for both the basic and extended versions of our graph colouring model. On the other hand, the pair-swap operator (Definition 4.4) is not suitable here because, for these graphs, swapping the colours of two nonadjacent vertices is equivalent to swapping the rounds of a pair of matches with no common team. Such moves will never maintain the feasibility of a round-robin and therefore should not be considered.

Recall that the number of vertices affected by a Kempe chain interchange can vary. For basic (non-extended) round-robin colouring problems involving t teams, the largest possible move involves t vertices (i.e., two colour classes, with $t/2$ vertices in each). In Fig. 8.10 we illustrate what we have found to be typical-shaped distri-

Fig. 8.10 Distribution of
differently sized Kempe
chains for $t = 10, 20$, and 40
for SRRs and DRRs

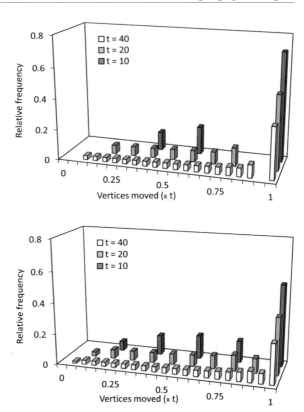

butions of the differently sized Kempe chains with single and double round-robins. These examples were gained by generating initial solutions with our graph colouring algorithms and then performing random walks of 10^6 neighbourhood moves. We see that, in the case of double round-robins, the smallest moves involve exactly two vertices, which only occurs when a chain is formed containing the complementary match vertices (t_i, t_j) and (t_j, t_i). In this case, the Kempe chain move is equivalent to the operator N_1 (Table 8.1) and it occurs with a probability $\frac{1}{k-1}$ (obviously moves of size 2 do not occur with single round-robins because a match does not have a corresponding reverse fixture with which to be swapped). Meanwhile, the most probable move in both cases is a total Kempe chain interchange (i.e., involving all vertices of the two associated colours). Moves of this size are equivalent to a corresponding move in N_4 and occur when all vertices in the two colours form a connected component. Such moves appear to be quite probable due to the relatively high edge densities of the graphs. Importantly, however, we see that for larger values of t the majority of moves are of sizes between these two extremes, resulting in moves that are beyond those achievable with neighbourhood operators N_1 and N_4.

We may also choose to perform random walks in this way from schedules generated by the circle, Greedy, or canonical algorithms. However, due to the structured way in which these go about constructing a schedule, many different values of t result in single round-robins in which *all* applications of the Kempe chain interchange operator are of size t. Such solutions are usually termed *perfect* one-factorisations and are known to be produced by the circle, Greedy, and canonical algorithms for any value of t for which $t - 1$ is a prime number [28]. Such features are undesirable as they do not allow the Kempe chain interchange operator to produce moves beyond what can already be achieved using N_1 and N_4, limiting the number of solutions accessible via the operator. We should note, however, that we found that this problem could be circumnavigated in some cases by applying neighbourhood operator N_5 from Table 8.1 to the solution. It seems that, unlike the other operators detailed in this table, N_5 has the potential of breaking up the structural properties of these solutions, allowing the Kempe chain distributions to assume their more "natural" shapes as seen in Fig. 8.10. However, we still found cases where this situation was not remedied.[3]

One of the main reasons why an analysis of move sizes is relevant here is because of the effects that the size of a move can have on the cost of a solution at different stages of the optimisation process. On the one hand, "large" moves can facilitate the exploration of wide expanses of the solution space and can provide useful mechanisms for escaping local optima. On the other hand, when relatively good candidate solutions are being considered, large moves will also be disruptive, usually worsening the quality of a solution as opposed to improving it. These effects are demonstrated in Fig. 8.11 where we illustrate the relationship between the size of a move and the resultant change in an arbitrary cost function. In the top chart graph, the Kempe chain interchange operator has been repeatedly applied to a solution that was randomly produced by one of our graph colouring algorithms. Note that larger moves here tend to give rise to greater variance in cost, but that many moves lead to improvements. In contrast, in the bottom chart the effects of the Kempe chain interchange operator on a relatively "good" solution (which has a cost of approximately a quarter of the previous one) are demonstrated. Here, larger moves again feature a larger variance in cost, but we also witness a statistically significant medium positive correlation ($r = 0.46$), demonstrating that larger moves tend to be associated with larger decreases in solution quality. Di Gaspero and Schaerf [23] have also noted the latter phenomenon (albeit with different neighbourhoods and a different cost function) and have suggested a modification to their neighbourhood search algorithm whereby any move seen to be above a specific size is automatically rejected, with no evaluation taking place. Because such moves lead to a degradation in quality and will therefore be rejected in the majority of cases, they find that their algorithm's performance over time is increased by skipping these mostly unnecessary evaluations. On the flip side, of course, such a strategy also eliminates the possibility of "larger" moves occurring which could diversify the search in a useful way.

---

[3]Specifically for SRRs and values of t less than fifty, this was seen to occur with $t = 12, 14, 20, 30$, and 38.

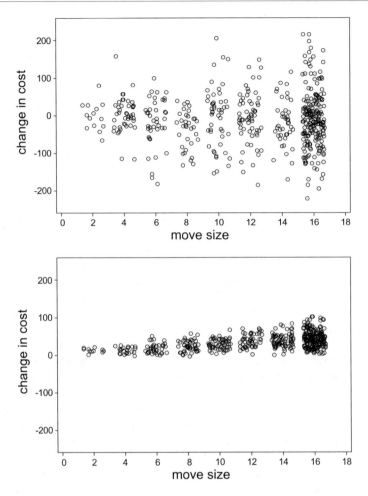

Fig. 8.11 Demonstrating how Kempe chain interchanges of different sizes influence the change in cost of a randomly generated solution (top) and a "good" solution (bottom). In both cases a double round-robin with $t = 16$ teams was considered using cost function c_2 defined in Sect. 8.6.1 (negative changes thus reflect an improvement)

8.6 Case Study: Welsh Premiership Rugby

In this section, we now present a real-world application of the graph colouring-based techniques introduced in this chapter. This problem was provided to us by the Welsh Rugby Union (WRU), based at the Millennium Stadium in Cardiff, which is the governing body for all rugby competitions, national and international, in Wales. The particular problem that we are concerned with is the *Principality Premiership* league, which is the highest level domestic league in the country.

The Principality Premiership problem involves t teams playing in a compact double round-robin tournament. Several round-specific (hard) constraints are also stipulated, all of which are mandatory in their satisfaction.

- *Hard Constraint A.* Some pairs of teams in the league share a home stadium. Therefore when one of these teams plays at home, the other team must play away.
- *Hard Constraint B.* Some teams in the league also share their stadia with teams from other leagues and sports. These stadia are therefore unavailable in certain rounds. (In practice, the other sports teams using these venues have their matches scheduled before the Principality Premiership teams, and so unavailable rounds are known in advance.)
- *Hard Constraint C.* Matches involving regional rivals (so-called "derby matches") need to be preassigned to two specific rounds in the league, corresponding to those falling on the Christmas and Easter weekends.

The league administrators also specify two soft constraints. First, they express a preference for keeping reverse fixtures (i.e., matches (t_i, t_j) and (t_j, t_i)) at least five rounds apart and, if possible, for reverse fixtures to appear in opposite "halves" of the schedule (they do not consider the stricter requirement of "mirroring" to be important, however). Second, they also express a need for all teams to have good home/away patterns, which means avoiding breaks wherever possible.

8.6.1 Solution Methods

In this section, we describe two algorithms for this scheduling problem. Both of these use the strategy of first producing a feasible solution, followed by a period of optimisation via neighbourhood search in which feasibility (i.e., validity, compactness, and adherence to all hard constraints) is maintained. Specific details of these methods together with a comparison are given in the next three subsections.

In both cases, initial feasible solutions are produced by encoding all of the hard constraints using the extended graph colouring model from Sect. 8.4, with one of our graph colouring algorithms then being applied. For Hard Constraint A, if a pair of teams t_i and t_j is specified as sharing a stadium then edges are simply added between all match-vertices corresponding to home matches of these teams. For Hard Constraint B, if a venue is specified as unavailable in a particular round, then edges are added between all match-vertices denoting home matches of the venue's team(s) and the associated round-vertex. Finally, for Hard Constraint C, edges are also added between vertices corresponding to derby matches and all round-vertices except those representing derby weekends.

Details of the specific problem instance faced at the WRU are given in bold in Table 8.2. To aid our analysis we also generated (artificially) a further nine instances of comparable size and difficulty, details of which are also given in the table. As it turned out, we found that it was quite straightforward to find a feasible solution to the WRU problem using the graph colouring algorithms from Chap. 5. For our

Table 8.2 Summary of the sports scheduling problem instances used. The entry in bold refers to the real-world WRU problem. These problems can be downloaded from http://www.rhydlewis.eu/resources/PrincipalityPremProbs.zip

#	Teams t	Vertices n	Graph Density	A^a	B^b	C^c
1	12	154	0.268	0	2 {5, 5}	3
2	12	154	0.292	1	3 {6, 8, 10}	6
3	12	154	0.308	2	4 {3, 6, 8, 10}	6
4	14	208	0.236	0	2 {4, 5}	4
5	**14**	**208**	**0.260**	**1**	**3 {8, 10, 10}**	**7**
6	14	208	0.271	2	5 {3, 6, 8, 10, 10}	7
7	16	270	0.219	1	3 {4, 5, 6}	5
8	16	270	0.237	2	5 {3, 6, 8, 10, 10}	8
9	18	340	0.194	1	3 {4, 5, 6}	6
10	18	340	0.212	2	6 {4, 5, 6, 7, 10, 10}	9

[a] Number of pairs of teams sharing a stadium
[b] Number of teams sharing a stadium with teams from another league/sport. The number of match-unavailability constraints for each of these teams is given in { }'s.
[c] Number of local derby pairings

experiments, we, therefore, ensured that all artificially generated problems also feature at least one feasible solution. However, the minimum number of soft-constraint violations achievable in these problems is not known.

The soft constraints of this problem are captured in two cost functions, c_1 and c_2. Both of these need to be minimised.

- *Spread Cost* (c_1): Here, a penalty of 1 is added each time a match (t_i, t_j) and its return fixture (t_j, t_i) are scheduled in rounds r_p and r_q, such that $|r_p - r_q| \leq 5$. In addition, a penalty of $\frac{1}{1/2 \times t(t-1)}$ is also added each time matches (t_i, t_j) and (t_j, t_i) are scheduled to occur in the same half of the schedule.
- *Break Cost* (c_2): Here, the home/away pattern of each team is analysed in turn and penalties of b^l are incurred for each occurrence of l consecutive breaks. In other words, if a team is required to play two home matches (or away matches) in succession, this is considered as one break and incurs a penalty of b^1. If a team has three consecutive home matches (or away matches), this is considered as two consecutive breaks and results in a penalty of b^2 being added, and so on.

The term $\frac{1}{1/2 \times t(t-1)}$ is used as part of c_1 to ensure that the total penalty due to match pairs occurring in the same half is never greater than 1, thus placing a greater emphasis on keeping matches and their return fixtures at least five rounds apart. The penalty unit of b^l in cost function c_2 is also used to help discourage long breaks from occurring in the schedule. In our case, we use $b = 2$: thus a penalty of 2 is incurred for single breaks, 4 for double breaks, 8 for triple breaks, and so on.

Fig. 8.12 The Multi-stage
algorithm. The procedure
takes as input a feasible
solution S provided by a
graph colouring algorithm

Multi-stage Algorithm (S)
(1) **while** (**not** stopping condition) **do**
(2)   Apply perturbation to S
(3)   Reduce cost c_1 in S via random descent
(4)   Reduce cost c_2 in S via simulated annealing
(5)   Update list L of non-dominated solutions

It is notable that, because the cost functions c_1 and c_2 measure different characteristics, use different penalty units, and feature different growth rates, they are in some sense incommensurable. For this reason, it is appropriate to use the concept of *dominance* to distinguish between solutions. This is defined as follows:

> **Definition 8.3** Let S_1 and S_2 be two feasible solutions. S_1 is said to *dominate* S_2 if and only if:
>
> - $c_1(S_1) \leq c_1(S_2)$ and $c_2(S_1) < c_2(S_2)$; or
> - $c_1(S_1) < c_1(S_2)$ and $c_2(S_1) \leq c_2(S_2)$.

In this definition, it is assumed that both cost functions are being minimised. Note that this definition can also be extended to more than two cost functions if required.

If S_1 does not dominate S_2, and S_2 does not dominate S_1, then S_1 and S_2 are said to be *incomparable*. The output to both algorithms is then a list L of mutually incomparable solutions that are not dominated by any other solutions encountered during the search.

Note that the concept of dominance is commonly used in the field of multiobjective optimisation where, in addition to being incommensurable, cost functions are often in conflict with one another (that is, an improvement in one cost will tend to invoke the worsening of another). It is unclear whether the two cost functions used here are necessarily in conflict, however.

8.6.1.1   The Multi-stage Approach

Our first method for the WRU problem operates in a series of stages, with each stage being concerned with minimising just one of the cost functions. A description of this method is given in Fig. 8.12.

The algorithm starts by taking an arbitrary feasible solution S produced by a graph colouring method. This solution is then "perturbed" by performing a series of randomly selected Kempe chain interchanges, paying no heed to either cost function. Next, a random descent procedure is applied that attempts to make reductions to the spread cost c_1. This is done by repeatedly selecting a random Kempe chain at each iteration and performing the interchange only if the resultant spread cost is less than or equal to the current spread cost. On completion of the random descent procedure, attempts are then made to reduce the break cost c_2 of the current schedule *without* *increasing the current spread cost.* This is achieved using a phase of simulated annealing with a restricted neighbourhood operator where only matches and their reverse fixtures (i.e., (t_i, t_j) and (t_j, t_i)) can be swapped. Note that the latter moves can, on occasion, violate some of the additional hard constraints of this problem, and so in these cases, such moves are rejected automatically. Also, note that moves in this restricted neighbourhood do not alter the spread cost of the schedule and therefore do not undo any of the work carried out in the previous random descent stage. On completion of Step (4), the best solution S^* found during this round of simulated annealing is used to update L. Specifically, if S^* is seen to dominate any solutions in L, then these solutions are removed from L and S^* is added to L. The entire process is then repeated.

Our choice of random descent for reducing c_1 arises simply because in initial experiments we observed that, in isolation, the associated soft constraints seemed quite easy to satisfy. Thus a simple descent procedure seems effective for making quick and significant gains in quality (for all instances spread costs of less than 1, and often 0, were nearly always achieved within our imposed cut-off point of 10,000 evaluations). In addition to this, we also noticed that only short execution times were needed for the simulated annealing stage due to the relatively small solution space resulting from the restricted neighbourhood operator, which meant that the search would tend to converge quite quickly at a local optimum. In preliminary experiments we also found that if we lengthened the simulated annealing process by allowing the temperature variable to be reset (thus allowing the search to escape these optima), then the very same optimum would be achieved after another period of search, perhaps suggesting that the convergence points in these searches are the true optima in these particular spaces.[4]

Finally, our use of a perturbation operator in the multi-stage algorithm is intended to encourage diversification in the search. In this case, a balance needs to be struck by applying enough changes to the current solution to cause the search to enter a different part of the solution space, but not applying too many changes so that the operator becomes nothing more than a random restart mechanism. In our case, we chose to simply apply the Kempe chain operator five times in succession, which proved sufficient for our purposes.

8.6.1.2 A Multiobjective Optimisation Approach

In contrast to the multi-stage approach, our second method attempts to eliminate violations of both types of soft-constraint simultaneously. This is done by combining both cost functions into a single weighted objective function $f(S) = w_1 \times c_1(S) + w_2 \times c_2(S)$, used in conjunction with the Kempe chain neighbourhood operator.

---

[4]In all cases we used an initial temperature $t = 20$, a cooling rate of $\alpha = 0.99$, and $z = \frac{n}{2}$ (refer to the simulated annealing algorithm in Fig. 4.12). The annealing process ended when no move was accepted for 20 successive temperatures. Such parameters were decided upon in preliminary testing and were not seen to be critical in dictating algorithm performance.

Multiobjective Algorithm $(S, \Delta w)$
(1) Set reference costs x_1 and x_2
(2) Set initial weights using $w_i = \frac{c_i(S)}{x_i}$ for $i \in \{1,2\}$
(3) Calculate weighted cost of solution, $f(S) = w_1 \times c_1(S) + w_2 \times c_2(S)$
(4) $B \leftarrow f(S)$
(5) **while** (**not** stopping condition) **do**
(6)   Form new solution S' by applying a Kempe chain interchange to S
(7)   **if** $(f(S') \leq f(S)$ **or** $(f(S') \leq B)$ **then**
(8) $S \leftarrow S'$
(9)   Update list L of non-dominated solutions using S
(10)   Find i corresponding to $\max_{i \in \{1,2\}} \left\{ \frac{c_1(S)}{x_1}, \frac{c_2(S)}{x_2} \right\}$
(11)   Increase weight $w_i \leftarrow w_i(1 + \Delta w)$

Fig. 8.13 Multiobjective algorithm with variable weights [29]. In all reported experiments, a setting of $\Delta w = 10^{-6}$ was used. The input S is a feasible solution provided by a suitable graph colouring algorithm

An obvious issue with the objective function f is that suitable values need to be assigned to the weights. Such assignments can, of course, have large effects on the performance of an algorithm, but they are not always easy to determine as they depend on many factors such as the size and type of problem instance, the nature of the individual cost functions, the user requirements, and the amount of available run time. To deal with this issue, we adopt a multiobjective optimisation technique of Petrovic and Bykov [29]. The strategy of this approach is to alter weights dynamically during the search based on the quality of solutions found so far, thus directing the search into specific regions of the solution space. This is achieved by providing two *reference costs* to the algorithm, x_1 and x_2. Using these values, we can then imagine a reference point (x_1, x_2) being plotted in a two-dimensional Cartesian space, with a straight *reference line* then being drawn from the origin $(0, 0)$ and through the reference point (see Fig. 8.14). During the search, all solutions encountered are then also represented as points in this Cartesian space and, at each iteration, the weights are adjusted automatically to encourage the search to move towards the origin while remaining close to the reference line. It is hoped that eventually solutions will be produced that feature costs less than the original reference costs.

A pseudocode description of this approach is given in Fig. 8.13. Note that the weight update mechanism used here (Step (11)) means that weights are gradually increased during the run. Since, according to Step (7), changes to solutions are only permitted if (a) they improve the cost, or (b) if the weighted cost is kept below a constant B, this implies that worsening moves become increasingly less likely during execution. In this respect, the search process is similar in nature to simulated annealing.

8.6.1.3 Experimental Analysis

To compare the performance of the two approaches, we performed 100 runs of each algorithm on each of the ten problem instances shown in Table 8.2. In all cases a cut-off point of 20,000,000 evaluations (of either cost function) was used, resulting in run times of approximately five to ten minutes.[5]

Note that, unlike with the multi-stage approach, reference costs x_1 and x_2 need to be supplied to the multiobjective approach. In our case we chose suitable reference costs by performing one run using the multi-stage approach. We then used the costs of the resultant solution as the reference costs for the multiobjective algorithm. In runs where more than one non-dominated solution was produced, solutions in L were sorted according to their costs, and the costs of the median solution were used. Note that because problem instance #5 is based on real-world data, in this case, we were also able to use the costs of the (manually produced) league schedule used in the competition. However, runs using the reference costs from both sources turned out to be similar in practice, and so we only report results achieved using the former method here.

We now examine the general behaviour of each method. Figure 8.14 displays the costs of a sample of solutions encountered in an example run with each algorithm. From the initial solution at point $(22.4, 686)$ we see that the multi-stage approach quickly finds solutions with very low values of cost c_1. It then spends the remainder of the run attempting to make reductions to c_2, which is the cause of the bunching of the points in the left of the graph. Due to the position of the reference line (the dotted line in the figure), the multiobjective approach follows a similar pattern to this, though initial progress is considerably slower. However, as the search nears the reference line, the algorithm then considers c_1 to have dropped to an appropriate level and so weight w_2 (used in evaluation function f (Fig. 8.13)) begins to increase such that reductions to cost c_2 are also sought. In this latter stage, unlike in the multi-stage approach, improvements to both cost functions are being made simultaneously. As both algorithms' searches converge, we see from the projection (inset) that the multiobjective approach produces the best solutions in this case.

A summary of the results over the ten test problems is provided in Table 8.3. The "best" column here shows the costs of the best solution found across all runs of both algorithms, which we use for comparative purposes. Note that in all problem instances except #2, only one best result has been produced—that is, we do not see any obvious conflict in the two cost functions. The values in the column labelled "maximum" are used for normalisation purposes and represent the highest cost values seen in any solution returned by the two algorithms. For both algorithms, the table then displays two statistics: the mean size of the solution lists L, and the distances between each of the solutions in L and the "best" solutions. This latter performance measure is based on the metric suggested by Deb et al. [30], which assesses the performance of a multiobjective algorithm by considering the distance between the costs of each solution in L and the costs of some optimal (or near-optimal) solution to the problem.

---

[5] Using a 3.0 GHz Windows 7 PC with 3.87 GB RAM.

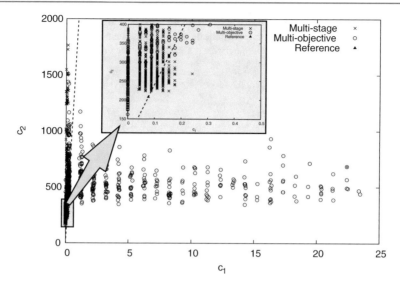

Fig. 8.14 Costs of solutions encountered by the multi-stage and multiobjective approaches in one run with problem instance #5. Solution costs were recorded every 10,000 evaluations. The dotted line (in both the main graph and the projection) represents the reference line used by the multiobjective approach and is drawn from the origin and through the reference point

Because the costs of the global optima are not known for these instances, we choose to use the values given in the "best" column as approximations to this. In our case distances are calculated as follows. First, the costs of all solutions returned by the algorithms, in addition to the costs of the "best" solutions, are normalised to values in [0, 1] by dividing by the maximum cost values specified in the table. Next, for each solution, the (Euclidean) distance between the normalised best costs and all normalised solution costs is calculated. The mean, standard deviation and median of all distances returned for each algorithm are recorded in the table.

The results in Table 8.3 can be split into two cases. The first involves the larger problem instances #3 to #10. Here, we see that the multiobjective approach consistently produces better results than the multi-stage approach, which is reflected in the lower means, medians, and deviations in the Distance from Best column. Note that the best results for these instances have also come from the multiobjective approach. We also see that the multi-stage approach has produced larger solution lists for these problem instances, which could be useful if a user wanted to be presented with a choice of solutions, though the solutions in these lists are of lower quality in general. Also note that the differences between the mean and median values here reveal that the distributions of distances with the multi-stage approach feature larger amounts of positive skew, reflecting the fact that this method produces solutions of very low quality on occasion. For the smaller problem instances #1 and #2 we see that these patterns are more or less reversed from those of the multi-stage approach, producing solutions with costs that are consistently closer (or equal to) to the best-known

Table 8.3 Summary of results achieved in 100 runs of the multi-stage and multiobjective algorithms. An asterisk (*) in the "best" column indicates that the solution with the associated costs was found using the multi-stage approach (otherwise it was found by the multiobjective approach)

#	Best	Maximum		Multi-stage			Multiobjective		
			$\lvert L \rvert$	Distance from Best		$\lvert L \rvert$	Distance from Best	
	(c_1, c_2)	c_1	c_2		Mean±SD	Med.		Mean±SD	Med.
1	$(0, 104)^*$	4.21	352	1.81	0.15 ± 0.22	0.04	2.57	0.27 ± 0.10	0.27
2	$(0, 134)^*$	11.3	394	2.81	0.06 ± 0.05	0.05	3.82	0.20 ± 0.11	0.17
	$(2.15, 132)^*$								
3	$(0, 102)$	8.24	560	4.32	0.24 ± 0.17	0.17	3.32	0.13 ± 0.10	0.12
4	$(0, 112)$	2.17	186	2.72	0.28 ± 0.09	0.26	1.00	0.14 ± 0.06	0.13
5	$(0, 134)$	3.29	294	3.61	0.34 ± 0.11	0.31	1.00	0.11 ± 0.04	0.11
6	$(0, 148)$	4.26	444	4.44	0.32 ± 0.11	0.30	1.04	0.08 ± 0.05	0.08
7	$(0, 156)$	2.18	816	4.51	0.19 ± 0.11	0.16	1.04	0.05 ± 0.04	0.04
8	$(0, 186)$	3.27	786	4.71	0.26 ± 0.12	0.22	1.53	0.08 ± 0.06	0.06
9	$(0, 204)$	1.26	940	4.32	0.21 ± 0.10	0.19	1.08	0.05 ± 0.06	0.04
10	$(0, 234)$	2.29	1398	4.63	0.20 ± 0.10	0.16	1.30	0.09 ± 0.05	0.07

costs. One feature to note in this case is the relatively large difference between the mean and median with the multi-stage approach for instance #1, where we saw about 60% of produced solutions being very close to the best, and the remainder having much larger distances. Finally, note that for all problem instances, the nonparametric Mann–Whitney test indicates that the distances of each algorithm are significantly different with significance level $\leq 0.01\%$.

In summary, the results in Table 8.3 suggest that the strategy used by the multi-stage algorithm of employing many rounds of short intensive searches seems more fitting for smaller, less constrained instances, but for larger instances, including the real-world problem instance, better solutions are achieved by using the multiobjective approach where longer, less intensive searches are performed.

8.7 Chapter Summary and Discussion

In this chapter, we have shown that round-robin schedules can be successfully constructed using graph colouring principles, often in the presence of many additional hard constraints. In Sect. 8.6 we exploited this link with graph colouring by proposing two algorithms for a real-world sports scheduling problem. In the case of the real-world problem instance (#5), we found that more than 98% of all solutions generated by our multiobjective approach dominated the solution that was manually produced by the WRU's league administrators. On the other hand, for the multi-stage approach, this figure was just 0.02%. We should, however, interpret these statistics with care, firstly because the manually produced solution was actually for a slightly different problem (the exact specifications of which we were unable to obtain from the league organisers), and secondly because our specific cost functions were not previously used by the league organisers for evaluating their solutions.

One further neighbourhood operator that might be used with these problems (and indeed any graph colouring problem) is an extension to the Kempe chain interchange operator known as the *s-chain* interchange operator. Let $\mathcal{S} = \{S_1, \ldots, S_k\}$ be a feasible graph colouring solution and let v be an arbitrary vertex in $\mathcal{S}$ coloured with colour j_1. Furthermore, let $j_2, \ldots, j_s$ be a sequence of distinct colours taken from the set $\{1, \ldots, k\} - \{j_1\}$. An s-chain is constructed by first identifying all vertices adjacent to v that are coloured with colour j_2. From these, adjacent vertices coloured with j_3 are then identified, and from these adjacent vertices with colour j_4, and so on. When considering vertices with colour j_s, adjacent vertices with colour j_1 are sought.

As an example, using the graph from Fig. 7.3, an s-chain using $s = 3$, $v = v_2$ and colours $j_1 = 2$, $j_2 = 4$, and $j_3 = 1$ can be seen to contain the vertices $\{v_2, v_3, v_4, v_5, v_7, v_8\}$. Through similar reasoning to that of Kempe chains (Theorem 4.1), it is simple to show that we can take the vertices of an s-chain and interchange their colours using the mapping $j_1 \leftarrow j_2, j_2 \leftarrow j_3, \ldots, j_{s-1} \leftarrow j_s, j_s \leftarrow j_1$ such that feasibility of the solution is maintained. Note also that s-chains are equivalent to Kempe chains when $s = 2$. In our experiments with round-robin schedules, we also tested the effects of the s-chain interchange operator for $s \geq 3$; however, because of the relatively high levels of connectivity between different colours in these graphs, we observed that over 99% of moves contained the maximum of $s(t/2)$ vertices. In other words, almost all moves simply resulted in moves that are also achievable through combinations of N_4. s-chains may show more promise in other applications, however, particularly those involving sparser graphs.

References

1. Dinitz J, Garnick D, McKay B (1994) There are 526,915,620 nonisomorphic one-factorizations of K$_{12}$. J Comb Des 2(2):273–285
2. de Werra D (1988) Some models of graphs for scheduling sports competitions. Discret Appl Math 21:47–65
3. Trick M (2001) A schedule-then-break approach to sports timetables. In: Burke E, Erben W (eds) Practice and theory of automated timetabling (PATAT) III. LNCS, vol 2079. Springer, pp 242–253
4. Elf M, Junger M, Rinaldi G (2003) Minizing breaks by maximizing cuts. Oper Res Lett 31(5):343–349
5. Miyashiro R, Matsui T (2006) Semidefinite programming based approaches to the break minimization problem. Comput Oper Res 33(7):1975–1992
6. Miyashiro R, Matsui T (2005) A polynomial-time algorithm to find an equitable home-away assignment. Oper Res Lett 33:235–241
7. Russell R, Leung J (1994) Devising a cost effective schedule for a baseball league. Oper Res 42(4):614–625

8. Nemhauser G, Trick M (1998) Scheduling a major college basketball conference. Oper Res 46:1–8

9. Anderson I (1991) Kirkman and GK_{2n}. Bull Inst Combin Appl 3:111–112

10. Bartsch T, Drexl A, Kroger S (2006) Scheduling the professional soccer leagues of Austria and Germany. Comput Oper Res 33(7):1907–1937

11. della Croce F, Oliveri D (2006) Scheduling the Italian football league: an ILP-based approach. Comput Oper Res 33(7):1963–1974

12. Wright M (2006) Scheduling fixtures for Basketball New Zealand. Comput Oper Res 33(7):1875–1893

13. della Croce F, Tadei R, Asioli P (1999) Scheduling a round-robin tennis tournament under courts and players unavailability constraints. Ann Oper Res 92:349–361

14. Wright M (1994) Timetabling county cricket fixtures using a form of Tabu search. J Oper Res Soc 47(7):758–770

15. Fleurent C, Ferland J (1993) Allocating games for the NHL using integer programming. Oper Res 41(4):649–654

16. Kendall G, Knust S, Ribeiro C, Urrutia S (2010) Scheduling in sports, an annotated bibliography. Comput Oper Res 37(1):1–19

17. Russell K (1980) Balancing carry-over effects in round-robin tournaments. Biometrika 67(1):127–131

18. Henz M, Muller T, Theil S (2004) Global constraints for round robin tournament scheduling. Eur J Oper Res 153:92–101

19. Easton K, Nemhauser G, Trick M (2001) The traveling tournament problem: description and benchmarks. In Walsh T (ed) Principles and practice of constraint programming. LNCS, vol 2239. Springer, pp 580–585

20. Easton K, Nemhauser G, Trick M (2003) Solving the traveling tournament problem: a combined integer programming and constraint programming approach. In Burke E, De Causmaecker P (eds) Practice and theory of automated timetabling (PATAT) IV. LNCS, vol 2740. Springer, pp 100–109

21. Anagnostopoulos A, Michel L, van Hentenryck P, Vergados Y (2006) A simulated annealing approach to the traveling tournament problem. J Sched 9(2):177–193

22. Lim A, Rodrigues B, Zhang X (2006) A simulated annealing and hill-climbing algorithm for the traveling tournament problem. Eur J Oper Res 174(3):1459–1478

23. Di Gaspero L, Schaerf A (2007) A composite-neighborhood Tabu search approach to the traveling tournament problem. J Heurist 13(2):189–207

24. Ribeiro C, Urrutia S (2007) Heuristics for the mirrored travelling tournament problem. Eur J Oper Res 179(3):775–787

25. Lambrechts E, Ficker M, Goossens D, Spieksma F (2018) Round-robin tournaments generated by the circle method have maximum carry-over. Math Program 172:277–302

26. Lewis R, Thompson J (2010) On the application of graph colouring techniques in round-robin sports scheduling. Comput Oper Res 38(1):190–204

27. Di Gaspero L, Schaerf A (2006) Neighborhood portfolio approach for local search applied to timetabling problems. J Math Model Algorithms 5(1):65–89

28. Januario T, Urrutia S, de Werra D (2016) Sports scheduling search space connectivity: a riffle shuffle driven approach. Discret Appl Math 211:113–120

29. Petrovic S, Bykov Y (2003) A multiobjective optimisation approach for exam timetabling based on trajectories. In Burke E, De Causmaecker P (eds) Practice and theory of automated timetabling (PATAT) IV. LNCS, vol 2740. Springer, pp 181–194

30. Deb K, Pratap A, Agarwal S, Meyarivan T (2000) A fast elitist multi-objective genetic algorithm: NSGA-II. IEEE Trans Evol Comput 6(2):182–197

Designing University Timetables

9

In this chapter, our case study looks at how graph colouring concepts can be used in the process of constructing high-quality timetables for universities and other types of educational establishments. As we will see, this problem area can contain a whole host of different constraints, which will often make problems very difficult to tackle. That said, most timetabling problems contain an underlying graph colouring problem, allowing us to use many of the concepts developed in previous chapters.

The first section of this chapter will look at university timetabling from a broad perspective, discussing among other things the various constraints that might be imposed on the problem. Section 9.2 onwards will then conduct a detailed analysis of a well-known timetabling formulation that has been the subject of various articles in the literature. As we will see, powerful algorithms derived from graph colouring principles can be developed for this problem, though careful modifications also need to be made to allow the methods to cope with the various other constraints that this problem involves.

9.1 Problem Background

In the context of higher education institutions, a timetable can be thought of as an assignment of events (such as lectures, tutorials, or exams) to a finite number of rooms and timeslots following a set of constraints, some of which will be mandatory, and others that may be optional. It is suggested by Corne et al. [1] that timetabling constraints can be categorised into five main classes.

© The Author(s), under exclusive license to Springer Nature Switzerland AG 2021   247
R. M. R. Lewis, *Guide to Graph Colouring*, Texts in Computer Science,
https://doi.org/10.1007/978-3-030-81054-2_9

- *Unary Constraints.* These involve just one event, such as the constraint "event a must not take place on a Tuesday", or the constraint "event a must occur in timeslot b".
- *Binary Constraints.* These concern *pairs* of events, such as the constraint "event a must take place before event b", or the *event clash* constraint, which specifies pairs of events that cannot be held at the same time in the timetable.
- *Capacity Constraints.* These are governed by room capacities. For example "All events should be assigned to a room that has a sufficient capacity".
- *Event Spread Constraints.* These concern requirements involving the "spreading-out" or "grouping-together" of events within the timetable to ease student/teacher workload, and/or to agree with a university's timetabling policy.
- *Agent Constraints.* These are imposed to promote the requirements and/or preferences of the people who will use the timetables, such as the constraint "lecturer a likes to teach event b on Mondays", or "lecturer c must have three free mornings per week".

Like many problems in operational research, a convention in automated timetabling is to group constraints into two classes: *hard* constraints and *soft* constraints. Hard constraints have a higher priority than soft, and their satisfaction is usually mandatory. Indeed, timetables will normally only be considered "feasible" if *all* of the imposed hard constraints have been satisfied. Soft constraints, meanwhile, are those that we want to obey *if possible*; they describe the criteria for a timetable to be "good" with regard to the timetabling policies of the university concerned, as well as the experiences of the people who will have to use it.

According to McCollum et al. [2], the problem of constructing university timetables can be divided into two categories: exam timetabling problems and course timetabling problems. It is also suggested that course timetabling problems can be further divided into two subcategories: "post enrolment-based course timetabling", where the constraints of the problem are specified by student enrolment data, and "curriculum-based course timetabling", where constraints are based on curricula specified by the university. Müller and Rudova [3] have also shown that these subcategories are closely related, demonstrating how instances of the latter can be transformed into those of the former in many cases.

We have seen in Sects. 1.1.2 and 5.7.5 that a fundamental constraint in university timetabling is the "event-clash" binary constraint. This specifies that if a person (or some other resource of which there is only one) is required to be present in a pair of events, then these must not be assigned to the same timeslot, as such an assignment will result in this person/resource having to be in two places at once. This constraint can be found in almost all university timetabling problems and allows us to draw parallels with the graph colouring problem by considering the events as vertices, clashing events as adjacent vertices, and the timeslots as colours. Beyond this near-universal constraint, however, timetabling problem formulations will usually vary widely from place to place because different universities will usually have their own needs and timetabling policies (and therefore set of constraints) that they need to satisfy (see Fig. 9.1). That said, it is still the case that nearly all timetabling prob-

Fig. 9.1 When constructing a timetable, meeting the needs of all concerned may not always be possible

Hi Dave - Professor Jones says that he can teach Monday mornings, but he'll need to finish early on Wednesdays, and will need three free hours on Fridays to walk his dog.

lems do feature an underlying graph colouring problem in some form or another in their definitions, and many timetabling algorithms do use various pieces of heuristic information extracted from the graph colouring problem as a driving force in their search for solutions.

Despite this wide range of constraints, it is known that university timetabling problems are $\mathcal{NP}$-complete in almost all variants. Cooper and Kingston [4], for example, have proved that $\mathcal{NP}$-completeness exists for a number of different problem interpretations that can arise in practice. They achieve this by providing polynomial transformations from various $\mathcal{NP}$-complete problems, including bin packing, three-dimensional matching as well as graph colouring itself.

9.1.1 Designing and Comparing Algorithms

The field of university timetabling has seen many solution approaches proposed over the years, including methods based on constructive heuristics, mathematical programming, branch and bound, and metaheuristics. (See, for example, the surveys of Carter et al. [5], Burke et al. [6], Schaerf [7], Lewis [8].) The latter survey has suggested that metaheuristic approaches for university timetabling can be classified into three categories as follows:

- *One-stage Optimisation Algorithms*. Here, the satisfaction of the hard constraints and soft constraints is attempted simultaneously, usually using a single objective function in which violations of hard constraints are penalised more heavily than violations of soft constraints. If desired, these weights can be altered during a run.

- *Two-stage Optimisation Algorithms.* In this case, the hard constraints are first satisfied to form a feasible solution. Attempts are then made to eliminate violations of the soft constraints by navigating the space of feasible solutions. Similar schemes to this have been used in the case studies from Chaps. 7 and 8.
- *Algorithms That Allow Relaxations.* Here, violations of the hard constraints are disallowed from the outset by relaxing some features of the problem. Attempts are then made to try and satisfy the soft constraints, while also considering the task of eliminating these relaxations. These relaxations could include allowing certain events to be left out of the timetable, or using additional timeslots or rooms.

The wide variety of constraints, coupled with the fact that each higher education institution will usually have its own timetabling policies, means that timetabling problem formulations have always tended to vary quite widely in the literature. While making the problem area very rich, one drawback has been the lack of opportunity for accurate comparison of algorithms. Since the early 2000s, this situation has been mitigated to a certain extent with the organisation of a series of timetabling competitions and the release of publicly available problem instances. In 2007, for example, the Second International Timetabling Competition (ITC2007) was organised by a group of timetabling researchers from different European Universities, which considered the three types of timetabling problems mentioned above: exam timetabling, post enrolment-based course timetabling, and curriculum-based timetabling. The competition operated by releasing problem instances into the public domain, with entrants then designing algorithms to try and solve these. Entrants' algorithms were then compared under strict time limits according to specific evaluation criteria.[1]

9.1.2   Chapter Outline

In this chapter, we will examine the post enrolment-based course timetabling problem used for ITC2007. This formulation models the real-world situation where students are given a choice of lectures that they wish to attend, with the timetable then being constructed according to these choices. The next section contains a formal definition of this problem, with Sect. 9.3 then containing a short review of the most noteworthy algorithms. We then go on to describe a high-performance graph colouring-based method in Sects. 9.4 and 9.5. The final results of our algorithm are given in Sect. 9.6, with a discussion and conclusions then being presented in Sect. 9.7.

---

[1] http://www.cs.qub.ac.uk/itc2007/.

9.2   Problem Definition and Preprocessing

As mentioned, the post enrolment-based course timetabling problem was introduced for use with the Second International Timetabling Competition, run in 2007. The problem involves seven types of hard constraint whose satisfaction is mandatory, and three soft constraints, whose satisfaction is desirable, but not essential. The problem involves assigning a set of events to 45 timeslots (5 days, with nine timeslots per day) according to these constraints.

The hard constraints for the problem are as follows. First, for each event there is a set of students who are enrolled to attend. Events should be assigned to timeslots such that no student is required to attend more than one event in any one timeslot. Next, each event also requires a set of room features (e.g., a certain number of seats, specialist teaching equipment, etc.), which will only be provided by certain rooms. Consequently, each event needs to be assigned to a suitable room that exhibits the room features that it requires. The double booking of rooms is also disallowed. Hard constraints are also imposed stating that some events cannot be taught in certain timeslots. Finally, precedence constraints—stating that some events need to be scheduled before or after others—are also stipulated.

More formally, a problem instance comprises a set of events $e = \{e_1, \ldots, e_n\}$, a set of timeslots $t = \{t_1, \ldots, t_k\}$ (where $k = 45$), a set of students $s = \{s_1, \ldots, s_{|s|}\}$, a set of rooms $r = \{r_1, \ldots, r_{|r|}\}$, and a set of room features $f = \{f_1, \ldots, f_{|f|}\}$. Each room $r_i \in r$ is also allocated a capacity $c(r_i)$ reflecting the number of seats it contains.

The relationships between the above sets are defined by five problem matrices: an *attends* matrix $\mathbf{P}^{(1)}_{|s| \times n}$, where

$$P^{(1)}_{ij} = \begin{cases} 1 & \text{if student } s_i \text{ is due to attend event } e_j \\ 0 & \text{otherwise,} \end{cases}$$

a *room features* matrix $\mathbf{P}^{(2)}_{|r| \times |f|}$, where

$$P^{(2)}_{ij} = \begin{cases} 1 & \text{if room } r_i \text{ has feature } f_j \\ 0 & \text{otherwise,} \end{cases}$$

an *event features* matrix $\mathbf{P}^{(3)}_{n \times |f|}$, where

$$P^{(3)}_{ij} = \begin{cases} 1 & \text{if event } e_i \text{ requires feature } f_j \\ 0 & \text{otherwise,} \end{cases}$$

an *event availability* matrix $\mathbf{P}^{(4)}_{n \times k}$ in which

$$P^{(4)}_{ij} = \begin{cases} 1 & \text{if event } e_i \text{ can be assigned to timeslot } t_j \\ 0 & \text{otherwise,} \end{cases}$$

and finally a *precedence* matrix $\mathbf{P}^{(5)}_{n \times n}$, where

$$
P^{(5)}_{ij} = \begin{cases} 1 & \text{if event } e_i \text{ should be assigned to an earlier timeslot than event } e_j \\ -1 & \text{if event } e_i \text{ should be assigned to a later timeslot than event } e_j \\ 0 & \text{otherwise.} \end{cases}
$$

For the precedence matrix above, note that two conditions are necessary for the relationships to be consistent: (a) $P^{(5)}_{ij} = 1$ if and only if $P^{(5)}_{ji} = -1$, and (b) $P^{(5)}_{ij} = 0$ if and only if $P^{(5)}_{ji} = 0$. We can also observe the transitivity of this relationship:

$$
\left(\exists e_i, e_j, e_l \in e : \left(P^{(5)}_{ij} = 1 \wedge P^{(5)}_{jl} = 1 \right) \right) \Rightarrow P^{(5)}_{il} = 1 \tag{9.1}
$$

In some of the competition problem instances, this transitivity is not fully expressed; however, observing it enables further 1's and -1's to be added to $\mathbf{P}^{(5)}$ during pre-processing, allowing the relationships to be more explicitly stated.

Given the above five matrices, we are also able to calculate two further matrices that allow fast detection of hard constraint violations. The first of these is a *room suitability* matrix $\mathbf{R}_{n \times |r|}$ defined as

$$
R_{ij} = \begin{cases} 1 & \text{if } \left(\sum_{l=1}^{|s|} P^{(1)}_{li} \le c(r_j) \right) \wedge \left(\nexists f_l \in f : \left(P^{(3)}_{il} = 1 \wedge P^{(2)}_{jl} = 0 \right) \right) \\ 0 & \text{otherwise.} \end{cases} \tag{9.2}
$$

The second is then a *conflicts* matrix $\mathbf{C}_{n \times n}$, defined:

$$
C_{ij} = \begin{cases} 1 & \text{if } \left(\exists s_l \in s : \left(P^{(1)}_{li} = 1 \wedge P^{(1)}_{lj} = 1 \right) \right) \\ & \vee \left(\left(\exists r_l \in r : (R_{il} = 1 \wedge R_{jl} = 1) \right) \wedge \left(\sum_{l=1}^{|r|} R_{il} = 1 \right) \wedge \left(\sum_{l=1}^{|r|} R_{jl} = 1 \right) \right) \\ & \vee \left(P^{(5)}_{ij} \ne 0 \right) \\ & \vee \left(\nexists t_l \in t : \left(P^{(4)}_{il} = 1 \wedge P^{(4)}_{jl} = 1 \right) \right) \\ 0 & \text{otherwise.} \end{cases} \tag{9.3}
$$

The matrix $\mathbf{R}$ therefore specifies the rooms that are suitable for each event (that is, rooms that are large enough for all attending students and that have all the required features). The $\mathbf{C}$ matrix, meanwhile, is a symmetrical matrix ($C_{ij} = C_{ji}$) that specifies pairs of events that cannot be assigned to the same timeslot (i.e., those that conflict). According to Eq. (9.3), this will be the case if two events e_i and e_j share a common student, require the same individual room, are subject to a precedence relation, or have mutually exclusive subsets of timeslots for which they are available.

Note that the matrix $\mathbf{C}$ is analogous to the adjacency matrix of a graph $G = (V, E)$ with n vertices, highlighting the similarities between this timetabling problem and the graph colouring problem. However, unlike the graph colouring problem, in this case the ordering of the timeslots (colour classes) is also an important property of a solution. Consequently, a solution is represented by an ordered set of sets $\mathcal{S} =$

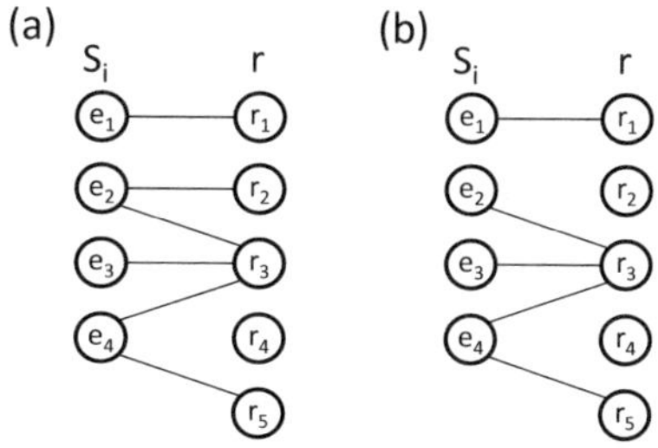

Fig. 9.2 Example bipartite graphs with and without a maximum bipartite matching of size $2|S_i|$. In Part (a), events e_1, e_2, e_3, and e_4 can be assigned to rooms r_1, r_2, r_3, and r_5, respectively

$(S_1, \ldots, S_{k=45})$ and is subject to the satisfaction of the following hard constraints.

$$\bigcup_{i=1}^{k} S_i \subseteq e \tag{9.4}$$

$$S_i \cap S_j = \emptyset \qquad (1 \leq i \neq j \leq k) \tag{9.5}$$

$$\forall e_j, e_l \in S_i, \; C_{jl} = 0 \qquad (1 \leq i \leq k) \tag{9.6}$$

$$\forall e_j \in S_i, \; P_{ji}^{(4)} = 1 \qquad (1 \leq i \leq k) \tag{9.7}$$

$$\forall e_j \in S_i, \; e_l \in S_{q<i}, \; P_{jl}^{(5)} \neq 1 \qquad (1 \leq i \leq k) \tag{9.8}$$

$$\forall e_j \in S_i, \; e_l \in S_{q>i}, \; P_{jl}^{(5)} \neq -1 \qquad (1 \leq i \leq k) \tag{9.9}$$

$$S_i \in \mathcal{M} \qquad (1 \leq i \leq k). \tag{9.10}$$

Constraints (9.4) and (9.5) state that S should partition the event set e (or a subset of e) into an ordered set of sets, labelled $S_1, \ldots, S_k$. Each set $S_i \in S$ contains the events that are assigned to timeslot t_i in the timetable. Equation (9.6) stipulates that no pair of conflicting events should be assigned to the same set $S_i \in S$ (the graph colouring constraint), while (9.7) states that each event should be assigned to a set $S_i \in S$ whose corresponding timeslot t_i is deemed available according to matrix $\mathbf{P}^{(4)}$. Constraints (9.8) and (9.9) then impose the precedence requirements of the problem.

Finally, (9.10) is concerned with ensuring that the events assigned to a set $S_i \in S$ can each be assigned to a suitable room from the room set r. To achieve this, it is necessary to solve a maximum bipartite matching problem. Specifically, let $G = (S_i, r, E)$ be a bipartite graph with vertex sets S_i and r, and an edge set $E = \{\{e_j \in S_i, r_l \in r\} : R_{jl} = 1\}$. Given G, the set S_i is a member of $\mathcal{M}$ if and only if there exists a maximum bipartite matching of G comprising $|S_i|$ edges. In this case, the room constraints for this timeslot are satisfied.

Figure 9.2 shows two examples of these ideas using $|S_i| = 4$ events and $|r| = 5$ rooms. In Fig. 9.2a, a matching exists—for example, event e_1 can be assigned to room r_1, e_2 to r_2, e_3 to r_3, and e_4 to r_5. On the other hand, a matching does not exist in Fig. 9.2b, meaning $S_i \notin \mathcal{M}$ in this case. Matching problems on bipartite graphs can be solved in polynomial time using, for example, the $\mathcal{O}(m\sqrt{n})$ Hopcroft–Karp algorithm.

In the competition's interpretation of this problem, a solution S is considered *valid* if and only if all of the constraints (9.4)–(9.10) are satisfied. The quality of a valid solution is then gauged by a *distance to feasibility* (DTF) measure, calculated as the sum of the sizes of all events not contained in the solution:

$$\text{DTF} = \sum_{e_i \in S'} \sum_{j=1}^{|s|} P_{ij}^{(1)}, \tag{9.11}$$

where $S' = e - (\bigcup_{i=1}^{k} S_i)$. If the solution S is valid and has a DTF of zero (implying $\bigcup_{i=1}^{k} S_i = e$ and $S' = \emptyset$) then it is considered *feasible* since all of the events have been feasibly timetabled. The set of feasible solutions is thus a subset of the set of valid solutions.

9.2.1 Soft Constraints

As mentioned, in addition to finding a solution that obeys all of the hard constraints, three soft constraints are also considered with this problem.

- *SC1*. Students should not be required to attend an event in the last timeslot of each day (i.e., timeslots 9, 18, 27, 36, or 45);
- *SC2*. Students should not have to attend events in three or more successive timeslots occurring in the same day; and,
- *SC3*. Students should not be required to attend just one event in a day.

The extent to which these constraints are violated is measured by a soft constraints cost (SCC), which is worked out in the following way. For SC1, if a student attends an event assigned to an end-of-day timeslot, this is counted as one penalty point. Naturally, if x students attend this class, this counts as x penalty points. For SC2, if a student attends three events in a row we count this as one penalty point. If a student has four events in a row we count this as two, and so on. Note that students assigned to events occurring in consecutive timeslots over two separate days are not counted as violations. Finally, each time we encounter a student with a single event on a day, we count this as one penalty point (two for 2 days with single events, etc.). The SCC is simply the total of these three values.

More formally, the SCC can be calculated using two matrices: $\mathbf{X}_{|s| \times 45}$, which tells us the timeslots for which each student is attending an event, and $\mathbf{Y}_{|s| \times 5}$, which specifies whether or not a student is required to attend just one event in each of the 5 days.

$$X_{ij} = \begin{cases} 1 & \text{if } \exists e_l \in S_j : P_{il}^{(1)} = 1 \\ 0 & \text{otherwise,} \end{cases} \tag{9.12}$$

$$Y_{ij} = \begin{cases} 1 & \text{if } \sum_{l=1}^{9} X_{i,9(j-1)+l} = 1 \\ 0 & \text{otherwise.} \end{cases} \tag{9.13}$$

Using these matrices, the SCC is calculated as follows:

$$\text{SCC} = \sum_{i=1}^{|s|} \sum_{j=1}^{5} \left(\left(X_{i,9j}\right) + \left(\sum_{l=1}^{7} \prod_{q=0}^{2} X_{i,9(j-1)+l+q}\right) + \left(Y_{i,j}\right) \right). \tag{9.14}$$

Here, the three terms summed in the outer parentheses of Eq. 9.14 define the number of violations of SC1, SC2, and SC3 (respectively) for each student on each day of the timetable.

9.2.2   Problem Complexity

Having defined the post enrolment-based course timetabling problem, we are now in a position to state its complexity.

Theorem 9.1 *The post enrolment-based course timetabling problem is $\mathcal{NP}$-hard.*

Proof. Let $\mathbf{C}_{n \times n}$ be our symmetric conflicts matrix as defined above, filled arbitrarily. In addition, let the following conditions hold:

$$|r| \geq n \tag{9.15}$$

$$R_{ij} = 1 \qquad\qquad \forall e_i \in e,\ r_j \in r \tag{9.16}$$

$$P_{ij}^{(4)} = 1 \qquad\qquad \forall e_i \in e,\ t_j \in t \tag{9.17}$$

$$P_{ij}^{(5)} = 0 \qquad\qquad \forall e_i, e_j \in e. \tag{9.18}$$

Here, there is an excess number of rooms which are suitable for all events ((9.15) and (9.16)), there are no event availability constraints (9.17), and no precedence constraints (9.18). In this special case we are therefore only concerned with satisfying Constraints (9.4)–(9.6) while minimising the DTF. Determining the existence of a feasible solution using k timeslots is therefore equivalent to the $\mathcal{NP}$-complete graph k-colouring problem.   □

From a different perspective, Cambazard et al. [9] have also shown that, in the absence of all hard constraints (9.6)–(9.10) and soft constraints SC1 and SC2, the problem of satisfying SC3 (i.e., minimising the number of occurrences of students sitting a single event in a day) is equivalent to the $\mathcal{NP}$-hard set covering problem.

9.2.3   Evaluation and Benchmarking

From the above descriptions, we see that a timetable's quality is described by two values: the distance to feasibility (DTF) and the soft constraint cost (SCC). According to the competition criteria, when comparing solutions the one with the lowest DTF is deemed the best timetable, reflecting the increased importance of the hard constraints over the soft constraints. However, when two or more solutions' DTFs are equal, the winner is deemed the solution among these that has the lowest SCC.

There are 24 benchmark instances available for this problem. These were generated so that all are known to feature at least one perfect solution (that is, a solution with DTF = 0 and SCC = 0). For comparative purposes, a benchmark timing program is also available on the competition website that allocates a strict time limit for each machine that it is executed on (based on its hardware and operating system). This allows researchers to use approximately the same amount of computational effort when testing their implementations, allowing more accurate comparisons.

9.3 Previous Approaches to This Problem

One of the first studies into the post enrolment-based timetabling problem (in this form) was carried out by Rossi-Doria et al. [10], who used it as a test problem for comparing five different metaheuristics, namely evolutionary algorithms, simulated annealing, iterated local search, ant colony optimisation, and tabu search. Two interesting observations were offered in their work:

- "The performance of a metaheuristic with respect to satisfying hard constraints and soft constraints may be different".
- "Our results suggest that a hybrid algorithm consisting of at least two phases, one for taking care of feasibility, the other taking care of minimising the number of soft constraint violations [without revI i olating any of the hard constraints in the process], is a promising direction" Rossi-Doria et al. [10].

These conclusions have since proven to be quite salient, with several successful algorithms following this suggested two-stage methodology. This includes the winning entry of ITC2007 itself, due to Cambazard et al. [11], which uses tabu search together with an intensification procedure to achieve feasibility, with simulated annealing then being used to satisfy the soft constraints.

Since the running of ITC2007, several papers have been published that have equalled or improved upon the results of the competition. Cambazard et al. [9] have shown how the results of their two-stage competition entry can be improved by relaxing Constraint (9.10) such that a timeslot t_i is considered feasible whenever $|S_i| < |r|$. The rationale for this relaxation is that it will "increase the solution density of the underlying search space", though a repair operator is also needed to make sure that the timeslots satisfy Constraint (9.10) at the end of execution. Cambazard et al. [9] have also examined constraint programming-based approaches and a large neighbourhood search (LNS) scheme, and find that their best results can be found when using simulated annealing together with the LNS operator for reinvigorating the search from time to time.

Other successful algorithms for this problem have followed the one-stage optimisation scheme by attempting to reduce violations of hard and soft constraints simultaneously. Ceschia et al. [12], for example, treat this problem as a single objective optimisation problem in which the space of valid *and invalid* solutions is explored. Specifically, they allow violations of Constraints (9.6), (9.8), and (9.9) within a solution, and use the number of students affected by such violations, together with the DTF, to form an infeasibility measure. This is then multiplied by a weighting coefficient w and added to the SCC to form the objective function. Simulated annealing is then used to optimise this objective function and, surprisingly, after extensive parameter tuning $w = 1$ is found to provide their best results.

Nothegger et al. [13] have also attempted to optimise the DTF and SCC simultaneously, making use of ant colony optimisation to explore the space of valid solutions. Here, the DTF and SCC measures are used to update the algorithm's pheromone matrices so that favourable assignments of events to rooms and timeslots will occur with higher probability in later iterations of the algorithm. Nothegger et al. also show that the results of their algorithm can be improved by adding a local search-based improvement method and by parallelising the algorithm.

Jat and Yang [14] have also used a weighted sum objective function in their hybrid evolutionary algorithm/tabu search approach, though their results do not appear as strong as those of the previous two papers. Similarly, van den Broek and Hurkens [15] have also used a weighted sum objective function in their deterministic algorithm based on column generation techniques.

From the above studies, it is clear that the density and connectivity of the underlying solution space is an important issue in the performance of a neighbourhood search algorithm for this problem. In particular, if connectivity is low then movements in the solution space will be more restricted, perhaps making improvements in the objective function more difficult to achieve. From the research discussed above, it is noticeable that some of the best approaches for this problem have attempted to mitigate this issue by relaxing some of the hard constraints and/or by allowing events to be kept out of the timetable. However, such methods also require mechanisms for coping with these relaxations, such as repair operators (which may ultimately require large alterations to be made to a solution), or by introducing terms into the objective function (which will require appropriate weighting coefficients to be determined, perhaps via tuning). On the other hand, a two-stage approach of the type discussed by Rossi-Doria et al. [10] will not need these features, though because feasibility must be maintained when the SCC is being optimised, the underlying solution space may be more sparsely connected, perhaps making good levels of optimisation more difficult to achieve. We will focus on the issue of connectivity in Sect. 9.5 onwards.

9.4   Algorithm Description: Stage One

Before looking at the task of eliminating soft constraint violations, it is first necessary to produce a valid solution that minimises the DTF measure (Eq. (9.11)). Previous strategies for this task have typically involved inserting all events into the timetable, and then rearranging them to remove violations of the hard constraints [9, 11, 12, 16].

Table 9.1 Heuristics used for producing an initial solution in Stage 1. Here, a "valid place" is defined as a room/timeslot pair that an event can be assigned to without violating Constraints (9.4)–(9.10)

Rule	Description
h_1	Choose the unplaced event with the smallest number of valid places in the timetable to which it can be assigned
h_2	Choose the unplaced event e_i that conflicts with the most other events (i.e., that maximises $\sum_{j=1}^{n} C_{ij}$)
h_3	Choose an event randomly
h_4	Choose the place that is valid for the least number of other unplaced events in U
h_5	Choose the valid timeslot containing the fewest events
h_6	Choose a place randomly

In contrast, we suggest using a method by which events are permitted to remain outside of the timetable, meaning that spaces within the timetable are not "blocked" by events causing hard constraint violations. To do this, we exploit the similarity between this problem and the graph colouring problem by using an adaptation of the PARTIALCOL algorithm from Sect. 5.2.

To begin, an initial solution is constructed by taking events one by one and assigning them to timeslots such that none of the hard constraints are violated. Events that cannot be assigned without breaking a hard constraint are kept to one side and are dealt with at the end of this process. To try and maximise the number of events inserted, a set of high-performance heuristics originally proposed by Lewis [17] and based on the DSATUR algorithm (Sect. 3.3) is used. At each step heuristic rule h_1 (Table 9.1) is used to select an event, with ties being broken using h_2, and then h_3 (if necessary). The selected event is then inserted into the timetable according to rule h_4, breaking ties with h_5 and further ties with h_6.

Completion of this constructive phase results in a valid solution S obeying Constraints (9.4)–(9.10). However, if $S' \neq \emptyset$ (where $S' = e - (\bigcup_{i=1}^{k} S_i)$) is the set of unplaced events), then the PARTIALCOL algorithm will need to be invoked.

As with the original algorithm, this method operates using tabu search with the simple cost function $|S'|$. During a run, the neighbourhood operator moves events between S' and timeslots in S while maintaining the validity of the solution. Given an event $e_i \in S'$ and timeslot $S_j \in S$, checks are first made to see if e_i can be assigned to S_j without violating Constraints (9.7)–(9.9). If the assignment of e_i violates one of these constraints, the move is rejected; else, all events e_l in S_j that conflict with e_i (according to the conflicts matrix $\mathbf{C}$) are transferred from S_j into S' and an attempt is made to insert event e_i into S_j, perhaps using a maximum matching algorithm. If this is not possible (i.e., Constraint (9.10) cannot be satisfied) then all changes are again reset; otherwise, the move is considered valid. Upon performing this change, all moves involving the reassignment of event(s) e_l to timeslot S_j are considered tabu for some iterations. Here, we use the same tabu tenure as that of the TABUCOL algorithm seen in Chap. 5. Hence, the tenure is defined as a random variable proportional to the current cost, $\lfloor 0.6 \times |S'| + x \rfloor$, where x is an integer uniformly selected from the set $\{0, 1, \ldots, 9\}$.

Similarly to the original PARTIALCOL algorithm, at each iteration the entire neighbourhood of $(|S'| \times k)$ moves is examined, and the move that is chosen is the one that invokes the largest decrease (or failing that, the smallest increase) in the cost of any valid, non-tabu move. Ties are broken randomly, and tabu moves are also permitted if they are seen to improve on the best solution found so far. From time to time, there may also be no valid non-tabu moves available from a particular solution, in which case a randomly selected event is transferred from S into S', before the process is continued as above.

9.4.1 Results

Table 9.2 contains the results of our PARTIALCOL algorithm and compares them to those reported by Cambazard et al. [9].[2] We report the percentage of runs in which each instance has been solved (i.e., where a DTF of zero has been achieved), and the average time that this took (calculated only from the solved runs).[3] We see that the success rates for the two approaches are similar, with all except one instance being solved in 100% of cases (instance #10 in Cambazard et al.'s case, instance #11 in ours). However, except for instance #11, the time required by PARTIALCOL is considerably less, with an average reduction of 97.4% in CPU time achieved across the 15 remaining instances.

Curiously, when using our PARTIALCOL algorithm with instance #11, most of the runs were solved very quickly. However, in a small number of cases, the algorithm seemed to quickly navigate to a point at which a small number of events remained unplaced and where no further improvements could be made, suggesting the search was caught in a conspicuous valley in the cost landscape. To remedy this situation we, therefore, added a diversification mechanism to the method which attempts to break out of such regions. We call this our improved PARTIALCOL algorithm and its results are also given in Table 9.2.

In the improved PARTIALCOL method, our diversification mechanism is used for making relatively large changes to the incumbent solution, allowing new regions of

---

[2]Our algorithm was implemented in C++, and all experiments were conducted on 3.0 GHz Windows 7 PCs with 3.87 GB RAM. The competition benchmarking program allocated 247 s on this equipment. The source code is available at http://www.rhydlewis.eu/resources/ttCodeResults.zip.

[3]For comparative purposes, the computation times stated by Cambazard et al. [9] have been altered in Table 9.2 to reflect the increased speed of our equipment. According to Cambazard et al. the competition benchmark program allocated them 324 s per run. Consequently, their original run times have been reduced by 23.8%. We should note, however, that when comparing algorithms in this way, discrepancies in results and times can also occur due to differences in the hardware, operating system, programming language, and compiler options used. Our use of the competition benchmark program attempts to reduce discrepancies caused by the first two factors, but cannot correct for differences arising due to the latter two.

Table 9.2 Comparison of results from the LS-colouring method of Cambazard et al. [9], and our PARTIALCOL and Improved PARTIALCOL algorithms (all figures taken from 100 runs per instance)

Instance #	1	2	3	4	5	6	7	8	9	10	11	12
Cambazard et al. [9]												
% Solved	100	100	100	100	100	100	100	100	100	**98**	100	100
Avg. time (s)	11.60	37.10	0.37	0.43	3.58	4.32	1.84	1.11	51.73	170.24	0.40	0.64
PARTIALCOL												
% Solved	100	100	100	100	100	100	100	100	100	100	**98**	100
Avg. time (s)	0.25	0.79	0.02	0.04	0.05	0.07	0.02	0.01	0.71	1.80	1.88	0.04
Improved PARTIALCOL												
% Solved	100	100	100	100	100	100	100	100	100	100	100	100
Avg. time (s)	0.25	0.79	0.02	0.02	0.06	0.08	0.03	0.01	0.68	2.03	0.03	0.04

Instance #	13	14	15	16	17	18	19	20	21	22	23	24
Cambazard et al. [9]												
% Solved	100	100	100	100	–	–	–	–	–	–	–	–
Avg. time (s)	8.86	7.97	0.80	0.55	–	–	–	–	–	–	–	–
PARTIALCOL												
% Solved	100	100	100	100	100	100	100	100	100	100	100	100
Avg. time (s)	0.08	0.11	0.01	0.01	0.00	0.02	0.74	0.01	0.07	3.77	1.33	0.17
Improved PARTIALCOL												
% Solved	100	100	100	100	100	100	100	100	100	100	100	100
Avg. time (s)	0.08	0.11	0.01	0.01	0.00	0.02	0.71	0.01	0.08	3.80	1.10	0.18

the solution space to be explored. It is called when the best solution found so far has not been improved for a set number of iterations. The mechanism operates by first randomly selecting a percentage of events in S and transferring them to the set of unplaced events S'. Next, alterations are made to S by performing a random walk using neighbourhood operator N_5 (to be described in Sect. 9.5). Finally, the tabu list is reset so that all potential moves are deemed non-tabu before PARTIALCOL continues to execute as before. For the results in Table 9.2, the diversification mechanism was called after 5000 non-improving iterations and extracted 10% of all events in S. A random walk of 100 neighbourhood moves was then performed, giving a $>95\%$ chance of all timeslots being altered by the neighbourhood operator (several other parameters were also tried here, though few differences in performance were observed). We see that the improved PARTIALCOL method has achieved feasibility in all runs in the sample, with the average time reduction remaining at 97.4% compared to the method of Cambazard et al. [9].

9.5   Algorithm Description: Stage Two

9.5.1   SA Cooling Scheme

In the second stage of this algorithm, we use simulated annealing (SA) to explore the space of valid/feasible solutions, and attempt to minimise the number of soft constraint violations measured by the SCC (Eq. (9.14)). This metaheuristic is applied similarly to that described in Chap. 4: starting at an initial temperature T_0, during execution the temperature variable is slowly reduced according to an update rule $T_{i+1} = \alpha T_i$, where the cooling rate $\alpha \in (0, 1)$. At each temperature T_i, a Markov chain is generated by performing n^2 applications of the neighbourhood operator. Moves that are seen to violate a hard constraint are immediately rejected. Moves that preserve feasibility but that increase the cost of the solution are accepted with probability $\exp(-|\delta|/T_i)$ (where δ is the change in cost), while moves that reduce or maintain the cost are always accepted. The initial temperature T_0 is calculated automatically by performing a small sample of neighbourhood moves and using the standard deviation of the cost over these moves [18].

Because this algorithm is intended to operate according to a time limit, a value for α is determined automatically so that the temperature is reduced as slowly as possible between T_0 and some end temperature T_{end}. This is achieved by allowing α to be modified during a run according to the length of time that each Markov chain takes to generate. Specifically, let μ^* denote the estimated number of Markov chains that will be completed in the remainder of the run, calculated by dividing the amount of remaining run time by the length of time the most recent Markov chain (operating at temperature T_i) took to generate. On completion of the ith Markov chain, a modified cooling rate can thus be calculated as

$$\alpha_{i+1} = (T_{\text{end}}/T_i)^{1/\mu^*} \tag{9.19}$$

The upshot is that the cooling rate will be altered slightly during a run, allowing the user-specified end temperature T_{end} to be reached at the time limit. Suitable values for T_{end}, the only parameter required for this phase, are examined in Sect. 9.6.

9.5.2 Neighbourhood Operators

We now define a number of different neighbourhood operators that can be used in conjunction with the SA algorithm for this problem. Let $N(S)$ be the set of candidate solutions in the neighbourhood of the incumbent solution S. Also, let $\mathbb{S}$ be the set of all valid solutions (i.e., $S \in \mathbb{S}$ if and only if Constraints (9.4)–(9.10) are satisfied). The relationship between the solution space and neighbourhood operator can now be defined by a graph $G = (\mathbb{S}, E)$ with vertex set $\mathbb{S}$ and edge set $E = \{\{S, S'\} : S, S' \in \mathbb{S} \wedge S' \in N(S)\}$. Note that all of the following neighbourhood operators are reversible, meaning that $S' \in N(S)$ if and only if $S \in N(S')$. Hence edges are expressed as unordered pairs. The various neighbourhood operators are now defined.

N_1: The first neighbourhood operator is based on those used by Lewis [17] and Nothegger et al. [13]. Consider a valid solution S represented as a matrix $\mathbf{Z}_{|r| \times k}$ in which rows represent rooms and columns represent timeslots. Each element of $\mathbf{Z}$ can be blank or can be occupied by exactly one event. If Z_{ij} is blank, then room r_i is *vacant* in timeslot t_j; if $Z_{ij} = e_l$, then event e_l is assigned to room r_i and timeslot t_j. N_1 operates by first randomly selecting an element $Z_{i_1 j_1}$ containing an arbitrary event e_l. A second element $Z_{i_2 j_2}$ is then randomly selected in a different timeslot ($j_1 \neq j_2$). If Z_{i_2, j_2} is blank, the operator attempts to transfer e_l from timeslot j_1 into any vacant room in timeslot j_2; if $Z_{i_2 j_2} = e_q$, then a swap is attempted in which e_l is moved into any vacant room in timeslot j_2, and e_q is moved into any vacant room in timeslot j_1. If such changes are seen to violate any of the hard constraints, they are rejected immediately; else they are kept and the new solution is evaluated according to Eq. (9.14).

N_2: This operates in the same manner as N_1. However, when seeking to insert an event into a timeslot, if no vacant, suitable room is available, a maximum matching algorithm is also executed to determine if a valid room allocation of the events can be found. A similar operator was used by Cambazard et al. [11] in their winning competition entry.

N_3: This is an extension of N_2. Specifically, if the proposed move in N_2 will result in a violation of Constraint (9.6), then a Kempe chain interchange is attempted (see Definition 4.3). An example of this process is shown in Fig. 9.3a. Imagine in this case that we have chosen to swap the events $e_5 \in S_i$ and $e_{10} \in S_j$. However, doing so will violate Constraint (9.6) because events e_5 and e_{11} conflict but would now both be assigned to timeslot S_j. In this case, we therefore construct the Kempe chain $\text{KEMPE}(e_5, i, j) = \{e_5, e_{10}, e_{11}\}$ which, when interchanged, guarantees the preservation of Constraint (9.6), as shown in Fig. 9.3b. Observe

that this neighbourhood operator also includes pair swaps (see Definition 4.4)—
for example, if we were to select events $e_4 \in S_i$ and $e_8 \in S_j$ from Fig. 9.3a.
Note, however, that as with the previous neighbourhood operators, applications
of N_3 may not preserve the satisfaction of the remaining hard constraints. Such
moves will again need to be rejected if this is the case.

N_4: This operator extends N_3 by using the idea of *double* Kempe chains, originally
proposed by Lü and Hao [19]. In many cases, a proposed Kempe chain interchange will be rejected because it will violate Constraint (9.10): that is, suitable
rooms will not be available for all of the events proposed for assignment to a particular timeslot. For example, in Figure 9.3a, the proposed Kempe interchange
involving events $\{e_1, e_2, e_3, e_6, e_7\}$ is guaranteed to violate Constraint (9.10)
because it will result in too many events in timeslot S_j for a feasible matching
to be possible. However, applying a second Kempe chain interchange at the
same time may result in feasibility being maintained, as illustrated in Fig. 9.3c.
In this operator, if a proposed single Kempe chain interchange is seen to violate
Constraint (9.10) only, then a random vertex from one of the two timeslots,
but from outside this chain, is randomly selected, and a second Kempe chain
is formed from it. If the proposed interchange of both Kempe chains does not
violate any of the hard constraints, then the move can be performed and the
new solution can be evaluated according to Eq. (9.14) as before.

N_5: Finally, N_5 defines a *multi*-Kempe chain operator. This generalises N_4 in that if
a proposed double Kempe chain interchange is seen to violate Constraint (9.10)
only, then triple Kempe chains, quadruple Kempe chains, and so on, can also
be investigated in the same manner. Note that when constructing these multiple
Kempe chains, a violation of any of the constraints (9.7)–(9.9) allows us to
reject the move immediately. However, if only Constraint (9.10) continues to
be violated, then eventually the considered Kempe chains will contain all events
in both timeslots, in which case the move becomes equivalent to swapping the
contents of the two timeslots. Trivially, in such a move Constraint (9.10) is
guaranteed to be satisfied.

From the above descriptions, it is clear that each successive neighbourhood operator requires more computation than its predecessor. Each operator also generalises
its predecessor—that is, $N_1(\mathcal{S}) \subseteq N_2(\mathcal{S}) \subseteq \cdots \subseteq N_5(\mathcal{S})$, $\forall \mathcal{S} \in \mathbb{S}$. From the perspective of the graph $G = (\mathbb{S}, E)$ defined above, this implies a greater connectivity of
the solution space since $E_1 \subseteq E_2 \subseteq \ldots \subseteq E_5$ (where $E_i = \{\{\mathcal{S}, \mathcal{S}'\} : \mathcal{S}' \in N_i(\mathcal{S})\}$
for $i = 1, \ldots, 5$). Note, though, that the set of vertices (solutions) $\mathbb{S}$ remains the
same under these different operators.

Finally, it is also worth mentioning that each of the above operators only ever
alters the contents of two timeslots in any particular move. In practice, this means
that we only need to consider the particular days and students affected by the move
when reevaluating the solution according to Eq. (9.14). This allows considerable
speed-up of the algorithm.

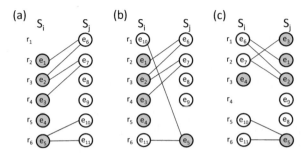

Fig. 9.3 Example moves using N_3 and N_4. Here, edges exist between pairs of vertices (events) e_l, e_q if and only if $C_{lq} = 1$. Part **a** shows two timeslots containing two Kempe chains, $\{e_1, e_2, e_3, e_6, e_7\}$ and $\{e_5, e_{10}, e_{11}\}$. Part **b** shows a result of interchanging the latter chain. Part **c** shows the result of interchanging both chains. Note that room allocations are determined via a matching algorithm and can therefore change during an interchange

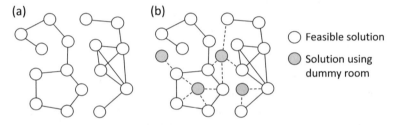

Fig. 9.4 Graphs depicting the connectivity of a solution space with **a** no dummy rooms, and **b** one or more dummy rooms

9.5.3   Dummy Rooms

An additional opportunity for altering the connectivity of the underlying solution space with this problem is through the use of *dummy rooms*. A dummy room is an extra room made available in all timeslots and defined as suitable for all events (i.e., it has an infinite seating capacity and possesses all available room features). Dummy rooms can be used with any of the previous neighbourhood operators, and multiple dummy rooms can also be applied if necessary. We therefore use the notation $N_i^{(j)}$ to denote the use of neighbourhood N_i using j dummy rooms (giving $|r| + j$ rooms in total). Similarly, we can use the notation $\mathbb{S}^{(j)}$ to denote the space of all solutions that obey all hard constraints, and where j dummy rooms are available. (For brevity, where a superscript is not used, we assume no dummy room is being used.) Dummy rooms can be viewed as a type of "placeholder" that are used to contain events not currently assigned to the "real" timetable. Transferring events in and out of dummy rooms might therefore be seen as similar to moving events in and out of the timetable.

Note that the use of dummy rooms increases the size of the set $\mathcal{M}$, making Constraint (9.10) easier to satisfy. This leads to the situation depicted in Fig. 9.4. Here, we observe that the presence of dummy rooms increases the number of vertices/solutions (i.e., $\mathbb{S}^{(j)} \subseteq \mathbb{S}^{(j+1)}$, $\forall j \geq 0$), with extra edges (dotted in the figure) being created between some of the original vertices and new vertices. As depicted, this could also allow previously disjoint components of the solution space to become connected.

Because they do not form part of the original problem, at the end of the optimisation process all dummy rooms will need to be removed. This means that any events assigned to these will contribute to the DTF measure. Because this is undesirable, in our case we attempt to discourage the assignment of events to the dummy rooms during evaluation by considering all events assigned to a dummy room as unplaced. We then use the cost function $w \times \text{DTF} + \text{SCP}$, where w is a weighting coefficient that will need to be set by the user. Additionally, when employing the maximum matching algorithm it also makes sense to ensure that the dummy room is only used when necessary—that is, if a feasible matching can be achieved without using dummy rooms, then this is the one that will be used.

9.5.4 Estimating Solution Space Connectivity

We have now seen various neighbourhood operators for this problem and made some observations on the connectivity of their underlying solution spaces, defined by the graph $G = (\mathbb{S}, E)$. Unfortunately, however, it is very difficult to gain a complete understanding of G's connectivity because it is simply too large. In particular, we are unlikely to be able to confirm whether G is connected or not, which would be useful information if we wanted to know whether an optimal solution could be reached from any other solution within the solution space.

One way to gain an indication of G's connectivity is to make use of what we call the *feasibility ratio*. This is defined as the proportion of proposed neighbourhood moves that are seen to not violate any of the hard constraints (i.e., that maintain validity/feasibility). A lower feasibility ratio suggests lower connectivity in G because, on average, more potential moves will be seen to violate a hard constraint, making movements within the solution space more restricted. A higher feasibility ratio will suggest a greater level of connectivity.

Figure 9.5 displays the feasibility ratios for neighbourhood operators $N_1, \ldots, N_5$ and also $N_5^{(1)}$ for all available problem instances. These mean figures were found by performing random walks of 50,000 feasible-preserving moves from a sample of 20 feasible solutions per instance. As expected, we see that the feasibility ratios increase for each successive neighbourhood operator, though the differences between N_3, N_4, and N_5 appear to be only marginal. We also observe quite a large range across the instances, with instance #10 appearing to exhibit the least connected solution space (with feasibility ratios ranging from just 0.0005 (N_1) to 0.004 ($N_5^{(1)}$)), and instance #17 having the highest levels of connectivity (0.04 (N_1) to 0.10 ($N_5^{(1)}$)). Standard deviations from these samples range between 0.000018 (N_2, #20) and 0.000806 (N_1, #23). These observations will help to explain the results in the following sections.

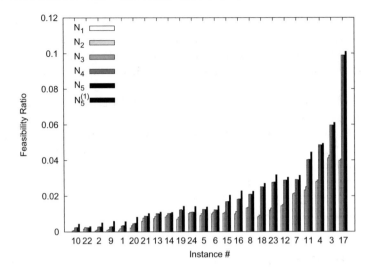

Fig. 9.5 Feasibility ratio for neighbourhood operators $N_1 \ldots N_5$ and also $N_5^{(1)}$ for all 24 problem instances

9.6   Experimental Results

9.6.1   Influence of the Neighbourhood Operators

We now examine the ability of each neighbourhood operator to reduce the soft constraint cost within the time limit specified by the competition benchmarking program (minus the time used for Stage 1). We also consider the effects of altering the end temperature of simulated annealing T_{end}, which is the only run-time parameter required in this stage. To measure performance, we compare our results to those achieved by the five finalists of the 2007 competition using the competition's ranking system. This involves calculating a "ranking score" for each algorithm, which is derived as follows.

Given x algorithms and a single problem instance, each algorithm is executed y times, giving xy results. These results are then ranked from 1 to xy, with ties receiving a rank equal to the average of the ranks they span. The mean of the ranks assigned to each algorithm is then calculated, giving the respective rank scores for the x algorithms on this instance. This process is then repeated on all instances, and the mean of all ranking scores for each algorithm is taken as its overall ranking score. A worked example of this process is shown in Table 9.3. We see that the best ranking score achievable for an algorithm on a particular instance is $(y+1)/2$, in which case its y results are better than all of the other algorithms' (as is the case with Algorithm A on instance #3 in the table). The worst possible ranking score, $(x-1)y+(y+1)/2$, has occurred with Algorithm C with instances #1, #2, and #3 in the table.

A full breakdown of the results and ranking scores of the five competition finalists can be found on the official website of ITC2007 at http://www.cs.qub.ac.uk/itc2007/.

Table 9.3 Worked example of how rank scores are calculated using, in this case, $y = 2$ runs of $x = 3$ algorithms on three problem instances. Results of each run are given by the DTF (in parentheses) and the SCC. Here, Algorithm A is deemed the winner and C the loser

Instance run	Results						Ranks			Rank scores			Mean
	#1		#2		#3		#1	#2	#3	#1	#2	#3	
	1	2	1	2	1	2							
Alg. A	(0) 0	(0) 10	(0) 1	(0) 5	(0) 0	(0) 2	1, 3	1.5, 3	1, 2	2	2.25	1.5	1.92
Alg. B	(0) 5	(0) 17	(0) 1	(0) 8	(0) 8	(0) 11	2, 4	1.5, 4	3, 4	3	2.75	3.5	3.08
Alg. C	(9) 3	(0) 19	(0) 18	(0) 16	(0) 12	(4) 0	6, 5	6, 5	5, 6	5.5	5.5	5.5	5.50

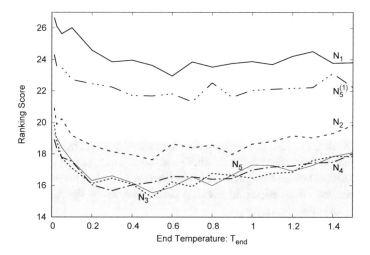

Fig. 9.6 Ranking scores achieved by the different neighbourhood operators using different end temperatures. The shaded area indicates the results that would have won the competition

In our case, we added results from ten runs of our algorithm to these published results, giving $x = 6$ and $y = 10$. A summary of the resultant ranking scores achieved by our algorithm with each neighbourhood operator over a range of different settings for T_{end} is given in Fig. 9.6. The shaded area of the figure indicates those settings where our algorithm would have won the competition (i.e., that have achieved a lower ranking score than the other five entries).

Figure 9.6 shows a clear difference in the performance of neighbourhood operators N_1 and N_2, illustrating the importance of the extra solution space connectivity provided by the maximum matching algorithm. Similarly, the results of N_3, N_4 and N_5 are better still, outperforming N_1 and N_2 across all of the values of T_{end} tested. However, there is very little difference between the performance of N_3, N_4 and N_5 themselves presumably because, for these particular problem instances, the behaviour and therefore feasibility ratios of these operators are very similar (as seen in Fig. 9.5). Moreover, we find that the extra expense of N_5 over N_3 and N_4 appears to have minimal effect, with N_5 producing less than 0.5% fewer Markov chains than N_3 over the course of the run on average. Of course, such similarities will not always be the case—they merely seem to be occurring with these particular problem instances because, in most cases, hard constraints are being broken (and the move rejected) before the inspection of more than one Kempe chain is deemed necessary.

Figure 9.6 also indicates that using dummy rooms does not seem to improve results across the instances. In initial experiments, we tested the use of one and two dummy rooms along with a range of different values for the weighting coefficient $w \in \{1, 2, 5, 10, 20, 200, \infty\}$ (some of these values were chosen due to their use in existing algorithms that employ weighted sum functions with this problem [12,13, 15]). Figure 9.6 reports the best of these: one dummy room with $w = 2$. For higher values of w, results were found to be inferior because the additional solutions in the solution space (shaded vertices in Fig. 9.4) would still be evaluated by the algorithm, but nearly always rejected due to their high cost. On the other hand, using a setting of $w = 1$ means that the penalty of assigning an event to a dummy room will be equal to the penalty of assigning the event to the last timeslot of a day (soft constraint SC1), meaning there is little distinction between the cost of infeasibility and the cost of soft constraint violations.

Note that we might consider the use of no dummy rooms as similar to using $w \approx \infty$, in that the algorithm will be unable to accept moves that involve moving an event into a dummy room (i.e., introducing infeasibility to the timetable). However, the difference is that, when using dummy rooms, such moves will still be evaluated by the algorithm before being rejected. On the other hand, without dummy rooms these unnecessary evaluations do not take place. This saves significant amounts of time during the course of a run.

As mentioned, a setting of $w = 2$, which seems the best compromise between these extremes, still produces inferior results on average compared to when using no dummy rooms. However, for problem instance #10 we found the opposite to be true, with significantly better results being produced when dummy rooms *are* used. From Fig. 9.5, we observe that instance #10 has the lowest feasibility ratio of all instances, and so the extra connectivity provided by the dummy rooms seems to be aiding the search in this case. On the other hand, the existence of a perfect solution here could mean that, while optimising the SCC, the search might also be being simultaneously guided towards regions of the solution space that are feasible. This matter will be discussed further in Sect. 9.7.

Finally, Fig. 9.7 illustrates, for the 24 problem instances, the relationship between the feasibility ratio and the proportion by which the SCC is reduced by the SA algorithm for two contrasting neighbourhood operators N_1 and N_5. We see that the points for N_5 are shifted upwards and rightwards compared to N_1, illustrating the larger feasibility ratios and higher performance of the operator. The general pattern in the figure suggests that higher feasibility ratios allow large decreases in cost during a run, while lower feasibility ratios can result in both large or small decreases, depending on the instance. Thus, while there is some relationship between the two variables, it seems that other factors also have an impact here, including the size

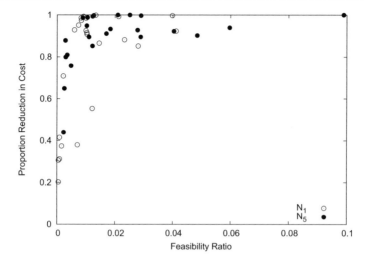

Fig. 9.7 Scatter plot showing the relationship between the feasibility ratio and the reduction in cost for the 24 competition instances, using neighbourhood operators N_1 and N_5

and shape of the cost landscape and the amount of computation needed for each application of the evaluation function.

9.6.2   Comparison to Published Results

In our next set of experiments, we compare the performance of our algorithm to the best results that were reported in the literature in the five years following the 2007 competition. Table 9.4 gives a breakdown of the results achieved by our method using $T_{end} = 0.5$ compared to the approaches of Cambazard et al. [9], Nothegger et al. [13] and van den Broek and Hurkens [15]. Note that the latter two papers only list results for the first 16 instances. In this table, all statistics are calculated from 100 runs on each instance except for van den Broek and Hurkens, whose algorithm is deterministic. All results were achieved strictly within the time limits specified by the competition benchmark program. Our experiments were performed using N_3, N_4, and N_5, though no significant difference was observed between the three operators' best, mean, or worst results. Consequently, we only present the results for one of these.[4]

Table 9.4 shows that, using N_4, perfect solutions have been achieved by our method in 17 of the 24 problem instances. A comparison to the 16 results reported by Cambazard et al. [9] indicates that our method's best, mean, and worst results are also significantly better than their corresponding results. Similarly, our best, mean, and

---

[4]For pairwise comparisons, Related Samples Wilcoxon Signed-Rank Tests were used; else Friedman Tests were used (significance level 0.05).

Table 9.4 Results from the literature, taken from samples of 100 runs per instance. Figures indicate the SCC achieved at the cut-off point defined by the competition benchmarking program. Numbers in parentheses indicate the % of runs where feasibility was found. No parentheses indicates that feasibility was achieved in all runs

#	Our method using N_4			Cambazard[a]			van den Broek[b]	Nothegger[c]	
	Best	Mean	Worst	Best	Mean	Worst	Result	Best	Mean
1	0	377.0	833	15	547	1072	1636	0	(54) 613
2	0	382.2	1934	9	403	1254	1634	0	(59) 556
3	122	181.8	240	174	254	465	355	110	680
4	18	319.4	444	249	361	666	644	53	580
5	0	7.5	60	0	26	154	525	13	92
6	0	22.8	229	0	16	133	640	0	(95) 212
7	0	5.5	11	1	8	32	0	0	4
8	0	0.6	59	0	0	0	241	0	61
9	0	514.4	1751	29	1167	1902	1889	0	(85) 202
10	0	1202.4	2215	2	(89) 1297	2637	1677	0	4
11	48	202.6	358	178	361	496	615	143	(99) 774
12	0	340.2	583	14	380	676	528	0	(86) 538
13	0	79.0	269	0	135	425	485	5	(94) 360
14	0	0.5	7	0	15	139	739	0	41
15	0	139.9	325	0	47	294	330	0	29
16	0	105.2	223	1	58	245	260	0	101
17	0	0.1	3	–	–	–	35	–	–
18	0	2.2	57	–	–	–	503	–	–
19	0	346.1	1222	–	–	–	963	–	–
20	557	724.5	881	–	–	–	1229	–	–
21	1	32.1	159	–	–	–	670	–	–
22	4	1790.1	2280	–	–	–	1956	–	–
23	0	514.1	1178	–	–	–	2368	–	–
24	18	328.2	818	–	–	–	945	–	–

[a]SA-colouring method [9, p. 122]
[b]Deterministic IP-based heuristic (one result per instance) [15, p. 451]
[c]Serial ACO algorithm [13, p. 334]

worst results are all seen to outperform the results of van den Broek and Hurkens [15]. Finally, no significant difference is observed between the best and mean results of our method compared to Nothegger et al. [13]; however, unlike our algorithm, they have failed to achieve feasibility in several cases.

9.6.3   Differing Time Limits

In this chapter's final set of experiments, we look at the effects of using different time limits with our algorithm. Until this point, experiments have been performed according to the time limit specified by the competition benchmark program; however, it is pertinent to ask whether the less expensive neighbourhood operators are more suitable when shorter time limits are used and whether further improvements can be achieved when the time limit is extended.

In Fig. 9.8 we show the relative performance of operators N_1, N_2, N_3, and N_5 using time limits of between 1 and 600 s, signifying very fast and very slow coolings, respectively. (N_4 is omitted here due to its close similarity with N_3 and N_5's results.) We see that even for very short time limits of less than five seconds, the more expensive neighbourhoods consistently produce superior solutions across the instances. We also see that when the time limit is extended beyond the benchmark and up to 600 s, the mean reduction in the soft cost rises from 89.1 to 94.6% (under N_5), indicating that superior results can also be gained with additional computing resources. This latter observation is consistent with that of Nothegger et al. [13], who were also able to improve the results of their algorithm, in their case via parallelisation.

9.7   Chapter Summary and Discussion

This chapter has considered the problem of constructing university timetables—a problem that can often involve a multitude of different constraints and requirements. Despite these variations, like the case studies seen in the previous two chapters, this problem usually contains an underlying graph colouring problem from which powerful algorithmic operators can then be derived.

Using this connection with graph colouring, we have proposed a robust two-stage algorithm for a well-known $\mathcal{NP}$-hard timetabling formulation known as the post enrolment-based timetabling problem. Stage 1 of this algorithm has proven to be very successful for finding feasibility with the considered problem instances with regard to both success rate and computation time. For Stage 2, we have then focussed on issues surrounding the connectivity of the solution space, and have seen that results generally improve when this connectivity is increased.

It is noticeable that many successful algorithms for the post enrolment-based timetabling problem have used simulated annealing as their main mechanism for reducing the number of soft constraint violations [9,16,20]. An obvious alternative to this metaheuristic is tabu search; however, this does not seem to have fared as favourably with this problem formulation in practice. One factor behind this apparent lack of performance could be due to the features noted in this chapter—that a decreased connectivity of the solution space tends to lead to fewer gains being made during the optimisation process.

To illustrate this, consider the situation shown in Fig. 9.9 where a solution space and neighbourhood operator is again defined as a graph $G = (\mathbb{S}, E)$. In Part (a) we

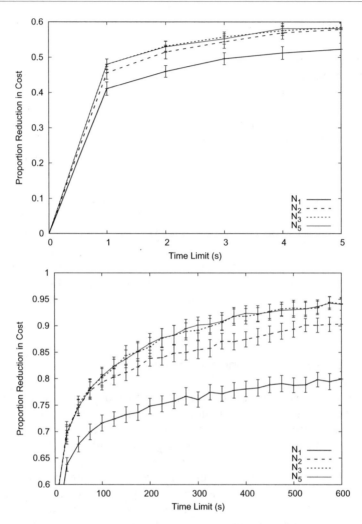

Fig. 9.8 Proportion decrease in SCC using differing time limits and differing neighbourhood operators. All points are taken from an average of ten runs on each problem instance (i.e., 240 runs). Error bars represent one standard error each side of the mean

show the effect of performing a neighbourhood move (i.e., changing the incumbent solution) with simulated annealing. In particular, we see that the connectivity of G does not change (though the probabilities of traversing the edges may change if the temperature parameter is subsequently updated). On the other hand, when the same move is performed using tabu search (Part (b)), several edges in G, including $\{S, S'\}$, will be made tabu for several iterations, effectively removing them from the graph for a time dictated by the tabu tenure. The exact edges that will be made tabu depends on the structure of the tabu list, and in typical applications, when an event e_i has been

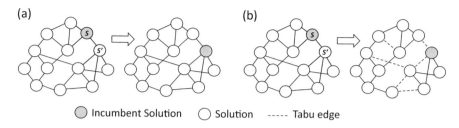

Fig. 9.9 Illustration of the effects of performing a neighbourhood move using **a** simulated annealing, and **b** tabu search

moved from timeslot S_j to a new timeslot, *all* moves that involve moving e_i back into S_j are made tabu. While the use of tabu moves helps to prevent cycling (which may regularly occur with SA), it therefore also has the effect of further reducing the connectivity of G. Over the course of a run, the cumulative effects of this phenomenon may put tabu search at a disadvantage with these particular problems.

In this chapter, we have also noted that an alternative approach to a two-stage algorithm is to use a one-stage optimisation algorithm in which the satisfaction of both hard and soft constraints is attempted simultaneously, as with the methods of Ceschia et al. [12] and Nothegger et al. [13]. As we have seen, despite the favourable performance of our two-stage algorithm overall, it does seem to struggle in comparison to these approaches for a small number of problem instances, particularly #10 and #22. According to Fig. 9.5, these instances exhibit the lowest feasibility ratios with our operators, seemingly suggesting that freedom of movement in the solution spaces is too restricted to allow adequate optimisation of the objective function.

On the other hand, it is also possible that the algorithms of Ceschia et al. [12] and Nothegger et al. [13] are being aided by the fact that perfect solutions to the 24 competition instances are known to exist—a feature that is unlikely to occur in real-world problem instances. For example, as mentioned in Sect. 9.3, Ceschia et al.'s algorithm is reported to produce its best results when optimisation is performed using an objective function in which hard and soft constraint violations are given equal weights. However, it could be that, by moving towards solutions with low SCCs, the search could also inadvertently be moving towards feasible regions of the solution space, simultaneously helping to satisfy the hard constraints along the way. This hypothesis was tested by Lewis [21], who compared the algorithm of this chapter to Ceschia et al.'s using a different suite of timetabling problems for which the existence of perfect solutions is not always known. These can be downloaded from http://www.rhydlewis.eu/hardTT. The results of their tests strongly support this hypothesis: for the 23 instances of this suite with no known perfect solution, the two-stage algorithm outperformed Ceschia et al.'s in 21 cases (91.3%), with stark differences in results. On the other hand, with the remaining 17 instances, Ceschia et al.'s approach produced better results in 12.5 cases (73.5%), suggesting that the existence of perfect solutions indeed benefits the algorithm.

As mentioned earlier, all of the problem instances used in this chapter are available online at http://www.cs.qub.ac.uk/itc2007/. Also, a full listing of this chapter's results, together with the C++ source code of the two-stage algorithm is available at http://www.rhydlewis.eu/resources/ttCodeResults.zip.

References

1. Corne D, Ross P, Fang H (1995) Evolving timetables. In: Chambers L (ed) The practical handbook of genetic algorithms, vol 1. CRC Press, pp 219–276
2. McCollum B, Schaerf A, Paechter B, McMullan P, Lewis R, Parkes A, Di Gaspero L, Qu R, Burke E (2010) Setting the research agenda in automated timetabling: the second international timetabling competition. INFORMS J Comput 22(1):120–130
3. Müller T, Rudova H (2012) Real life curriculum-based timetabling. In: Kjenstad D, Riise A, Nordlander T, McCollum B, Burke E (eds) Practice and theory of automated timetabling (PATAT 2012), pp 57–72
4. Cooper T, Kingston J (1996) The complexity of timetable construction problems. In: Burke E, Ross P (eds) Practice and theory of automated timetabling (PATAT) I. LNCS, vol 1153. Springer, pp 283–295
5. Carter M, Laporte G, Lee SY (1996) Examination timetabling: algorithmic strategies and applications. J Oper Res Soc 47:373–383
6. Burke E, Elliman D, Ford P, Weare R (1996) Examination timetabling in British universities: a survey. In: Burke E, Ross P (eds) Practice and theory of automated timetabling (PATAT) I. LNCS, vol 1153. Springer, pp 76–92
7. Schaerf A (1999) A survey of automated timetabling. Artif Intell Rev 13(2):87–127
8. Lewis R (2008) A survey of metaheuristic-based techniques for university timetabling problems. OR Spectr 30(1):167–190
9. Cambazard H, Hebrard E, O'Sullivan B, Papadopoulos A (2012) Local search and constraint programming for the post enrolment-based timetabling problem. Ann Oper Res 194:111–135
10. Rossi-Doria O, Samples M, Birattari M, Chiarandini M, Knowles J, Manfrin M, Mastrolilli M, Paquete L, Paechter B, Stützle T (2002) A comparison of the performance of different metaheuristics on the timetabling problem. In: Burke E, De Causmaecker P (eds) Practice and theory of automated timetabling (PATAT) IV. LNCS, vol 2740. Springer, pp 329–351
11. Cambazard H, Hebrard E, O'Sullivan B, Papadopoulos A (2008) Local search and constraint programming for the post enrolment-based course timetabling problem. In: Burke E, Gendreau M (eds) Practice and theory of automated timetabling (PATAT) VII
12. Ceschia S, Di Gaspero L, Schaerf A (2012) Design, engineering, and experimental analysis of a simulated annealing approach to the post-enrolment course timetabling problem. Comput Oper Res 39:1615–1624
13. Nothegger C, Mayer A, Chwatal A, Raidl G (2012) Solving the post enrolment course timetabling problem by ant colony optimization. Ann Oper Res 194:325–339
14. Jat S, Yang S (2011) A hybrid genetic algorithm and Tabu search approach for post enrolment course timetabling. J Sched 14:617–637
15. van den Broek J, Hurkens C (2012) An IP-based heuristic for the post enrolment course timetabling problem of the ITC2007. Ann Oper Res 194:439–454
16. Chiarandini M, Birattari M, Socha K, Rossi-Doria O (2006) An effective hybrid algorithm for university course timetabling. J Sched 9(5):403–432. ISSN 1094-6136
17. Lewis R (2012) A time-dependent metaheuristic algorithm for post enrolment-based course timetabling. Ann Oper Res 194(1):273–289

18. van Laarhoven P, Aarts E (1987) Simulated Annealing: Theory and Applications. Kluwer Academic Publishers
19. Lü Z, Hao J-K (2010b) Adaptive Tabu search for course timetabling. Eur J Oper Res 200(1):235–244
20. Kostuch P (2005) The university course timetabling problem with a 3-phase approach. In: Burke E, Trick M (eds) Practice and theory of automated timetabling (PATAT) V. LNCS, vol 3616. Springer, pp 109–125
21. Lewis R, Thompson J (2015) Analysing the effects of solution space connectivity with an effective metaheuristic for the course timetabling problem. Eur J Oper Res 240:637–648

Computing Resources

A

A.1 Algorithm User Guide

This section contains instructions on how to compile and use the implementations of the algorithms described in Chaps. 3 and 5 of this book. These can all be downloaded directly from:

> http://rhydlewis.eu/resources/gCol.zip

Once downloaded and unzipped, we see that the directory contains a number of subdirectories. Specifically, these are:

- **Constructive**: The GREEDY, Welsh-Powell, DSATUR, GREEDY-IS and RLF algorithms seen in Chap. 3.
- **PartialColAndTabuCol**: The PARTIALCOL and TABUCOL algorithms (Sects. 5.1 and 5.2).
- **HybridEA**: The hybrid evolutionary algorithm (Sect. 5.3).
- **AntCol**: The Ant Colony Optimisation-based algorithm for graph colouring (Sect. 5.4).
- **HillClimber**: The hill-climbing algorithm (Sect. 5.5).
- **BacktrackingDSatur**: The Backtracking algorithm based on the DSatur heuristic (Sect. 5.6).

All of these algorithms are programmed in C++. They have been successfully compiled in Windows using Microsoft Visual Studio and in Linux using **g++**. Both of these compilers are available for free online.

To compile and execute using Microsoft Visual Studio the following steps can be taken:

© The Editor(s) (if applicable) and The Author(s), under exclusive license to Springer   277
Nature Switzerland AG 2021
R. M. R. Lewis, *Guide to Graph Colouring*, Texts in Computer Science,
https://doi.org/10.1007/978-3-030-81054-2

1. Open Visual Studio and click **File**, then **New**, and then **Project from Existing Code**.
2. In the dialogue box, select **Visual C++** and click **Next**.
3. Select one of the subdirectories above, give the project a name and click **Next**.
4. Finally, select **Console Application Project** for the project type and then click **Finish**.

The source code for the chosen algorithm can then be viewed and executed from the Visual Studio application. Release mode should be used during compilation to allow the programs to execute at maximum speed.

Appropriate **makefiles** are included for compiling the source code using **g++**. Simply navigate to the correct directory at the command prompt, type **make** and hit return.

A.1.1 Usage

Once generated, the executable files (one per subdirectory) can be run from the command line. If the programs are called with no arguments, usage information is output to the screen. For example, suppose we are using the executable file **HillClimber**. Running this program with no arguments from the command line gives the following output:

```
1   Hill Climbing Algorithm for Graph Colouring
2
3   USAGE:
4   <InputFile>  (Required. File must be in DIMACS format)
5   -s <int>     (Stopping criteria expressed as number of
6                 constraint checks. Can be anything up to
7                 9x10^18. DEFAULT = 100,000,000.)
8   -I <int>     (Number of iterations of local search per cycle.
9                 DEFAULT = 1000)
10  -r <int>     (Random seed. DEFAULT = 1)
11  -T <int>     (Target number of colours. Algorithm halts if this
12                is reached. DEFAULT = 1.)
13  -v           (Verbosity. If present, output is sent to screen.
14                If -v is repeated, more output is given.)
15  ****
```

The input file specifies the graph we intend to colour. This is the only mandatory argument. Input files should be text files in the DIMACS format. Specifically,

- Initial lines in the text file will begin with the character **c**. These are used for comments but are otherwise ignored.
- After the comments, the next line in the file should start with the character **p**. This is followed by the text **edge**, which tells us that the graph is being specified by using a list of its edges. The number of vertices n and edges m are then given.
- Finally, a series of m lines beginning with the character **e** should be included. Each of these specifies a single edge of the graph by giving its two endpoints. Note that each edge **e u v** should appear exactly once in the input file. It is therefore not repeated as **e v u**.

It is assumed that files are well-formed and consistent—vertex labels are valid, exactly m edges are defined, self-loops are not present, and so on. Vertices are numbered from 1 to n.

For illustrative purposes, the following text specifies a wheel graph (comprising $n = 5$ vertices and $m = 8$ edges) using the DIMACS format. An example input file called **graph.txt** is also provided in each subdirectory.

```
1   c Example text file specifying a graph using the DIMACS format.
2   c
3   c This is a wheel graph comprising 5 vertices and 8 edges
4   c
5   p edge 5 8
6   e 1 2
7   e 1 4
8   e 1 5
9   e 2 5
10  e 2 3
11  e 3 4
12  e 3 5
13  e 4 5
```

The remaining arguments for each of the programs are optional and are allocated default values if left unspecified. Here are some example commands using the **HillClimber** executable:

```
1   HillClimber graph.txt
```

This will execute the algorithm on the problem given in the file **graph.txt**, using the default of 1000 iterations of local search per cycle and a random seed of 1. The algorithm will halt when 100,000,000 constraint checks have been performed. No output will be written to the screen. Another example command is

```
1   HillClimber graph.txt -r 6 -T 50 -v -s 500000000000
```

This run will be similar to the previous one but will use a random seed of six and will halt either when 500,000,000,000 constraint checks have been performed, or when a feasible solution using fifty or fewer colours has been found. The presence of **-v** in this command means that output will be written to the screen. Including **-v** more than once will increase the amount of output.

The arguments **-r** and **-v** are used with all of the algorithms supplied here. Similarly, **-T** and **-s** are used with all algorithms except for the single-parse constructive algorithms. Descriptions of arguments particular to just one algorithm are found by typing the name of the program with no arguments, as described above. Interpretations of the run-time parameters for the various algorithms can be found by consulting the algorithm descriptions in this book.

A.1.2 Output

When a run of any of the programs is completed, three files are created: **ceffort.txt** (computational effort), **teffort.txt** (time effort), and **solution.txt**. The first two files

specify how long (in terms of constraint checks and milliseconds, respectively) solutions with certain numbers of colours took to produce during the run. For example,
we might get the following computational effort file:

1	40	126186
2	39	427143
3	38	835996
4	37	1187086
5	36	1714932
6	35	2685661
7	34	6849302
8	33	X

This file is interpreted as follows: The first feasible solution observed used 40
colours, and this took 126,186 constraint checks to achieve. A solution with 39
colours was then found after 427,143 constraint checks, and so on. To find a solution
using 34 colours, a total of 6,849,302 constraint checks was required. Once a row
with an "X" is encountered, this indicates that no further improvements were made:
that is, no solution using fewer colours than that indicated in the previous row was
achieved. Therefore, in this example, the best solution found used 34 colours. For
consistency, the "X" is always present in a file, even if a specified target has been
met.

The file **teffort.txt** is interpreted in the same way as **ceffort.txt**, with the right-
hand column giving the time (in milliseconds) as opposed to the number of constraint
checks. Both of these files are useful for analysing algorithm speed and performance.
For example, the computational effort file above can be used to generate the following
plot:

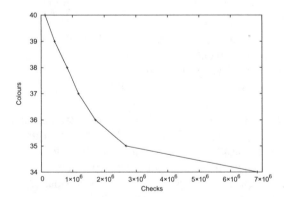

The file **solution.txt** contains the best feasible solution (i.e., the solution with the
fewest colours) that was achieved during the run. The first line of this file gives the
number of vertices n, and the remaining n lines then state the colour of each vertex,
using colour labels 0, 1, 2, ... For example, the following solution file

1	5
2	0

```
3  2
4  1
5  0
6  1
```

is interpreted as follows: There are five vertices; the first and fourth vertices are assigned to colour 0, the third and fifth vertices are assigned to colour 1, and the second vertex is assigned to colour 2. Hence, three colours are being used.

Finally, on completion, a single line is also appended to the file **resultsLog.log**. This contains the following pieces of information, separated by tabs:

1. The name of the algorithm;
2. The number of colours used in the best feasible solution observed during the run;
3. The total run time (in milliseconds);
4. The total number of constraint checks performed during the run.

If the run resulted in an error (due to unrecognised input parameters or an invalid input file), then an appropriate single-line error message is also appended to the file.

A.2 Graph Colouring in Sage

In this section, we give a brief demonstration of how Sage can be used to create, colour, and visualise graphs.

Sage is specialised software that allows the exploration of many aspects of mathematics, including combinatorics, graph theory, algebra, calculus, and number theory. It is both free to use and open source. To use Sage, commands can be typed into a Sage notebook. Blocks of commands are then executed by hitting **Shift+Enter** next to these commands, with output (if applicable) then being written back to the notebook.

Sage contains a whole host of elementary and specialised mathematical functions that are documented online at https://doc.sagemath.org/html/en/reference/index.html. Of particular interest to us here is the functionality surrounding graph colouring and graph visualisation. A full description of the graph colouring library for Sage can be found at https://doc.sagemath.org/html/en/reference/graphs/sage/graphs/graph_coloring.html.

The following text now shows some example commands from this library, together with the output that Sage produces. In our case, these commands have been typed into notebooks provided by the online tool *CoCalc*. This allows the editing and execution of Sage notebooks through a web browser and can be freely accessed online (go to https://cocalc.com/ and open a blank Sage notebook).

The following pieces of code each represent an individual block of executable Sage commands. Textual output produced by these commands is indicated by the » symbol.

To begin, it is first necessary to specify the names of the libraries that we intend to use in our Sage program. We therefore type:

```
1  from sage.graphs.graph_coloring import
       chromatic_number
2  from sage.graphs.graph_coloring import
       vertex_coloring
3  from sage.graphs.graph_coloring import
       number_of_n_colorings
4  from sage.graphs.graph_coloring import edge_coloring
```

This will allow us to access the various graph colouring functions used below.

We now use Sage to generate a small graph G. In this case, our graph has $n = 4$ vertices, $m = 5$ edges, and is defined by the following adjacency matrix

$$: A = \begin{pmatrix} 0 & 1 & 1 & 0 \\ 1 & 0 & 1 & 1 \\ 1 & 1 & 0 & 1 \\ 0 & 1 & 1 & 0 \end{pmatrix}.$$

The first Sage command below defines this matrix. The next command then transfers this information into a graph G. The final command draws G to the screen.

```
1  A = matrix([[0,1,1,0],[1,0,1,1],[1,1,0,1],[0,1,1,0]])
2  G = Graph(A)
3  G.show()
```

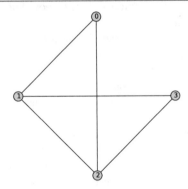

Note that, by default, Sage labels the vertices from 0 to $n - 1$ in this diagram as opposed to using indices 1 to n.

We will now produce an optimal colouring of this graph. The algorithms that Sage uses to obtain these solutions are based on integer programming techniques (see Chap. 4). They are therefore able to produce provably optimal solutions for small graphs. A colouring is produced via the following command (note the spelling of "coloring" as opposed to "colouring"):

```
1  vertex_coloring(G)
2  >> {[[2], [1], [3, 0]]}
```

The output produced by Sage tells us that G can be optimally coloured using three colours, with vertices 0 and 3 receiving the same colour. This solution is expressed as a partition of the vertices, which can be used to produce a visualisation of the colouring as follows:

```
1  S = vertex_coloring(G)
2  G.show(partition=S)
```

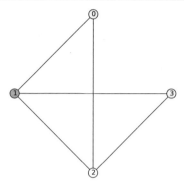

Here, the partition produced by vertex_coloring(G) is assigned to the variable S, which is then used as an additional argument in the G.show() command to produce the above visualisation.

We can also use the vertex_coloring command to test if a graph is k-colourable. For example, to test whether G is two-colourable, we get

```
1  vertex_coloring(G,2)
2  >> False
```

which tells us that a two-colouring is not possible for this graph. On the other hand, if we want to confirm whether G is four-colourable, we get

```
1  vertex_coloring(G,4)
2  >> {[[3, 0], [2], [1], []]}
```

which tells us that one way of four-colouring the G is to not use the fourth colour!

In addition to the above, commands are also available in Sage for determining the chromatic number

```
1  chromatic_number(G)
2  >> 3
```

and for calculating the number of different k-colourings. For example, with $k = 2$, we get

```
1  number_of_n_colorings(G,2)
2  >> 0
```

which is what we would expect since no two-colouring of G exists. On the other hand, for $k = 3$, we get

```
1  number_of_n_colorings(G,3)
2  >> 6
```

telling us that there are six different ways of feasibly assigning three colours to *G*.

Sage also provides commands for calculating edge colourings of a graph (see Sect. 6.2). For example, continuing with the graph *G* from above, we can use the `edge_coloring()` command to get

```
1  edge_coloring(G)
2  >> {[[(0, 1), (2, 3)], [(0, 2), (1, 3)], [(1, 2)]]}
```

This tells us that the chromatic index of *G* is 4, with edges {0, 1} and {2, 3} being assigned to one colour, {0, 2} and {1, 3} being assigned to a second, and {1, 2} being assigned to a third.

Sage also contains a collection of predefined graphs. This allows us to make use of common graph topologies without having to manually type out their adjacency matrices. A full list of these graphs is provided at https://doc.sagemath.org/html/en/reference/graphs/sage/graphs/graph_generators.html. For example, here are the commands for producing an optimal colouring of a dodecahedral graph. In this case, we have switched off vertex labelling to make the illustration clearer:

```
1  G = graphs.DodecahedralGraph()
2  S = vertex_coloring(G)
3  G.show(partition=S, vertex_labels=False)
```

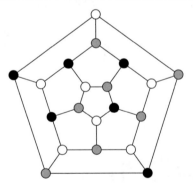

The following is an optimal colouring of the complete graph with ten vertices, K_{10}:

```
1  G = graphs.CompleteGraph(10)
2  S = vertex_coloring(G)
3  G.show(partition=S, vertex_labels=False)
```

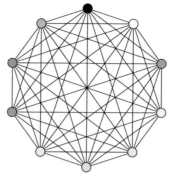

So-called lollipop graphs are defined by a path of n_1 vertices representing the "stick" and a complete graph K_{n_2} to represent the "head". Here is an example colouring using a lollipop graph with $n_1 = 4$ and $n_2 = 8$:

```
1  G = graphs.LollipopGraph(8,4)
2  S = vertex_coloring(G)
3  G.show(partition=S, vertex_labels=False)
```

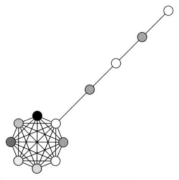

Our next graph, the "flower snark", is optimally coloured as follows:

```
1  G = graphs.FlowerSnark()
2  S = vertex_coloring(G)
3  G.show(partition=S, vertex_labels=False)
```

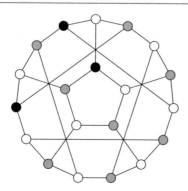

As we can see,, this graph is three-colourable. However, we can confirm that it is not planar using the following command:

```
1  G.is_planar()
2  >> False
```

Finally, Sage also allows us to define random graphs $G_{n,p}$ that have n vertices and edge probabilities p (see Definition 3.15). Here is an example with $n = 50$ and $p = 0.05$:

```
1  G = graphs.RandomGNP(50,0.05)
2  S = vertex_coloring(G)
3  G.show(partition=S, vertex_labels=False)
```

It can be seen that this particular graph is three-colourable, although the default layout of this graph is not very helpful because the connected component on the left is tightly clustered. If desired, we can change this layout so that vertices are shown in a circle:

```
1  G.show(vertex_labels=False, layout='circular',
         partition=S)
```

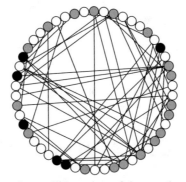

This, arguably, gives a clearer illustration of the graph.

A.3 Graph Colouring in Python with NetworkX

In this section,, we show how to create and colour graphs using NetworkX. NetworkX is a Python package used for the creation, manipulation, and study of graphs. An open-source distribution of Python including NetworkX can be downloaded for free from https://www.anaconda.com/download/. Versions for all major operating systems are available.

The listing below gives a short Python program and uses NetworkX to create, analyse, colour, and visualise a graph. Note that Python uses indentation to indicate a block of code (it replaces curly braces, semi-colons, etc., used in other programming languages). The hash character indicates comments.

```
1   import networkx as nx
2   import matplotlib.pyplot as plt
3   from networkx.algorithms import approximation as approx
4
5   G = nx.binomial_graph(20, 0.2)
6   nx.draw_networkx(G, pos=nx.circular_layout(G))
7   plt.show()
8
9   A = set(nx.articulation_points(G))
10  print("The articulation points of G are", A)
11
12  B = set(nx.bridges(G))
13  print("The bridges of G are", B)
14
15  omega = nx.graph_clique_number(G)
16  print("The clique number of G is", omega)
17
18  C = approx.max_clique(G)
19  print("A large clique in G comprises vertices", C)
20
21  I = approx.maximum_independent_set(G)
22  print("A large independent set in G comprises vertices", I)
23
24  def drawcoloring(G, S, pos):
25      #Function for drawing a coloring of G
26      cols = ["black","gray","gainsboro","darkgray","lightgray"]
27      k = max(S.values()) + 1
28      assert k <= len(cols)
29      for i in range(len(cols)):
30          nx.draw_networkx_nodes(G, pos, [v for v in G.nodes()
                    if S[v] == i], node_color=cols[i])
31      nx.draw_networkx_edges(G, pos)
32      plt.show()
33
34  #Colour the vertices of G using the greedy algorithm with a
        random permutation of the vertices
35  S = nx.coloring.greedy_color(G, strategy="random_sequential")
36  print("Here is a", max(S.values()) + 1 ,"coloring of G:", S)
37  drawcoloring(G, S, nx.circular_layout(G))
38
39  #Colour the vertices of G using the DSATUR algorithm
40  S = nx.coloring.greedy_color(G, strategy="DSATUR")
41  print("Here is a", max(S.values()) + 1 ,"coloring of G:", S)
42  drawcoloring(G, S, nx.circular_layout(G))
```

In this example, Lines 1–3 first import the libraries that we need. Line 5 then uses the command nx.binomial_graph(20, 0.2) to create a random graph G with twenty vertices and an edge probability of 0.2. Similar commands can be used in NetworkX to generate other topologies such as cycles, scale-free graphs, d-regular graphs, and so on. Lines 6 and 7 then draw this graph to the screen using a circular layout of the vertices. This output is shown in Fig. A.1. Note that, by default, vertices are labelled from zero upwards.

Having created an example graph, the commands on Lines 9–22 output some of its properties. The problem of identifying cut vertices (articulation points) and bridges in a graph can be solved in $\mathcal{O}(m)$ time, where m is the number of edges. The commands on Lines 9 and 12 are therefore fast to execute. On the other hand, identifying the clique number in a graph is $\mathcal{NP}$-hard; the algorithm employed by NetworkX is exact and therefore has an exponential running time. Finally, the commands on Lines 18 and 21 employ approximation algorithms to identify a large clique and a large independent set in the graph.

Lines 35 and 40 show some of the available commands for feasibly colouring a graph. The first example uses the GREEDY algorithm with a random permutation of the vertices, the second uses DSATUR. The output of these commands is a Python dictionary in which each vertex is assigned to a colour, labelled from zero upwards. In our example, these dictionaries are used to draw the resultant colourings to the screen via the user-defined function drawcolouring(), shown in Lines 24–32.

The full output of this program is given below. Note that the application of GREEDY has resulted in a four-colouring in this case while DSATUR has given a three-colouring. These solutions are visualised in Fig. A.2.

```
1   The articulation points of G are {8}
2   The bridges of G are {(8, 16)}
3   The clique number of G is 3
4   A large clique in G comprises vertices {0, 18, 10}
5   A large independent set in G comprises vertices {0, 1, 2, 3,
        6, 11, 16, 17}
6   Here is a 4 coloring of G: {3: 0, 16: 0, 2: 0, 19: 0, 14: 1,
        15: 0, 17: 0, 9: 2, 4: 1, 8: 1, 6: 3, 0: 0, 12: 1, 11:
        0, 1: 2, 13: 3, 7: 1, 10: 2, 5: 1, 18: 3}
7   Here is a 3 coloring of G: {5: 0, 6: 1, 10: 1, 18: 2, 0: 0,
        2: 2, 14: 0, 9: 2, 17: 1, 7: 2, 12: 1, 15: 2, 1: 1, 19:
        2, 4: 0, 8: 0, 13: 2, 3: 1, 16: 1, 11: 0}
```

Fig. A.1 NetworkX visualisation of a random graph with $n = 20$ vertices and an edge probability of 0.2

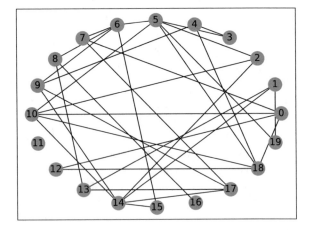

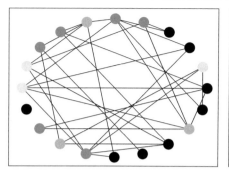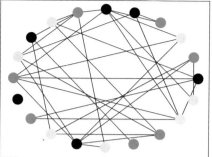

Fig. A.2 Colourings of our example graph due to the GREEDY (left) and DSATUR (right) algorithms

A.4 Generating Planar Graphs in Python

In this section, we provide the Python code used for randomly generating planar graphs. This program requires three parameters from the user: the number of vertices n, the number of edges m, and a random seed (Lines 6–10). The program first generates random coordinates for n points in the unit square (Line 13). A Delaunay triangulation is then generated from these points and this is converted into a corresponding planar graph T (Lines 14–24). The task is to now identify a random subset of T's edges so that the resultant graph G has n vertices, m edges, and is connected. To do this, G is first set as a minimum spanning tree of T (Line 30). Additional edges from T are then copied to G until the required number of edges is reached (Lines 33–41). The command `nx.graph_clique_number(G)` is also used in this example to calculate the size of the largest clique in G. For planar graphs, this command executes quickly.

An illustration of this process is shown in Fig. A.3. The complete code listing is as follows:

```
1   import networkx as nx
2   import random
3   from scipy.spatial import Delaunay
4
5   #Get the input variables, check their validity, and seed the
        random number generator
6   n = int(input("Enter n >> "))
7   m = int(input("Enter m >> "))
8   s = int(input("Enter seed >> "))
9   assert n > 0 and m >= n - 1 and m <= 3 * n - 6, "Illegal
        parameter combination."
10  random.seed(s)
11
12  #Generate a list P of n points in the unit square and form a
        Delaunay triangulation
13  P = [(random.random(), random.random()) for i in range(n)]
14  T = Delaunay(P)
15  tri = T.simplices.copy()
16
17  #Convert the triangulation to a simple graph T
18  T = nx.Graph()
19  for v in range(n):
```

```
20       T.add_node(v)
21  for e in tri:
22       T.add_edge(e[0], e[1])
23       T.add_edge(e[0], e[2])
24       T.add_edge(e[1], e[2])
25
26  #Check the triangulation gives enough edges
27  assert T.number_of_edges() >= m, "Cannot form planar graph
        with this many edges."
28
29  #Build the graph G. It starts as a spanning tree of T
30  G = nx.minimum_spanning_tree(T);
31
32  #Construct a list L of edges in T that might be added to G
33  L = []
34  for e in T.edges():
35       if e not in G.edges():
36            L.append((e[0], e[1]))
37
38  #Take a random sample of L and add these edges to G.
39  S = random.sample(L, m - (n - 1))
40  for e in S:
41       G.add_edge(e[0], e[1])
42
43  #Draw the graph to the screen
44  nx.draw_networkx(G, pos=P, with_labels=False, node_size=30)
45
46  #Write the graph to the file 'graph.txt'
47  f = open("graph.txt", "w+")
48  f.write("c Undirected planar graph\n")
49  f.write("c Seed = " + str(s) + "\n")
50  f.write("c Clique number = " + str(nx.graph_clique_number(G))
        + "\n")
51  f.write("p edge " + str(G.number_of_nodes()) + " " + str(G.
        number_of_edges()) + "\n")
52  for e in G.edges():
53       f.write("e " + str(e[0] + 1) + " " + str(e[1] + 1) + "\n")
54  f.close()
```

A.5 Graph Colouring with Commercial IP Software

The following code demonstrates how the graph colouring problem can be specified using integer programming methods and then solved using off-the-shelf optimisation software. This example gives the implementation used in the experiments of Sect. 4.1.2.4 and is coded in the Xpress-Mosel language, which comes as part of the FICO Xpress Optimisation Suite. Comments in the code are preceded by exclamation marks.

```
1  model GCOL
2
3  !Gain access to the Xpress-Optimizer solver and timer
4  uses "mmxprs", "mmsystem"
5
6  !Specify a hard time limit of 60 seconds
7  setparam("XPRS_MAXTIME",-60)
8
9  !Start the timer
```

```
10   starttime:=gettime
11
12   !Define input file
13   fopen("graph.txt",F_INPUT)
14
15   !Define integers used in the program
16   declarations
17       n,m,v1,v2: integer
18   end-declarations
19
20   !Read the number of vertices and edges from the input file
21   read(n,m)
22   writeln("n = ",n,", m = ",m)
23
24   !Declare the decision variables and make them binary
25   declarations
26       X: array(1..n,1..n) of mpvar
27       Y: array(1..n) of mpvar
28   end-declarations
29   forall (i in 1..n) do
30       forall (j in 1..n) do
31           X(i,j) is_binary
32       end-do
33       Y(i) is_binary
34   end-do
```

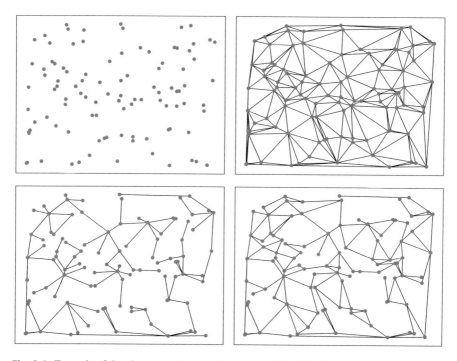

Fig. A.3 Example of the planar graph generation process. Top-left shows $n = 100$ points placed at random coordinates in the unit square; top-right then shows a Delaunay triangulation of these points. Bottom-left shows a minimum spanning tree of this triangulation, while bottom-right shows the final planar graph G, formed by adding further edges

```
35
36   !Read in all of the edges and define the constraints that
         ensure that (a) adjacent vertices are assigned to
         different colours, and (b) Y(i) is a set to 1 only if a
         vertex is assigned to colour i
37   write("E = {")
38   forall (j in 1..m) do
39       read(v1,v2)
40       forall (i in 1..n) do
41           X(v1,i) + X(v2,i) <= Y(i)
42       end-do
43       write("{",v1,",",v2,"}")
44   end-do
45   writeln("}")
46
47   !Specify that each vertex is to be assigned to exactly one
         colour
48   forall (i in 1..n) do
49       sum(j in 1..n) X(i,j) = 1
50   end-do
51
52   !Eliminate solution symmetries
53   forall (i in 1..n) do
54       forall (j in i+1..n) do
55           X(i,j) = 0
56       end-do
57   end-do
58   forall (i in 2..n) do
59       forall (j in 2..i-1) do
60           X(i,j) <= sum(l in j-1..i-1) X(l,j-1)
61       end-do
62   end-do
63
64   !Specify the objective function
65   objfn := sum(i in 1..n) Y(i)
66
67   !Run the model
68   writeln
69   writeln("Running model...")
70   minimise(objfn)
71   writeln("...Run ended")
72
73   !Write the output to the screen
74   writeln
75   writeln("Total time  = ",gettime-starttime," secs")
76   writeln("Upper bound = ",getobjval," (number of colours in
         best observed solution)")
77   writeln("Lower bound = ",getparam("XPRS_BESTBOUND"))
78   writeln
79
80   writeln("X = ")
81   forall (i in 1..n) do
82       forall (j in 1..n) do
83           write(getsol(X(i,j))," ")
84       end-do
85       writeln
86   end-do
87   writeln
88   writeln("Y = ")
89   forall (j in 1..n) do
90       write(getsol(Y(j))," ")
91   end-do
92   writeln
93   writeln
94   writeln("Solution = ")
```

```
95   forall (i in 1..n) do
96       write("c(v_",i,") = ")
97       forall (j in 1..n) do
98           if(getsol(X(i,j))=1) then
99               writeln(j)
100          end-if
101      end-do
102  end-do
103
104  end-model
```

The above program starts by reading in a graph colouring problem from a text file, called **graph.txt** in this case. The objective function and constraints of the problem are then specified, before the optimisation process itself is invoked using the `minimize(objfn)` command on Line 70. In this implementation, the optimisation process is terminated once a provably optimal solution has been found, or when a time limit has been reached (specified as sixty seconds here). The upper and lower bounds are then written to the screen, together with the total run time. If the lower and upper bounds are equal, then the provably optimal solution has been found. If this is not the case, then the best-observed integer solution found within the time limit (if, indeed, one has been found) is output. The number of colours used in this integer solution corresponds to the upper bound.

Here is some example input that can be read in by the above program. The first two lines give the number of vertices and edges, n and m, respectively. The m edges then follow, one per line. This particular example corresponds to the graph shown in Fig. 4.5.

```
1    8
2    12
3    1 2
4    1 3
5    1 4
6    2 5
7    2 6
8    2 8
9    3 4
10   3 7
11   4 7
12   5 8
13   6 8
14   7 8
```

On completion of the program, the following output is produced:

```
1    n = 8, m = 12
2    E = {{1,2}{1,3}{1,4}{2,5}{2,6}{2,8}{3,4}{3,7}{4,7}{5,8}{6,8}
         {7,8}}
3
4    Running model...
5    ...Run ended
6
7    Total time   = 0.025 secs
8    Upper bound = 3 (number of colours in best observed solution
         )
```

```
 9  | Lower  bound  =  3
10  |
11  | X  =
12  | 1  0  0  0  0  0  0  0
13  | 0  1  0  0  0  0  0  0
14  | 0  0  1  0  0  0  0  0
15  | 0  1  0  0  0  0  0  0
16  | 1  0  0  0  0  0  0  0
17  | 1  0  0  0  0  0  0  0
18  | 1  0  0  0  0  0  0  0
19  | 0  0  1  0  0  0  0  0
20  |
21  | Y  =
22  | 1  1  1  0  0  0  0  0
23  |
24  | Solution  =
25  | c ( v_1 )  =  1
26  | c ( v_2 )  =  2
27  | c ( v_3 )  =  3
28  | c ( v_4 )  =  2
29  | c ( v_5 )  =  1
30  | c ( v_6 )  =  1
31  | c ( v_7 )  =  1
32  | c ( v_8 )  =  3
```

It can be seen that the upper and lower bounds reported here are equal. A provably optimal solution has therefore been found.

A.6 Useful Web Links

Here are some further web resources related to graph colouring. A page of resources maintained by Joseph Culberson featuring, most notably, a collection of problem generators and C code for the algorithms presented by Culberson and Luo [1] can be found at:

> http://webdocs.cs.ualberta.ca/~joe/Coloring/

An excellent bibliography on the graph colouring problem can also be found at:

> http://www.imada.sdu.dk/~marco/gcp/

A large set of graph colouring problem instances has been collected by the Center for Discrete Mathematics and Theoretical Computer Science (DIMACS) as part of their DIMACS Implementation Challenge series. These can be downloaded at:

> http://mat.gsia.cmu.edu/COLOR/instances.html

These problem instances have been used in a large number of graph colouring-based papers and are written in the DIMACS graph format, a specification of which can be found in the following (postscript) document:

http://mat.gsia.cmu.edu/COLOR/general/ccformat.ps

or at

http://prolland.free.fr/works/research/dsat/dimacs.html

The fun graph colouring game *CoLoRaTiOn*, which is suitable for both adults and children, can be downloaded from:

http://vispo.com/software/

The goal in this game is to achieve a feasible colouring within a certain number of moves. The difficulty of each puzzle depends on several factors, including its topology, whether you can see all of the edges, the number of vertices, and the number of available colours.

Finally, C++ code for the random Sudoku problem instance generator used in Sect. 6.4.2 of this book can be downloaded from:

http://rhydlewis.eu/resources/sudokuGeneratorMetaheuristics.zip

A Sudoku-to-graph-colouring problem converter can also be found at:

http://rhydlewis.eu/resources/sudokuToGCol.zip

When compiled, this program reads in a single Sudoku problem (from a text file) and converts it into the equivalent graph colouring problem in the DIMACS format mentioned above.

Reference

1. Culberson J, Luo F (1996) Exploring the k-colorable landscape with iterated greedy. In: Cliques, coloring, and satisfiability: second DIMACS implementation challenge, vol 26. American Mathematical Society, pp 245–284

Table of Notation

B

The following table lists the main notation used throughout this book. Further information can be found by consulting the index and the relevant definitions.

Notation	Description
Graph properties	
$G = (V, E)$	Graph G comprising a vertex set V and an edge set E
n	Number of vertices in a graph, $n = |V|$
m	Number of edges in a graph, $m = |E|$
$\Gamma(v)$	Set of vertices adjacent to a vertex $v \in V$ (i.e., the neighbourhood of v)
$\deg(v)$	Degree (number of neighbours) of a vertex v. $\deg(v) = |\Gamma(v)|$
$\Delta(G)$	Maximum degree of any vertex in the graph G
$c(v)$	Colour label of a vertex $v \in V$
$\text{sat}(v)$	Saturation degree of a vertex v. Used with the DSATUR algorithm
KEMPE(v, i, j)	Set of vertices in the Kempe chain generated by $v \in V$ and colours $i = c(v)$ and j
$\chi(G)$	Chromatic number of a graph G
$\alpha(G)$	Independence number of a graph G
$\omega(G)$	Clique number of a graph G
$L(G)$	Line graph of a graph G
$\chi'(G)$	Chromatic index of a graph G
$\chi_L(G)$	Choice number of a graph G
$\chi_e(G)$	Equitable chromatic number of a graph G
General mathematical notation	
$\mathcal{O}(.)$	Big-O notation for describing the order of growth of a function
$|S|$	Set cardinality (number of elements in the set S)
$\lg n$	Binary logarithm. $\lg n = \log_2 n$
$n!$	Factorial function. $n! = n \times (n-1) \times (n-2) \times \ldots \times 2 \times 1$
$^n P_k$	Permutation function. Number of ways of picking an ordered set of k elements from a set of n elements. $^n P_k = \frac{n!}{(n-k)!}$

(continued)

R. M. R. Lewis, *Guide to Graph Colouring*, Texts in Computer Science,

$\binom{n}{k}$	Binomial coefficient. Number of ways of picking an unordered set of k elements from a set of n elements. $\binom{n}{k} = \frac{n!}{k!(n-k)!}$
B_n	The nth Bell number
$\left\{{n \atop k}\right\}$	Stirling numbers of the second kind
Complexity theory	
$\mathcal{P}$	Set of decision problems solvable in polynomial time
$\mathcal{NP}$	Set of decision problems verifiable in polynomial time
$\Pi_1 \propto \Pi_2$	A polynomial-time reduction exists from decision problem Π_1 to Π_2

Bibliography

1. Beineke L, Wilson J (eds) (2015) Topics in chromatic graph theory. Encyclopedia of mathematics and its applications (no. 156). Cambridge University Press
2. Jensen T, Toft B (1994) Graph coloring problems, 1st edn. Wiley-Interscience
3. Lewis R, Thompson J, Mumford C, Gillard J (2012) A wide-ranging computational comparison of high-performance graph colouring algorithms. Comput Oper Res 39(9):1933–1950

Index

A

Adjacency list, 14
Adjacency matrix, 14
Adjacent (vertices), 10
Algorithm user guide, 277–281
Ant colony optimisation, 118–121, 257
AntCol algorithm, 118–121
  empirical performance, 124–145
Appel, Kenneth, 9, 161, 165
Approximation algorithms, 75
Approximation ratio, 75
Articulation point, *see* Cut vertex
Aspiration criterion (tabu search), 114

B

Backtracking algorithm, 78–81, 122–124
  empirical performance, 124–145
Barabási-Albert method, 134
Bell number B_n, 21
Betweenness centrality, 153
Big O notation $\mathcal{O}$
  definition, 19
  examples, 19
Binary heap, 67
Bipartite graph, 34, 43, 56, 57, 60, 149, 158,
    169
Block, 51
Boolean satisfiability, *see* Satisfiability problem
Branch-and-bound algorithm, 81–86
  empirical performance, 88–91
Breadth-first search, 25
Bridge, 156
  identification using NetworkX, 288
Brooks' bound, *see* Chromatic number $\chi(G)$

C

Canonical round-robin algorithm, 222, 234
Cayley, Arthur, 163
Checks (counting), 15
Choice number $\chi_L(G)$, 188
Chordal graph, 48
Chromatic index $\chi'(G)$, 166
Chromatic number $\chi(G)$
  Berge's bound, 75
  Bollobás' bound, 74
  Brooks' bound, 50–53
  definition, 11
  Hoffman's bound, 75
  interval graphs, 47–48
  lower bounds, 45–48
  Reed's bound, 71
  Reed's conjecture, 71
  upper bounds, 49–54
  Welsh–Powell bound, 53–54
Chromatic polynomial, 196–199
Circle method, 168, 222, 234
Clash, 10
Clique
  counting in NetworkX, 92
  definition, 11
  identification using NetworkX, 288
  use with backtracking, 80
  use with branch-and-bound, 88
  use with multicolouring, 196
Clique number $\omega(G)$, *see also* Maximum
    clique problem, 45
  identification with NetworkX, 289
CoLoRaTiOn game, 295
Colour class, 11

© The Editor(s) (if applicable) and The Author(s), under exclusive license to Springer   299
Nature Switzerland AG 2021
R. M. R. Lewis, *Guide to Graph Colouring*, Texts in Computer Science,

Column generation, 91–96
Compact round-robin, 222
Complete colouring, 10
Complete graph, 34, 284
Complete improper k-colourings, 102–105
Component, 50
Connected graph, 40, 50
Constraint checks, *see* Checks (counting)
Constructing timetables, *see* Timetabling
Contraction (of vertices), 171
Cook's Theorem, 29
Cut vertex, *see also* Vertex separating set
   definition, 51
   identification using NetworkX, 288
Cycle, 40
Cyclegraph, 35

D
Decentralised graph colouring, 182–185
Decision problem, 23–25
Degree deg(v), 40
Delauney triangulation, 289
DIMACS
   instance download, 294
   instance format, 278
Disconnected graph, 40
Distance (between vertices), 40
Dodecahedral graph, 284
Domination (multiobjective optimisation), 239
DSATUR algorithm, 54–58
   bipartite graphs, 57
   complexity, 66–67
   cycle graphs, 57
   empirical performance, 68–71
   equitable colouring, 191
   pseudocode, 55
   wheel graphs, 57
Dual graph, 158–160
Dummy room, 264
Dynamic graph colouring, 187

E
Edge colouring, 166–170
Edge set E, *see* Graph
Empty graph, 34
Endpoints, 10
Equitable chromatic number $\chi_e(G)$, 189
Equitable graph colouring, 189–193
Eulerian graph, 160–161
Euler, Leonhard, 156, 172
Euler's characteristic, 156

Evaporation rate, *see* Ant colony optimisation
Event-clash constraint, 4, 248
Evolutionary algorithm, *see also* HEA, 97–98
Evolutionary pressure, *see* Evolutionary
   algorithm
Exact algorithm, 77
Exam timetabling, *see* Timetabling
Experimental design, 172
Exponential growth rate, 19

F
Facebook, 3
Face colouring, 7–9, 156–166
Feasibility ratio, 265
Feasible colouring, 11
Feasible-only solution spaces, 97–101
Five colour theorem, 162
Flat graph
   algorithm performance on, 126–129
Flower snark graph, 285
Four colour theorem, 9, 159–166
Frequency assignment, 182–183, 186

G
GET- MAXIMAL- I- SET algorithm
   pseudocode, 60
Girth, 157
Graph
   centrality, 153
   connectivity, 51
   definition, 9
   density, 41
   energy, 153
   representation, 13–15
Graph colouring
   in NetworkX, 286–289
   in Sage, 281–286
   polynomially solvable cases, 33–37
   problem definition, 10
   proof of $\mathcal{NP}$-completeness, 30–33
   with weighted edges, 195–196
   with weighted vertices, 193–195
GREEDY algorithm, 41–45
   complexity, 65
   empirical performance, 68–71
   example with social networks, 4
   pseudocode, 41
   worst-case performance, *see* Grundy
      chromatic number
GREEDY-I-SET algorithm, 58–59
   complexity, 67–68

empirical performance, 68–71
    pseudocode, 60
Grid graph, 36
Grundy chromatic number, 186
Guthrie, Francis, 7, 8, 159

H
Hajnal's theorem, 190
Haken, Wolfgang, 9, 161, 165
HAM-CYCLE problem, 24
Hamilton, William, 163
Hamiltonian cycle problem, 23
    application in seating plans, 205
HC algorithm, 121–122
    empirical performance, 124–145
HEA, 117–118
    alternative local search operator, 151–152
    alternative recombination operators,
        149–151
    empirical performance, 124–145
    Greedy partition crossover (GPX), 117
    maintaining diversity, 146–149
Heawood, Percy, 164
Hill-climbing algorithm, *see* HC algorithm
Hybrid evolutionary algorithm, *see* HEA

I
Improper colouring, 10
Incident, 10
Independence number $\alpha(G)$, *see also*
        Maximum independent set problem
Independent set
    counting in NetworkX, 92
    definition, 11
    extraction, 108–109
    identification using NetworkX, 288
    in random graphs, 92
Induced subgraph, 40
Integer programming (IP), 81–91
    example implementation, 290–294
    use with WSP, 214–219
Integrality constraints, 82
Interference graph, 6
Interval graph, 6, 47–48
Intractable problem, *see* $\mathcal{NP}$-complete
Isomorphism (of graphs), 64–65
Iterated Greedy algorithm, 97

J
Jaccard distance measure, 146

K
k-CLIQUE problem, 24
k-COL problem, 25
k-connected graph, 51
k-DISTANCE problem, 24
k-I-SET problem, 24
k-partition problem, 208
Kempe chain
    definition, 100
    evaluating all moves, 210
    example application, 101
    total chain, 213, 234
    use in the HC algorithm, 121
    use with round-robin graphs, 233–235
    use with timetabling, 262
    use with WSP, 210
Kempe, Alfred, 163–164
Kirkman, Thomas, 168
König's line colouring theorem, 169
Kuratowski's theorem, 158

L
Latin squares, 172–178
    relationship to graph colouring, 173–174
Line graph $L(G)$, 166
Linear programming, *see* Integer programming
        (IP)
List colouring, 188
Literal, 29
Liveness analysis, 7
Lollipop graph, 285

M
Map colouring, *see* Face colouring
Maximal independent set, 59
Maximum clique problem, 23
Maximum independent set, 59
Maximum independent set problem, 23
Maximum matching
    in bipartite graphs, 253
    in general graphs, 194
Maximum weighted independent set problem,
        93
Metaheuristics, *see also* Ant colony optimi-
        sation, Evolutionary algorithm,
        Simulated annealing, Tabu search,
        96
Minimum distance problem, 23
Minimum set covering problem
    definition, 91
    IP formulation, 91

Multicolouring, 196
Mutation, *see* Evolutionary algorithm
Mycielskian graphs, 46

N
NetworkX, 286–290
    counting cliques, 92
    counting independent sets, 92
    generation of planar graphs, 289–290
    graph generation, 287
    graph visualisation, 287
Nonadjacent (vertices), 10
$\mathcal{NP}$-complete, 27–33
$\mathcal{NP}$-hard, 37
$\mathcal{NP}$ problem class, 25

O
One-factor, 222
One-factorisation, 222
    perfect, 235
Online colouring, 185–186
Optimal colouring, 11
OR-gadget, 31

P
$\mathcal{P}$ problem class, 25
$\mathcal{P} \neq \mathcal{NP}$, 37
Pair swap
    definition, 100
    evaluating all moves, 210
    use in the HC algorithm, 121
    use with WSP, 210
Partial colouring, 10
Partial proper k-colouring, 105–106
PARTIALCOL algorithm, 116
    empirical performance, 124–145
    use in timetabling, 257–261
Path, 40
    length, 40
Perfect elimination ordering, 47
Perfect graph, 71
Phase transition, 126, 176
Pheromone, *see* Ant colony optimisation
Pierce, Charles, 163
Planar graph, 7, 36, 156–166
    algorithm performance on, 129–134
    generation, 131, 289–290
Polynomial growth rate, 19
Polynomial-time reduction, 26
    examples, 27
Polynomial-time verification, 25

Precolouring, 171
Preferential attachment, 134
Principality premiership problem, 237
Proper colouring, 10
Pruning, 79
Pseudocode, 15
Python, *see* NetworkX

R
Random descent, 102, 239
Random graph
    algorithm performance on, 68–71, 88–91,
        125–126
    definition, 64
    degree distribution, 65, 135
    in NetworkX, 287
    in Sage, 286
Recombination, *see* Evolutionary algorithm
Recursive largest first, *see* RLF algorithm
Reducible configuration, 165
Reducing problem size, 107–109
Register allocation, 6–7
RLF algorithm, 59–62
    Bipartite graphs, 60
    complexity, 67–68
    cycle graphs, 62
    empirical performance, 68–71
    wheel graphs, 62
Round-robin graph
    algorithm performance on, 227, 230–232
Round-robin schedule, 168, 221–245
    adding constraints, 227–232
    as a graph colouring problem, 225–226
    carryover, 224, 227
    definition, 222
    home-away break, 222–223
    mirroring, 225
    neighbourhood operators, 233–235

S
s-chain, *see also* Kempe chain, 245
Sage, 281–286
    edge colouring, 284
    graph colouring, 282–284
    graph construction, 282
    graph visualisation, 283
    planarity testing, 286
    predefined graphs, 284–286
Satisfiability problem, 28–30
Saturation degree sat(v), 55
Scale-free graph

algorithm performance on, 134–137
  degree distribution, 135
Search tree, 78
Seating plans, 203–220
Set covering problem, *see* Minimum set
  covering problem
Short circuit testing, 179–182
Simple graph, *see* Graph
Simulated annealing, 103–104, 151, 240
  cooling schedule, 261
  pseudocode, 103
  with timetabling, 261–269
Six colour theorem, 161
Social networks, 2–4
  algorithm performance on, 143
  decentralised colouring, 185
Sorting problem, 23
Spanning tree, 289
Sports league, *see* Round-robin schedule
Star graph, 190
Steepest descent, 104
  pseudocode, 104
Stirling numbers of the second kind $\left\{ {n \atop k} \right\}$, 22
Subgraph, *see also* Induced subgraph, 40
Sudoku, 172–178
  logic solvable puzzles, 174, 178
  random puzzles, 176
  relationship to graph colouring, 173–175

T
Table plans, *see* Seating plans
Tabu search, 104
Tabu tenure, 115
TABUCOL algorithm, 115, 210
  empirical performance, 124–145
Taxi scheduling, 5–6

Team building, 2–4
3-CNF-SAT, 30
3-conjunctive normal form (3-CNF), 29
Timetable graph
  algorithm performance on, 137
Timetabling, 4–5, 195, 247–274
  constraints, 247–249
  distance to feasibility (DTF), 254
  evaluation and benchmarking, 255
  neighbourhood operators, 262–263
  problem complexity, 255
Travelling salesman problem, 23
Travelling tournament problem, 224–225
TSP problem, 24

U
Unavoidable set, 165
University timetabling, *see* Timetabling

V
Vertex separating set, 51
Vertex set V, *see* Graph
Vizing's theorem, 169

W
Wedding seating problem (WSP), 206
Weighted graph colouring, 193–196
Welsh–Powell algorithm, 54
  complexity, 65
  empirical performance, 68–71
Welsh–Powell bound, *see* Chromatic number
  $\chi(G)$
Welsh Rugby Union (WRU), 236
Wheel graph, 36
Wireless network, *see* Frequency assignment
WSP, *see* Wedding seating problem

Printed in the United States
by Baker & Taylor Publisher Services